Henrici-Olivé/Olivé
Coordination and Catalysis

Monographs in Modern Chemistry

Series Editor: Hans F. Ebel

The series is to be continued

G. HENRICI-OLIVÉ
S. OLIVÉ

COORDINATION AND CATALYSIS

Verlag Chemie · Weinheim · New York · 1977

Dr. Gisela Henrici-Olivé
Prof. Dr. Salvador Olivé

Monsanto Triangle Park Development Center, Inc.
P.O. Box 12274
Research Triangle Park, N.C. 27709
USA

CIP-Kurztitelaufnahme der Deutschen Bibliothek

Henrici-Olivé, Gisela
Coordination and catalysis / G. Henrici-Olivé;
S. Olivé. – 1. Aufl. – Weinheim, New York:
Verlag Chemie, 1976.
  (Monographs in modern chemistry; Vol. 9)
  ISBN 3-527-25686-5
NE: Olivé, Salvador:

Printed by Schwetzinger Verlagsdruckerei GmbH, D-6830 Schwetzingen
Bookbinder: Georg Kränkl, D-6148 Heppenheim

Printed in West Germany

Preface

Catalysis with coordination compounds of transition metals has become, during the past decade, a real forum of encounter for various disciplines of chemistry. Preparative organic chemistry and polymer chemistry are perhaps primarily interested from the point of view of the reaction products, organometallic and inorganic coordination chemistry from the side of the catalytically active transition metal center; the methods of physical chemistry are often extremely helpful in elucidating the mechanism of a catalytic process and, last not least, the conceptions and computational results of theoretical chemistry may guide the catalysis chemist to a better insight into the interactions of catalyst and substrates, at a molecular level.

In particular, catalysis with well defined, soluble transition metal complexes (homogeneous catalysis) plays an eminent role, not only as the medium of very important industrial and laboratory processes such as hydroformylation, hydrogenation, dimerization, oligomerization and polymerization of olefins, etc., but also as a model for heterogeneous processes, the mechanism of which is inherently more difficult to elucidate directly, and even as a model for certain natural enzymatic processes.

The present book deals with such well defined transition metal complexes and their use in catalysis. Some comments concerning the layout of the book may be appropriate. The title "Coordination and Catalysis" indicates a certain division into two parts. The interdisciplinary character of the subject matter has the consequence that the practical chemist, interested in a certain catalytic reaction, very often is not familiar enough with the methods and the special technical jargon used by his colleagues of the contiguous disciplines. Such lack of knowledge may result in a serious handicap during literature studies, and may impede a fruitful cooperation between the various interested groups. The first six chapters of the book are designed to bridge this gap. They provide the basic background of coordination chemistry as far as it is useful or even indispensable for a chemist working with transition metal complexes. In presenting this theoretical background, we kept in mind that it is more important for the practical catalysis chemist to understand the significance of the information available from a specific theory than to be able to apply its often sophisticated mathematics for his own computations.

The chapters "Atomic Orbitals" and "Transition Metal Ions" shall remind the reader of his basic knowledge of atomic structure, with particular emphasis on d orbitals and transition metals. Readers with a background in introductory quantum chemistry will be able to cover these two chapters rather quickly, but we hope that for others they might fill some voids. Informations that should be known from other sources are sometimes treated only very briefly. More space is given to certain details which, from the authors' experience, are often not clearly understood. Thus we have tried to answer such elementary questions as: What is the physical significance of the four-lobe representations of d orbitals, so trustfully used by chemists? Why has one of the d orbitals a different·shape? What does the labeling of the d orbitals ($d_{x^2-y^2}$, d_{xz}, etc.) mean? Special care has also been taken to explain the determination of the Russell-Saunders terms of a given d^n configuration.

The following three chapters are dedicated to the three theories which are the basis for all theoretical and semitheoretical work in the field: group theory, ligand field theory and molecular orbital theory. Most modern research publications relative to the matter are written in the language of one or the other of these theories. The concepts of group theory are presented in this book in a rather pragmatic, application oriented fashion. The trained reader may pardon the complete lack of mathematical rigour, but we have endeavoured to treat this valuable device as a tool rather than as a science. In particular the understanding of ligand field theory is greatly facilitated by some basic knowledge of group theory. And ligand field theory, this lucid account of the influences of ligands on the electronic levels of a central transition metal ion in a complex, is indispensable for the comprehension of the electronic spectra and the magnetic properties of such a complex, as well as of its behaviour in catalysis. Group theory is also one of the elements of the molecular orbital theory of transition metal complexes which, in our context, is used to account for all effects and consequences of covalent bonding between the metal center and its ligands.

In the whole of this first part of the book, we have tried to emphasize the relation to practical application, which sometimes gets lost in more sophisticated treatises. We hope that our often deliberately simplifying treatment will stimulate the interested reader to proceed to the more advanced texts which are listed at the end of each chapter (with increasing degree of difficulty, if there are more than one on the same topic).

The application part of the book starts with the chapter "Some Aspects of Catalysis with Transition Metal Complexes", where the present knowledge (and sometimes opinion) concerning the general course of a transition metal catalyzed reaction is discussed in the light of coordination chemistry. This field of work is doubtlessly still in a state of flux, where new hypotheses and suggestions are brought continuously into consideration. Under such conditions it is perhaps inevitable that the chosen presentation occasionally overemphasizes the authors' view.

The following chapters on particular catalytic reactions are organized in such a way as to minimize overlap. Thus, although the accent is generally on mechanistic aspects, and in particular on the processes occurring within the coordination sphere of the transition metal center, the division is according to substrates rather than to metal centers. Reactions of olefins, conjugated diolefins, carbon monoxide, oxygen, and nitrogen are treated, each in a chapter. The first three of these chapters are rigorously restricted to real catalysis, i.e. omitting all stoichiometric reactions of ligands in the coordination sphere of a metal. Sufficient technically important catalytic reactions are available to serve as illustrations for the principles to be discussed. In the last two chapters certain concessions in this respect appeared appropriate. The activation of molecular oxygen and nitrogen are processes produced by nature under mildest conditions, mostly in the presence of a transition metal carrying enzyme. Chemistry is still at the very beginning of learning to understand, and hopefully one day to copy, the work of Nature.

Obviously it was not possible, nor our intent, to include in the application part of the book reference to all papers in the literature concerning the vast field of homogeneous catalysis with transition metal complexes. Very often more recent papers are cited which in most cases refer to the older original papers. This was done generally when the treated material was presented more didactically in the more recent work. The suggested "additional reading" at the end of each of the later chapters refers mostly to review papers which permit the reader to broaden his knowledge on certain catalytic reactions treated in the chapter. Here again, a somewhat subjective selection appears inevitable.

Finally, we would like to indicate for which classes of readers the book has been conceived. Proceeding from the authors' teaching experience at the Swiss Federal Institute of Technology and at the University of Mainz, Germany, the book is addressed, on one hand, to advanced students who wish to specialize in this interesting field. On the other hand, however, we have undertaken a serious effort to make the book suitable for self-study, considering those chemists who completed their studies several years ago and who wish to keep up with modern developments. The expert research chemist working in the field, last not least, will hopefully find sufficient stimulation in the selective compilation of recent research work.

It is a great pleasure to acknowledge the valuable help of our friends and collegues, who read parts of the book and offered many clarifying suggestions: Ph. Teyssié, University of Liège, Belgium; J. Halpern, University of Chicago; L. Markó, University of Veszprem, Hungary; B. Bogdanović, Max-Planck-Institute, Mülheim, Germany; D. A. v. Bézard and E. Spitzer-Wien, Swiss Federal Institute of Technology; Ch. Jungo, University of Fribourg, Switzerland. Furthermore, we greatly appreciate the effort of D.R.M. Walton, University of Sussex, England, in polishing our English.

We should also like to thank the publishers of the following journals for permission to reproduce various figures in the book: Journal of the American Chemical Society; Inorganic Chemistry; Inorganic and Nuclear Chemistry Letters; our sincere thanks are also due to the authors of the papers from which these figures were chosen.

This book was written at Monsanto Research S.A., Zürich, Switzerland. We gratefully acknowledge the encouragement by its president, H. H. Zeiss.

Contents

1. Introduction

Catalysis with soluble transition metal complexes has attracted very great interest over the past few decades both because of the novelty of much of the chemistry involved and because of its potential and, as already proven in numerous examples, practical applications. The starting point of this development was probably the discovery by O. Roelen [1] in 1938 of the reaction of olefins with carbon monoxide and hydrogen to form aldehydes (the "oxo process" or hydroformylation), where a soluble cobalt carbonyl complex served as catalyst. Many other homogeneous reactions were subsequently discovered, some of the most important of which are the oxidation of ethylene to acetaldehyde on a palladium complex ("Wacker process") [2], the carbonylation of methanol with a rhodium species [3], the cyclo-oligomerization of conjugated diolefins on nickel [4], and the dimerization [5], oligomerization [6] and polymerization [7] of olefins with soluble Ziegler catalysts.

From a technical point of view, this development was challenged by the availability, at reasonable prices, of olefins after the displacement of coal by natural gas and oil as the most important feedstocks for the chemical industry. In the nineteen forties, coal-based acetylene was still a major raw material in chemical processing (especially in Germany), useful because of its high reactivity. Gradually it has been replaced by ethylene and other olefinic compounds which are less expensive, but where superior catalysts have to compensate for lower reactivity. Another impulse came from a general trend toward milder reaction conditions. The more "classical" processes using heterogeneous metal oxide or metal catalysts are often termed "capital-intensive", meaning that high capital has to be invested per ton of product. This is mainly due to the high pressures and temperatures usually required, with the often disagreeable consequence of reduced selectivity. Moreover, relatively low conversions per pass frequently necessitate the separation of product from highly diluted mixtures, and recycling of the unreacted components.

Homogeneous catalysts, on the other hand, usually work at low pressure and temperature, with high efficiency, and sometimes with an amazing selectivity. Thus, for instance, the manufacture of acetaldehyde from ethylene and oxygen in an aqueous catalyst system by the Wacker process operates at the boiling temperature of water at pressures of only a few atmospheres. This process has displaced the older vapor-phase oxidation of ethyl alcohol in the temperature range 375 to 500 °C. An example of selectivity is the cyclooligomerization of butadiene which can be oriented towards cyclooctadiene or to cyclododecatriene, by only minor variations of the nickel catalyst (20 °C, normal pressure).

Nevertheless, soluble transition metal complexes are by no means expected to replace the heterogeneous metal oxide and metal catalysts in all domains. Their robustness and easy regeneration guarantee them a secure future in catalytic chemistry. However, experiences obtained with soluble catalyst species have also had an impact on heterogeneous catalysis. Most solid catalysts are non-stoichiometric, ill-defined materials; moreover, the fact that the reactions take place at a gas-solid interface implies certain physical complications, such as rate control by mass transfer, influence of pore volume, *etc.* These conditions are not favorable for a mechanistic interpretation of heterogeneous processes. Work with well-defined soluble complexes has contributed greatly to the understanding of the basic principles of transition metal catalysis which may be assumed to be common to heterogeneous and homogeneous processes. The Fischer-Tropsch reaction, for instance, which comprises the formation of hydrocarbons from carbon monoxide and hydrogen on heterogeneous cobalt or iron catalysts, and which was very important during the coal-based feedstocks period, is to-day at least partly understood in terms of individual steps recognized in the course of the homogeneous hydroformylation of olefins [8]. The Fischer-Tropsch, and related reactions may revive when the oil reserves decline. The related synthesis of methanol from carbon monoxide and hydrogen is of even more significant interest, as a step in one of the routes from methane (earth gas component) to the easier transportable methanol, which may become an important export capacity from the Middle East in the nineteen eighties [9]. Evidently the discovery of a soluble system for the synthesis of methanol from carbon monoxide and hydrogen would constitute a great step towards its mechanistic elucidation. On the other hand, the ideas gained in the work with soluble complexes have, in certain cases, even stimulated systematic variations of heterogeneous catalysts [10].

Although the discovery of the first soluble catalysts was, presumably, more or less accidental, modern catalysis research appears to be progressing towards an intelligent synthesis of contributions from many sources and from different disciplines of chemistry. In particular, it has benefited greatly from two major scientific developments of the past decades. On one hand, transition metals attracted growing interest in preparative coordination chemistry as a great number of new compounds became available, while on the other hand important advances in theoretical inorganic chemistry (particularly in ligand field theory) influenced the thinking of catalysis chemists. As a result, more and more attention was devoted to the significance of coordination sphere symmetry of a catalytic transition metal center and of the distribution of electron density therein. It was recognized that the metal centers are in a sense polyfunctional, possessing a multisite capability for forming several electron pair bonds in clearly defined geometric juxtaposition, thus predisposing reactants within the framework of a complex towards specific interaction. The influence of certain atoms or groups of atoms attached to a metal center (the "ligands") on the activity of the latter was observed and traced back to the fact that the ligands may either donate electron density to, or withdraw it from the metal.

Two different approaches have led to the present situation of knowledge and concepts. The first, of a phenomenological nature, was aimed at the effectiveness of catalysts and consists essentially of varying parameters and measuring activities. This procedure, with prominent industrial support, has been very successful in several occasions. Thus, the production of aldehydes, acids, esters, etc., by the oxo process and related reactions was able to make "big money" during the many years until a well-argued suggestion concerning the mechanism was advanced in 1961 [11]. The same criterion holds good for low pressure polymerization of ethylene with heterogeneous titanium/aluminum catalysts discovered by Ziegler in 1952 [12], by which millions of tons of polyethylene were procuced world-wide, prior to 1964 when a theoretical interpretation of the process was offered [13], and before the use of analogous soluble catalysts permitted a mechanistic interpretation [7].

The second path of research, located more within the University sphere, is concerned more with the manner in which a catalyzed reaction proceeds. This, the more basic approach, consists of the determination of the sequence of the elementary reactions within the co-ordination sphere of the metal center in the course of a catalytic cycle, measurement of rate and equilibrium constants for these individual steps, isolation or spectroscopic identification of intermediates, and determination of the valency state and coordination number of the active metal center, *etc.*

Much progress has also been achieved by an imaginative transfer of ideas and experience from one catalytic process to another. An illustration is provided by the Ziegler catalysts. The original discovery referred to a combination of an aluminum alkyl and a transition metal salt (in particular triethylaluminum and titanium tetrachloride) as catalyst for the low pressure polymerization of ethylene. While initially the aluminum component was considered as the catalyst, and the transition metal as a "cocatalyst", careful fundamental work in several laboratories, with related soluble systems, led to the conclusion that the transition metal is the active center, and that the aluminum alkyl can carry out one or more of several tasks: It can alkylate the transition metal center, forming an active metal-carbon bond. If the metal alkyl is unstable, and is homolytically cleaved, the aluminum alkyl provokes indirectly reduction of the transition metal ion, thus bringing the latter eventually into a valency state required for catalysis. Finally, the aluminum alkyl can form a complex with the transition metal center, operating as an activating ligand. This multipurpose action of aluminum alkyls in Ziegler systems once recognized, it was a logical corollary to use these systems not only for polymerization, but also in other cases where the *in situ* formation of low valent and/or alkylated transition metal species was required for catalysis. In this way, Ziegler systems have found ample application in many catalytic reactions, such as the dimerization of olefins and of conjugated diolefins, metathesis, and even hydrogenation of olefins.

From these few introductory remarks it will be evident that our treatment of the phenomenon catalysis by coordination compounds of transition metals will emphasise those processes occurring within the coordination sphere of the metal center more than the technical aspects of the catalytic reactions. In this sense, we considered it useful to initiate the book with a basic consideration of the electronic structure of transition metal ions, and with the theoretical concepts underlying their coordination chemistry which have contributed greatly to the understanding of transition metal catalysis. The most essential concepts stem from group theory, ligand field theory and molecular orbital theory. With regard to the latter two it should, however, always be born in mind that they represent approximations, and that our exact knowledge of bonding in coordination compounds (and of chemical bonding in general) is far from being complete. Nevertheless, these theories have provided us with extremely useful interpretations of certain aspects of the bonding in transition metal complexes, and of properties of the complexes depending thereupon (*e.g.* spin pairing of electrons in d orbitals and magnetism; directed valency and symmetry; electron distribution and activity; *etc.*).

References

[1] O. Roelen, Angew. Chem., *60*, 62 (1948).   [2] J. Smidt, W. Hafner, R. Jira, R. Sieber, J. Sedlmeier, and A. Sabel, Angew. Chem. internat. Edit., *1*, 80 (1962).   [3] J. F. Roth, J. H. Craddock, A. Hershman, and F. E. Paulik, Chem. Technol., *1971*, 600.   [4] G. Wilke, Angew. Chem. internat. Edit., *2*, 105 (1963).   [5] G. Lefebvre and Y. Chauvin, in: R. Ugo, Ed., *Aspects of Homogeneous Catalysis*, Carlo Manfredi, Editore, Milano 1970, vol. *1*, p. 108.   [6] G. Henrici-Olivé and S. Olivé, Advan. Polym. Sci., *15*, 1 (1974).   [7] See, e.g. G. Henrici-Olivé, and S. Olivé, *Polymerisation*, Verlag Chemie, Weinheim 1969.   [8] G. Henrici-Olivé and S. Olivé, Angew. Chem., internat. Edit., *15*, 123 (1976).   [9] L. Löhmer, Chem. Industrie *27*, 396 (1975).   [10] See, e.g. G. Henrici-Olivé and S. Olivé, Angew. Chem. internat. Edit., *12*, 153, 754 (1973).
[11] R. F. Heck and D. S. Breslow, J. Amer. Chem. Soc., *83*, 4023 (1961).   [12] K. Ziegler, E. Holzkamp, H. Breil, and H. Martin, Angew. Chem., *67*, 541 (1955).   [13] P. Cossee, J. Catal., *3*, 80 (1964).

2. Atomic Orbitals

In this section we shall summarize briefly the quantum mechanical treatment of the hydrogen atom. This will remind the reader of the origin of our working knowledge about the geometry and directional properties of atomic orbitals.

2.1. Wave Equation and Wave Functions

In 1926 the chemist's electron was displaced by the quantum mechanicist's electron. The chemist's electron was a small particle, moving in a defined circular or elliptic orbit around the nucleus of an atom (Bohr-Sommerfeld model). Quantum mechanics tells us that the electron is comparable to a standing wave which can be described by a wave equation. The starting point for this development was de Broglie's wave theory of matter (1924), which was based on theoretical as well as metaphysical considerations. At that time, it had become evident that light, apparently well-defined as a wave process since Fresnel, can also be described as a corpuscle. The belief in a general harmony in Nature led de Broglie to the ingenious assumption that a wave-particle dualism would also apply to matter. He formulated the relationship between wave length λ, mass m, and velocity v of a particle:

$$\lambda = \frac{h}{m \cdot v} \tag{2.1}$$

where h is Planck's constant. The formulation of a wave equation for the electron by Schrödinger (1926) was a logical corollary of these ideas. Experimentally, the wave-particle dualism manifests itself through the observation of both diffraction patterns and interferences, with particles (electrons, protons, etc.) as well as with electromagnetic radiation (light, X-rays).

Fortunately some of the properties of the electron, interesting for the chemist, are such that they still may be rationalized by the classical concept, and in view of the wave-particle dualism we are allowed to do so whenever it appears convenient. Other properties, however,

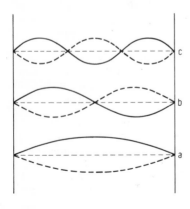

Fig. 2.1. The classical standing wave; fundamental wave (a), and first (b) and second (c) overtones.

can only be comprehended within the mathematical framework of wave mechanics. (The conventional four lobe shape of the d orbitals is an example.)

Certain formal analogies between the equation describing a mechanical vibrating system and the Schrödinger wave equation of an electron may perhaps help to familiarize the reader with the mathematical formalism of quantum chemistry. A standing wave can be generated, for example, by fixing a violin string at both ends and plucking it. Due to the fixed boundary conditions, only certain vibrations are possible in such a system (see Fig. 2.1).

Fig. 2.1(a) shows the fundamental wave, which has no node, *i.e.*, there is no point of zero displacement of the string. The first, second, ... overtones have one, two, ... nodes [Fig. 2.1(b) and (c)]. At one side of the node, a segment of the string moves up, while another segment moves down on the other side. Calling the upward motion notionally plus and the downward motion minus, there is a chance of sign, as the wave passes through a nodal point. The wave equation for a mechanical standing wave as represented in Fig. 2.1 is a second order differential equation:

$$\frac{\mathrm{d}^2 f(x)}{\mathrm{d}x^2} + \frac{4\,\pi^2}{\lambda^2}\, f(x) = 0 \tag{2.2}$$

where the "amplitude function" $f(x)$ describes the amplitude of the wave as a function of the distance, x, along the wave; λ is the wave length. The expression $\mathrm{d}^2/\mathrm{d}x^2$ is an operator, *i.e.* it gives the instruction to differentiate twice with respect to x whatever follows, in this case the function $f(x)$. Such a vibrating system, with fixed boundary conditions, is a classical example of an "eigenvalue problem". Only certain wave lengths (or frequencies) are acceptable as solutions (*cf.* Fig. 2.1); they are called eigenfrequencies. The corresponding amplitude functions are called eigenfunctions.

2.1.1. The Schrödinger Wave Equation

Somewhat analogously, we may imagine the electron as a three dimensional standing wave, confined to the limited space of an atom, with fixed boundary conditions, and so being another, necessarily much more complicated, eigenvalue problem. Schrödinger has formulated the corresponding wave equation. Although this equation must be understood as one of the basic postulates of quantum mechanics which cannot be derived rigorously, we can turn to analogies with Eq. (2.2). (It is these analogies that have given the name "wave mechanics" to the Schrödinger method[*].)

Since the electron in an atom does not move merely along one axis, but extends in all directions (taking the nucleus as the origin), the relevant mathematical description must necessarily be a three dimensional function $\Psi(x, y, z)$, abbreviated Ψ. Correspondingly, the operator $\mathrm{d}^2/\mathrm{d}x^2$ in Eq. (2.1) must be replaced by a three dimensional operator, the Laplacian operator ∇^2 (pronounced "del squared"):

---

[*] Note that Schrödinger's "wave mechanics" is only one of several quantum mechanical descriptions of matter, the other most important ones being Heisenberg's "matrix mechanics" and Dirac's "operator method". Schrödinger's approach appears to be the most adequate to interpret chemical phenomena however.

$$V^2 = \frac{\partial^2}{\partial x^2} + \frac{\partial^2}{\partial y^2} + \frac{\partial^2}{\partial z^2}$$

With the aid of de Broglie's equation (2.1), the wave length λ in Eq. (2.2) may be replaced by the mass and velocity of the electron. Furthermore, the overall energy of a system, E, is equal to the sum of the kinetic energy $mv^2/2$, and the potential energy, V, leading to:

$$\lambda^2 = \frac{h^2}{m^2 v^2} = \frac{h^2}{2\,m(E-V)} \tag{2.3}$$

Taking this into account, we now can see the formal similarity between Eq. (2.2), describing a classical standing wave, and the Schrödinger wave equation, Eq. (2.4), describing the movement of electrons in atomic systems (m_e = mass of the electron):

$$V^2\,\Psi + \frac{8\,\pi^2\,m_e}{h^2}\,(E-V)\,\Psi = 0 \tag{2.4)*)}$$

For subsequent applications it is convenient to rewrite Eq. (2.4):

$$\left(-\frac{h^2}{8\,\pi^2\,m_e}\,V^2 + V\right)\,\Psi = E\,\Psi \tag{2.5}$$

The whole expression within brackets now has the character of an operator; it indicates all operations that have to be carried out on the function Ψ and is called the Hamilton operator (or the Hamiltonian), $\mathcal{H}$. The abbreviated, and very commonly used form of Eq. (2.5) is then:

$$\mathcal{H}\,\Psi = E\,\Psi \tag{2.6}$$

The solution of the Schrödinger equation consists in finding functions Ψ which, after carrying out all operations required by the Hamiltonian, result in some constant multiple of Ψ. The factor E is the total energy of the electron in the particular state described by the wave function Ψ. Whereas Ψ is called an eigenfunction of the Hamilton operator, E is said to be its eigenvalue**).

2.1.2. The Meaning of the Wave Function

The problem "electron in an atom" has now become a mathematical question: to find the wave function Ψ which satisfies the Schrödinger wave equation. But what does Ψ really

---

* This equation, which refers only to stationary states, but not to atomic processes, is time-independent. The latter processes are governed by the "time dependent Schrödinger equation", see reference books at the end of the Chapter.

** To familiarize with the terminology, consider a simple eigenvalue problem; e^{2x} is an eigenfunction of the operator d/dx with an eigenvalue of 2, because $(d/dx)e^{2x} = 2\,e^{2x}$.

represent? The wave function is defined in such a way that $\Psi^2\,\mathrm{d}\tau$ is equal to the probability $\mathrm{d}P$ of finding the electron in the volume element $\mathrm{d}\tau \equiv \mathrm{d}x\,\mathrm{d}y\,\mathrm{d}z$:

$$\mathrm{d}P = \Psi^2\,\mathrm{d}\tau \tag{2.7}$$

Wave functions are sometimes complex functions (*cf.* Section 2.2.2); in these cases, the square of the wave function has to be replaced by the product of the function and its complex conjugate*[)]:

$$\mathrm{d}P = \Psi\Psi^*\,\mathrm{d}\tau \tag{2.8}$$

Eqs. (2.7) and (2.8) indicate that the wave function Ψ must be a "well behaved function", because only then is the probability of finding the electron in the volume element $\mathrm{d}\tau$ *well defined**[)]. Of course the value of the function can be zero in certain regions of space; at a large distance from the origin (nucleus) doubtlessly it will be zero. Furthermore, the overall probability of finding the electron represented by Ψ in space must be unity, hence Ψ must* be such that the integral of Ψ^2 over all space is unity:

$$P = \int \mathrm{d}P = \int \Psi^2\,\mathrm{d}\tau = 1 \tag{2.9}$$

It should be noted that, whenever the shorthand symbol $\mathrm{d}\tau$ for all coordinates is used, the integration is understood to be over all coordinates. In case of imaginary wave functions, again $\Psi\Psi^*$ replaces Ψ^2.

A wave function satisfying Eq. (2.9) is said to be normalized. Frequently a wave function, obtained as a solution of the Schrödinger equation, is not normalized. In this case it has to be multiplied by a number N (normalization constant) such that the integral becomes equal to unity.

The wave functions, which now have been related to the probability of finding the electron at a certain point in space, are called orbitals, reminiscent of the Bohr-Sommerfeld orbits of the classical electron.

2.1.3. The Wave Equation for the Hydrogen Atom

The Schrödinger equation has been solved exactly for the hydrogen atom. In this case the potential energy of the atomic system is given simply by the Coulomb attraction potential between the single electron of charge $-e$ and the nucleus of charge Ze, at the equilibrium distance r:

$$V = -Ze^2/r$$

For reasons, that will become clear in Chapter 3, we have introduced the atomic number Z which, of course, for hydrogen is unity.

---

* Complex functions contain the imaginary quantity $\sqrt{-1} = i$; the complex conjugate has i replaced by $-i$ (e.g. $\Psi = re^{im\varphi}$; $\Psi^* = re^{-im\varphi}$, i.e. $\Psi\Psi^* = r^2$).

** A "well behaved function" is continuous, single valued and quadratically integrable.

The special form of Eq. (2.5) for the electron in the hydrogen atom is then:

$$\left(-\frac{h^2}{8\,\pi^2 m_e}\, \nabla^2 - \frac{Ze^2}{r} \right)\, \Psi = E\Psi \tag{2.10}$$

where it is assumed that the nucleus does not move*).

It turns out that a great number of wave functions, with increasing values of E, satisfy Eq. (2.10) which means that these wave functions represent the single electron of the hydrogen atom in its ground state, and in its many excited states.

In principle, there is no problem in formulating the wave equation also for the higher atoms. Unfortunately, however, the mathematical complication in solving it becomes prohibitive for any atomic system with more than one electron. The main difficulty arises from the Coulombic repulsion between the electrons, which adds further terms to the expression for the potential energy. The way out of the dilemma is to assume that the wave functions (orbitals) representing the electrons of the higher atoms are similar to those of the hydrogen atom ("hydrogen-like orbitals"), and that deviations from the exact H-orbitals can be evaluated by suitable approximations. This general procedure evidently requires a detailed knowledge of the wave functions of the hydrogen atom.

2.2. The Orbitals of the Hydrogen Atom

The solution of the Schrödinger equation for the hydrogen atom, Eq. (2.10), involves the transformation from Cartesian coordinates x, y, z to polar coordinates r, ϑ, φ, and a separation of the variables such that three independent equations are obtained, each containing only one of the three variables. These operations are carried out by standard procedures which are described in most relevant textbooks [2, 3]. We are interested only in a qualitative discussion of the results. (For the relationship between Cartesian and polar coordinates see Fig. 2.2).

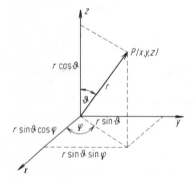

Fig. 2.2. Cartesian and polar coordinates. The Cartesian coordinate system is fixed according to the "right hand rule": thumb ($+x$), index ($+y$) and middle finger ($+z$) of the right hand are extended in three mutually perpendicular directions.

---

* If this assumption is not made, the same equation results except that the electronic mass m_e is replaced by the reduced mass of the system, $\mu = m_e M/(m_e + M)$, M being the mass of the nucleus. Since $M \gg m_e$, $\mu \simeq m_e$.

One can write the wave function in the general form

$$\Psi(r, \vartheta, \varphi) = R(r) \cdot \Theta(\vartheta) \cdot \Phi(\varphi) \tag{2.11}$$

where Ψ, which is a function of all three coordinates, is given as the product of three functions, each containing only one of the variables. Eq. (2.10) can then be replaced by a system of three independent differential equations:

$$\frac{d^2 \Phi(\varphi)}{d\varphi^2} + m^2 \Phi(\varphi) = 0 \tag{2.12}$$

$$\frac{1}{\sin \vartheta} \frac{d}{d\vartheta} \left(\sin \vartheta \frac{d\Theta(\vartheta)}{d\vartheta} \right) - \frac{m^2 \Theta(\vartheta)}{\sin^2 \vartheta} + \beta \Theta(\vartheta) = 0 \tag{2.13}$$

$$\frac{1}{r^2} \frac{d}{dr} \left(r^2 \frac{dR(r)}{dr} \right) - \frac{\beta}{r^2} R(r) + \frac{8 \pi^2 m_e}{h^2} \left(E + \frac{Ze^2}{r} \right) R(r) = 0 \tag{2.14}$$

The symbols m and β are "separation constants", constants which appear in the course of the separation process. Note that only Eq. (2.14) contains the energy E.
The differential equation (2.12) has a relatively simple solution:

$$\Phi = \frac{1}{\sqrt{2\pi}} e^{\pm im\varphi} \tag{2.15}$$

If Φ is a well-behaved function, m must be zero or a positive or negative integer. These particular restrictions indicate that m is the analogue of the magnetic quantum number of the Bohr-Sommerfeld model.
The solutions of Eqs. (2.13) and (2.14) involve more sophisticated mathematics, the results being:

$$\Theta = \sqrt{\frac{(2l+1)(l-|m|)!}{2(l+|m|)!}} \ P_l^{|m|} (\cos \vartheta) \tag{2.16}$$

and

$$R = -\sqrt{\left(\frac{2Z}{na_0} \right)^3 \frac{(n-l-1)!}{2n[(n+l)!]^3}} \left(\frac{2Zr}{na_0} \right)^l L_{n+l}^{2l+1} (2Zr/na_0) \ e^{-Zr/na_0} \tag{2.17}$$

Herein l is related to β in Eq. (2.13) and (2.14) by $\beta = l(l+1)$; a_0 is the atomic unit of length, $a_0 = h^2/4 \pi^2 m_e e^2$, which is the "Bohr radius", the radius of the innermost orbit of the electron in the classical Bohr model of the hydrogen atom. $P_l^{|m|}$ (Legendre function) and $L_{n+l}^{2l+1} (2Zr/na_0)$ (associated Laguerre polynominal) are two complicated mathematical functions which, however, reduce to relatively simple expressions for certain values of the parameters m and l.
In spite of the complicated nature of the solutions, Eqs. (2.15) to (2.17), some important features may be deduced in a simple way. The first refers to the appearance of further

quantum numbers, l and n. Since factorials are restricted to positive integers (and zero)*⁾, it may be seen from Eq. (2.16) that allowed values for l are 0, 1, 2, 3, ..., and that $|m|$ cannot be greater than l. Thus, the restriction on m has now become $m = 0, \pm 1, \pm 2, \pm 3, ... \pm l$. From Eq. (2.17) it follows that the parameter n cannot be zero (a fraction with denominator zero is infinite); the expression $(n - l - 1)!$ requires that n be an integer, and that the maximum value of l be $n - 1$.

The three quantum numbers and their restrictions are summarized in Table 2.1. Their names are taken from the Bohr-Sommerfeld model, where they were introduced *ad hoc* to reconcile a theoretical model with experiment (spectra). Interestingly, they appear now also, and with the same restrictions, in the wave mechanical treatment of the hydrogen atom.

Table 2.1. Quantum numbers.

Symbol	Name	Possible Values
n	Principal quantum number	$1 \leqslant n \leqslant \infty$
l	Azimuthal quantum number	$0 \leqslant l \leqslant n - 1$
m	Magnetic quantum number	$-l \leqslant m \leqslant l$

Introducing allowed sets of the three quantum numbers n, l and m into Eqs. (2.15), (2.16) and (2.17), and forming the product function according to (2.11), one obtains the allowed wave functions of the electron in the hydrogen atom – the atomic orbitals. But before viewing some of the orbitals in more detail, we should consider another interesting point evolving from a qualitative study of the separated differential equations (2.12) to (2.14), and their solutions (2.15) to (2.17).

2.2.1. Angular and Radial Part of the Wave Functions

Two of the functions, $\Theta (\vartheta)$ and $\Phi (\varphi)$, depend only on angles in space and on the quantum numbers l and m. In the corresponding differential equations there is no energy term. [See Eqs. (2.15) and (2.16).] These two functions are often treated together in the form:

$$Y (\vartheta, \varphi) = \Theta (\vartheta) \, \Phi (\varphi)$$

$Y(\vartheta, \varphi)$ is named the angular component of the wave function, or the spherical harmonics**⁾. This part determines the geometry of the orbitals, and therefore it is of primary concern for the treatment of directional bonding.

The spherical harmonics are given in Table 2.2, for $l \leqslant 2$. They are obtained by introducing allowed combinations of l and m [cf. (Table 2.1)] into Eqs. (2.15) and (2.16), and forming the product function $Y (\vartheta, \varphi)$.

---

* In fact this is not quite correct, since factorials of positive non-integers may be interpolated by means of the Gamma-Function, $\Gamma (n + 1) = n!$. In the present context, however, the argument is considered admissible. A more rigorous determination of the restrictions on the quantum numbers must be sought in more specialized literature.

** This name was given to solution of Eqs. (2.12), (2.13) long before the advent of quantum mechanics, in the mathematical treatment of spherical vibrating systems.

Table 2.2. Spherical harmonics $Y(\vartheta, \varphi) = \Theta(\vartheta)\,\Phi(\varphi)$ for $l \leqslant 2$.

l	m	$Y(\vartheta, \varphi)$
0	0	$\sqrt{\dfrac{1}{2\pi}}\,\sqrt{\dfrac{1}{2}}$
1	0	$\sqrt{\dfrac{1}{2\pi}}\,\sqrt{\dfrac{3}{2}}\,\cos\vartheta$
1	± 1	$\mp\sqrt{\dfrac{1}{2\pi}}\,\sqrt{\dfrac{3}{4}}\,\sin\vartheta\,e^{\pm i\varphi}$
2	0	$\sqrt{\dfrac{1}{2\pi}}\,\sqrt{\dfrac{5}{8}}\,(2\cos^2\vartheta - \sin^2\vartheta)$
2	± 1	$\mp\sqrt{\dfrac{1}{2\pi}}\,\sqrt{\dfrac{15}{4}}\,\cos\vartheta\,\sin\vartheta\,e^{\pm i\varphi}$
2	± 2	$\sqrt{\dfrac{1}{2\pi}}\,\sqrt{\dfrac{15}{16}}\,\sin^2\vartheta\,e^{\pm 2i\varphi}$

The other part of the wave function, $R(r)$, gives the electron distribution with respect to the distance from the nucleus and is referred to as the radial part. Table 2.3 shows the radial part of the hydrogen atom wave functions for $n \leqslant 2$. These functions are obtained by introducing allowed combinations of n and l into Eq. (2.17).

The radial part of the wave function is intimately connected to the energy of the electron in a particular orbital; E is an eigenvalue of Eq. (2.14). Introducing allowed function $R(r)$ into Eq. (2.14), this equation can be solved for E. The allowed values of the energy for the hydrogen atom are:

$$E_n = -\frac{2\pi^2 m_e Z^2 e^4}{n^2 h^2} \qquad\qquad n = 1, 2, 3\ldots \qquad\qquad (2.18)$$

Table 2.3. Radial component of the hydrogen atom wave functions, for $n \leqslant 3$.

n	l	$R(r)$
1	0	$2\,Z^{3/2}\,e^{-Zr}$
2	0	$\dfrac{1}{\sqrt{2}}\,Z^{3/2}\left(1 - \dfrac{1}{2}Zr\right)e^{-Zr/2}$
2	1	$\dfrac{1}{2\sqrt{6}}\,Z^{5/2}\,r\,e^{-Zr/2}$
3	0	$\dfrac{2}{3\sqrt{3}}\,Z^{3/2}\left(1 - \dfrac{2}{3}Zr + \dfrac{2}{27}Z^2 r^2\right)e^{-Zr/3}$
3	1	$\dfrac{8}{27\sqrt{6}}\,Z^{3/2}\left(Zr - \dfrac{1}{6}Z^2 r^2\right)e^{-Zr/3}$
3	2	$\dfrac{4}{81\sqrt{30}}\,Z^{7/2}\,r^2\,e^{-Zr/3}$

It will be noted that the energy depends only on the principal quantum number n. Orbitals having the same energy are said to be degenerate. Thus, all orbitals Ψ, having the same quantum number n, whatever their quantum numbers l and m might be, are degenerate. As will be seen later, this is only true for the hydrogen atom; for higher atoms E depends also on l, but not on m. It will also be noted that the energy is negative. There exist solutions for Eq. (2.14) resulting in positive values of E, but the corresponding wave functions have no physical sense, because they are finite everywhere in space. An allowed solution must obviously vanish at infinite distance from the nucleus, and this is only true for those radial wave functions which are connected with negative E. Consequently the energy of an electron at infinity with respect to the nucleus is taken as zero.

The radial part of the wave function has to be considered explicitly, whenever the energy levels of ground state and excited states are involved (particularly in the interpretation of spectra). However, it is quite fortunate that the mathematically less complicated geometric component may be treated separately, and that much information concerning the directional properties of the orbitals can be estimated from this part alone.

2.2.2. The "Real Orbitals" of the Hydrogen Atom

A wave function $\Psi(r, \vartheta, \varphi)$ is the product of radial part and spherical harmonics [cf. Eq. (2.11)]. From Table 2.2 it may be seen that most spherical harmonics (in fact all with $m \neq 0$) are complex functions, due to the $e^{\pm im\varphi}$ term in Eq. (2.15). Calculations are usually performed with these complex functions. For the purpose of vizualizing the directional properties, and of drawing illustrative pictures of the distribution of the electron in space, however, it is more convenient to express the angular part of the wave function in an alternative form which does not contain the imaginary quantity i. The new functions are obtained on the basis of a theorem of eigenvalue problems: if two or more linearly independent eigenfunctions of an operator $\mathscr{H}$ have the same eigenvalue (*i.e.* they are degenerate), then any linear combination of these functions is also an eigenfunction, with the same eigenvalue. The combinations (sums and differences) can be chosen such as to eliminate i. The resulting wave functions are generally referred to as the "real orbitals". It is seen easily that the most convenient combinations are:

$$\Psi_{n,l,m} \pm \Psi_{n,l,-m} = R_{n,l}\left(Y_{l,m} \pm Y_{l,-m}\right) \tag{2.19}$$

The procedure for calculating the real orbitals is exemplified for the sum of $Y_{2,\pm 2}$ (*i.e.* the spherical harmonics with $l = 2$, $m = 2$ and with $l = 2$, $m - 2$, *cf.* Table 2.2). Taking into account Euler's equation:

$$e^{\pm iy} = \cos y \pm i \sin y$$

one obtains:

$$Y_{2,2} = \sqrt{\frac{1}{2\pi}} \sqrt{\frac{15}{16}} \sin^2\vartheta \,(\cos 2\,\varphi + i \sin 2\,\varphi)$$

$$Y_{2,-2} = \sqrt{\frac{1}{2\pi}} \sqrt{\frac{15}{16}} \sin^2 \vartheta \, (\cos 2\varphi - i \sin 2\varphi)$$

(2.20)

$$Y_{2,2} + Y_{2,-2} = 2\sqrt{\frac{1}{2\pi}} \sqrt{\frac{15}{16}} \sin^2 \vartheta \cos 2\varphi = 2\sqrt{\frac{1}{2\pi}} \sqrt{\frac{15}{16}} \sin^2 \vartheta \, (\cos^2 \varphi - \sin^2 \varphi)$$

Remembering the relationships between Cartesian and polar coordinates (see Fig. 2.2):

$$x = r \sin \vartheta \cos \varphi, \quad y = r \sin \vartheta \sin \varphi, \quad z = r \cos \vartheta$$

(2.21)

the new angular function may also be expressed in terms of x, y, and z:

$$Y_{2,2} + Y_{2,-2} = 2\sqrt{\frac{1}{2\pi}} \sqrt{\frac{15}{16}} \frac{x^2 - y^2}{r^2}$$

Normalization*) adds another factor $1/\sqrt{2}$. The resulting real orbital is then:

$$\Psi = R_{n,2} \sqrt{\frac{1}{\pi}} \sqrt{\frac{15}{16}} \frac{x^2 - y^2}{r^2}$$

The difference $Y_{2,2} - Y_{2,-2}$ gives:

$$Y_{2,2} - Y_{2,-2} = 2i \sqrt{\frac{1}{2\pi}} \sqrt{\frac{15}{16}} \sin^2 \vartheta \sin 2\varphi = i \sqrt{\frac{8}{\pi}} \sqrt{\frac{15}{16}} \frac{xy}{r^2} \quad (2.22)$$

In this case, i is eliminated by the normalization factor $1/(i\sqrt{2})$; the resulting orbital is:

$$\Psi = R_{n,2} \sqrt{\frac{1}{\pi}} \sqrt{\frac{15}{4}} \frac{xy}{r^2}$$

Application of the same procedure to the other functions in Table 2.2 leads to the complete set of real orbitals for the hydrogen atom, given in Table 2.4 (through $n = 3$). The procedure adopted for the construction of the linear combinations [cf. Eq. (2.19)] ensures that their number is equal to the number of initial wave functions. Evidently the number of independent linear combinations can never exceed the number allowed by the quantum number restrictions. In principle, however, other sets of linear combinations are possible, and have in fact been calculated [1]. However, they are not generally used, because they are mathematically more complicated.

The possibility of describing a given set of orbitals in different ways may still appear rather confusing. In a very approximate way, one may vizualize the situation in the following manner: the unique and definite "electron cloud" of an atom is formally divided into different regions by mathematical procedures, so that different calculations may give different pat-

---

* The process of normalization will be treated in more detail in Section 6.1.4.

Table 2.4. The real orbitals of the hydrogen atom, for $1 \leqslant n \leqslant 3^{*)}$.

Quantum number n	l	Radial function $R(r)$	Angular function $Y(\vartheta, \varphi)$	Label
1	0	$2\,Z^{3/2}\,e^{-Zr}$	$\dfrac{1}{2\sqrt{\pi}}$	$1s$
2	0	$\dfrac{1}{\sqrt{2}}\,Z^{3/2}\left(1 - \dfrac{1}{2}\,Zr\right)e^{-Zr/2}$	$\dfrac{1}{2\sqrt{\pi}}$	$2s$
	1	$\dfrac{1}{2\sqrt{6}}\,Z^{5/2}\,r\,e^{-Zr/2}$	$\dfrac{\sqrt{3}}{2\sqrt{\pi}}\,\dfrac{x}{r}$	$2p_x$
			$\dfrac{\sqrt{3}}{2\sqrt{\pi}}\,\dfrac{z}{r}$	$2p_z$
			$\dfrac{\sqrt{3}}{2\sqrt{\pi}}\,\dfrac{y}{r}$	$2p_y$
3	0	$\dfrac{2}{3\sqrt{3}}\,Z^{3/2}\left(1 - \dfrac{2}{3}\,Zr + \dfrac{2}{27}\,Z^2r^2\right)e^{-Zr/3}$	$\dfrac{1}{2\sqrt{\pi}}$	$3s$
	1	$\dfrac{8}{27\sqrt{6}}\,Z^{3/2}\left(Zr - \dfrac{1}{6}\,Z^2r^2\right)e^{-Zr/3}$	$\dfrac{\sqrt{3}}{2\sqrt{\pi}}\,\dfrac{x}{r}$	$3p_x$
			$\dfrac{\sqrt{3}}{2\sqrt{\pi}}\,\dfrac{z}{r}$	$3p_z$
			$\dfrac{\sqrt{3}}{2\sqrt{\pi}}\,\dfrac{y}{r}$	$3p_y$
	2	$\dfrac{4}{81\sqrt{30}}\,Z^{7/2}\,r^2\,e^{-Zr/3}$	$\dfrac{\sqrt{15}}{4\sqrt{\pi}}\,\dfrac{x^2 - y^2}{r^2}$	$3d_{x^2-y^2}$
			$\dfrac{\sqrt{15}}{2\sqrt{\pi}}\,\dfrac{xz}{r^2}$	$3d_{xz}$
			$\dfrac{\sqrt{5}}{4\sqrt{\pi}}\,\dfrac{3z^2 - r^2}{r^2}$	$3d_{z^2}$
			$\dfrac{\sqrt{15}}{2\sqrt{\pi}}\,\dfrac{yz}{r^2}$	$3d_{yz}$
			$\dfrac{\sqrt{15}}{2\sqrt{\pi}}\,\dfrac{xy}{r^2}$	$3d_{xy}$

* r is in atomic units of length (*i.e.*, in units of a_0); R and Y are separately normalized.

terns. For the treatment of particular properties one or other description might be more advantageous.

The quantum number m does not appear in Table 2.4, since it is not correct to assign m values to the real orbitals, because they are linear combinations of functions belonging to different m.

The familiar labeling of the real orbits is given in the last column of Table 2.4. Each label consists of a number, a letter and an index. The number represents the principal quantum number n. The letter is related to the azimuthal quantum number in the following way:

$$l = \quad 0 \quad 1 \quad 2 \quad 3 \quad 4 \quad 5 \quad \ldots$$
$$ \quad s \quad p \quad d \quad f \quad g \quad h \quad \ldots$$

The first four letters are taken from spectroscopical notation (s = sharp, p = principal, d = diffuse and f = fundamental series in the emission spectra of alkali metal atoms); following f, the letters are in alphabetical order. The meaning of the index in the orbital labelling should by now be evident.

Orbitals belonging to the same principal quantum number n are named a shell, and those with equal n and l, a subshell. Each shell has one s orbital. Because of the allowed combinations of quantum numbers (Table 2.1) each shell with $n \geq 2$ has a subshell of three p orbitals, and for $n \geq 3$, each shell has also a subshell of five d orbitals.

Note that the angular part of the wave function is the same for all s orbitals, regardless of the principal quantum number n of the shell to which they belong. The same is true for all p_x, all p_y etc. The reason is that $Y(\vartheta, \varphi)$, in contrast to $R(r)$, does not depend on n [cf. Eqs. (2.15)–(2.17)]. Thus, for instance, a $2p_x$ and a $3p_x$ orbital have the same "angular distribution", but they differ through an overall geater distance of the electron from the nucleus in the case of $3p_x$, because of the radial part of the wave function (see Table 2.4).

2.2.3. Graphical Description of the Orbitals

We shall first discuss the graphical description of the angular part of the wave functions. For several reasons the spherical harmonics are treated separately. One reason is the difficulty encountered in representing a complicated function of three variables adequately on two-dimensional paper. Another, and more important reason arises because the angular part of the wave function is not restricted to the hydrogen atom, but is valid for all atoms. (Remember that the energy term, which complicates the problem with higher atoms, occurs only in Eq. (2.13), but not in (2.11) and (2.12).) Finally, the angular dependence is the only one important for a qualitative discussion of chemical bonding.

Like mechanical standing waves (Fig. 2.1), the spherical harmonics may have regions of positive and negative sign; the nodal points are replaced by nodal planes. Take the $d_{x^2-y^2}$ orbital as an example. Its Y function is given in Table 2.4 in Cartesian coordinates, but its angular dependency is best deduced from Eq. (2.20), making use of polar coordinates. The part $\sin^2 \vartheta$ is zero or positive for all ϑ; it is zero along the $\pm z$ axis, and has its maximum [$\sin^2 (\pi/2) = 1$] in the xy plane (refer to Fig. 2.2). The part $\cos 2\varphi$, on the other hand, changes sign four times, as φ goes from 0 to 2π. The four maxima of $|\cos 2\varphi|$ are evidently in the direction of the $\pm x$ axis and $\pm y$ axis. In the two planes bisecting these axes, $\cos 2\,\varphi$ is exactly zero; these are nodal planes.

Not the function itself, but the square of it is generally chosen for graphical representation. (As mentioned before, the square of the value of the function at a certain point gives the probability of finding the electron at that point.) In order to construct the graphical picture of the angular part of the $d_{x^2-y^2}$ orbital, calculate $(\sin^2 \vartheta \cos 2\,\varphi)^2$ for any combination of ϑ and φ in space, and draw the vector fixed by the couple ϑ, φ under consideration, with a length equal to $(\sin^2 \vartheta \cos 2\,\varphi)^2$. The ends of all possible vectors in space will describe a surface, enclosing the familiar four-lobe volume of a d orbital. From the discussion of the angular dependence of $d_{x^2-y^2}$ above, it should be easy to recognize this orbital in Fig. 2.3 (left-hand side sketch). Reasoning along the same lines leads from Eq. (2.22) to the right-hand side drawing in Fig. 2.3.

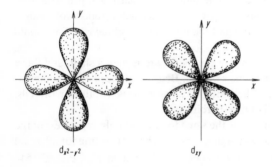

Fig. 2.3. Graphical description of the angular part of the wave functions, for $d_{x^2-y^2}$ and d_{xy}.

Corresponding drawings can be made for all other orbitals. It is, however, common practice to represent the orbitals in short-hand notation, by a cross-section through the boundary surface along the relevant axes, *cf*. Fig. 2.4.

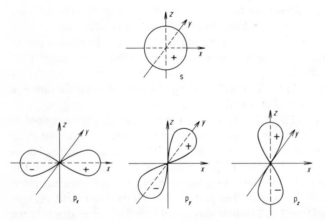

Fig. 2.4. Dual purpose sketches of atomic orbitals.

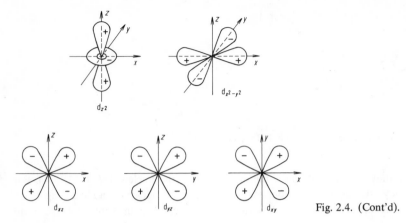

d_{z^2} $d_{z^2-y^2}$

d_{xz} d_{yz} d_{xy} Fig. 2.4. (Cont'd).

It should be recognized that these sketches are dual-purpose drawings. On the one hand they give a picture of the distribution of the electron density in space, on the other hand they give also the changes in sign of the wave function, the square of which is represented. The necessity in taking care of the changes of sign will become evident later.

Note that only s orbitals have electron density at the nucleus, all others having a node there. The geometry of the orbitals is determined by the quantum number l (s, p, d orbitals). The sets of orbitals belonging to the same l (three p orbitals, five d orbitals) differ only by orientation in space.

One of the d orbitals, d_{z^2}, apparently departs from this rule. It has a different shape, and it is not obvious from an inspection of the sketches in Fig. 2.4 that this orbital is equivalent to the other four d orbitals. To realize the equivalence, d_{z^2} may be considered as a linear combination of two orbitals, $d_{z^2-x^2}$ and $d_{z^2-y^2}$. Introducing $\sin^2 \varphi + \cos^2 \varphi = 1$ into the angular wave function $Y_{2,0}$ of Table 2.2, and applying Eqs. (2.21), one obtains (discarding numerical constants):

$$d_{z^2} \sim 2 \cos^2 \vartheta - \sin^2 \vartheta \sin^2 \varphi - \sin^2 \vartheta \cos^2 \varphi = \frac{z^2 - x^2}{r^2} + \frac{z^2 - y^2}{r^2}$$

The two orbitals are illustrated in Fig. 2.5. Both have positive lobes in the direction of the z axis; one has negative lobes along x, the other along y. Their combination is strongly positive in the z direction, and, to a smaller extent, negative in the xy plane. It must be stressed, however, that the two orbitals, $d_{z^2-x^2}$ and $d_{z^2-y^2}$, do not have separate existence because of the quantum number restrictions (there can be only one orbital with $l = 2$ and $m = 0$).

A set of five equally shaped d orbitals has also been calculated [1]. They lie along five equivalent directions in space which may be described as the body diagonals of a pentagonal antiprism. However, their disadvantages (mathematical bulkiness, difficulty of representation) are greater evidently than the satisfaction of having all five d orbitals equally shaped, and they are not generally used.

It should be born in mind that each wave function Ψ is the product of spherical harmonics and a radial wave function [Eq. (2.11)]. Electron density can be found only in those parts of space, where both functions are different from zero. The $R(r)$ are spherically symmetric

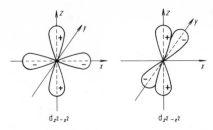

Fig. 2.5. Sketches of the (hypothetical) orbitals $d_{z^2-x^2}$ and $d_{z^2-y^2}$.

functions, since they depend only on r, the distance from the nucleus (*cf*. Table 2.4). For 1s, 2p and 3d they are simple e-functions, indicating that $R(r)$ decays with increasing r, but never becomes negative. The other $R(r)$ functions represented in Table 2.4 are more complicated; they have one or more changes of sign as r increases. This means that there are concentric spherical surfaces around the nucleus where the radial part of the wave function is zero (nodal surfaces). For 2s and 3p there is one nodal surface; for 3s there are two.

Thus, if we wish to adhere to the intuitive and practical sketches of the spherical harmonics (Fig. 2.4), we should bear in mind that the radial part of the wave function is responsible for a certain fine structure of electron density within the boundary surfaces of the spherical harmonics. This is most easily seen for the s orbitals, for which the angular function is $1/(2\sqrt{\pi})$ (*cf*. Table 2.4). Regardless of what values the angles ϑ and φ may take, the angular part of the wave function is a constant, and therefore spherically symmetric. Hence the circular drawing of an s orbital in Fig. 2.4 represents equally well a 1s, 2s or 3s orbital. However taking into account also the radial part of the wave function we realize that in a 2s orbital there are two, and in a 3s orbital there are three concentric zones of electron density, separated by nodal surfaces. A comparable fine structure concerning electron density is found also in the two lobes of the higher p orbitals (for $n \geq 3$) and in the four lobes of the higher d orbitals (for $n \geq 4$).

2.2.4. Orthogonality of the Orbitals

We can now introduce another important concept: all hydrogen orbitals are "orthogonal"[*], and that means that the integral over all space of the product of any two of the atomic orbitals, Ψ_m and Ψ_n, is zero:

$$\int \Psi_m \Psi_n d\tau = 0$$

At first sight it might appear difficult to realize how, for instance, $3d_{z^2}$ and $4d_{z^2}$ can be orthogonal, if one has in mind just the usual dual purpose sketches of the orbitals (Fig. 2.4). But note that the integral defining orthogonality contains the wave functions (not the squares thereof), and that the radial components of the wave functions cause positive and negative zones in each of the "lobes" of the orbitals.

If two wave functions Ψ_m and Ψ_n are orthogonal and normalized ("orthonormal"), we can write

---

[*] More general: eigenfunctions of any quantum mechanical operator are orthogonal.

$$\int \Psi_m \Psi_n \mathrm{d}\tau = \delta_{mn}$$

where δ_{mn} is the so-called "Kronecker delta" which is by definition equal to zero if $m \neq n$ and equal to unity if $m = n$.

2.3. Angular momentum

In order to introduce the concept of angular momenta related to the electron, we have recourse to the classical picture of the electron moving in an orbit (Fig. 2.6.). The angular momentum related to such a motion is classically defined as the vector product of the linear momentum $\vec{p}$ and the radius vector $\vec{r}$; it may be represented therefore as a vector $\vec{l}$ perpendicular to the plane of the orbit:

$$\vec{l} = \vec{p} \times \vec{r} = m\vec{r} \times \vec{v}$$

(m = mass, $\vec{v}$ = linear velocity of the particle).

The angular momentum has thus the dimension of an action (kg m^2 s^{-1} = J s). Planck's great discovery was that physical magnitudes having the dimension of an action cannot occur in nature in infinitely small amounts, but are always quantized, in multiples of the "quatum of action", $h = 6.625 \ 10^{-34}$ J s.

Fig. 2.6. The orbital angular momentum vector $\vec{l}$ of the "classical" electron.

The angular momentum related to the orbital motion of an electron is called orbital angular momentum. The quantum mechanical treatment of the problem, which is beyond the scope of this book, tells us that the magnitude of the orbital angular momentum vector, $|\vec{l}|$, is related to the azimuthal quantum number l by the equation[*]:

$$|\vec{l}| = \sqrt{l(l+1)} \ \frac{h}{2\pi} = \sqrt{l(l+1)} \ \hbar \qquad (2.23)$$

The quantum number l is therefore also called the angular momentum quantum number. Planck's constant divided by 2π is the atomic unit of angular momentum; it is often abbreviated as $\hbar$.

---

[*] To introduce the vectorial nature of orbital angular momentum, we used the classical picture. Its ultimate inadequacy becomes evident if we consider that an s orbital, which is most similar to the circular orbit in Fig. 2.6, actually has orbital angular momentum zero according to Eq. (2.23), with $l = 0$.

Furthermore, quantum mechanics states that only certain directions in space are allowed for the orbital angular momentum vector. These are such that the projection of the vector on a reference axis (*e.g.* the z axis) is restricted to the values $m\hbar$, where:

$$m = l, l-1, l-2, \ldots -l \tag{2.24}$$

We recognize the quantum number m (Table 2.1). It is interesting that the angular momentum itself, $\sqrt{l(l+1)}\ \hbar$, is larger than its maximum component in any direction, $l\hbar$. This is accordant with Heisenberg's uncertainty principle: if the component, say in the z direction, happened to be exactly equal to the angular momentum itself, we would know that the components in the x and y directions would be precisely zero. We would know then all three components precisely, in contradiction to the uncertainty principle which states that the product of the uncertainty of the momentum, and the uncertainty of the position of an electron must be at least of the same order of magnitude as Planck's constant.

The classical picture (Fig. 2.6) may also help us to understand that there is a magnetic dipole moment associated with the orbital motion of an electron (remember an electric current travelling in a loop of wire). According to quantum mechanics the magnetic moment is given by the equation:

$$\mu_l = |\vec{l}|\,\beta = \sqrt{l(l+1)}\,\beta \tag{2.25}$$

where $\beta = eh/4\pi\,m_e c$ is the atomic unit of magnetic moment, known as the Bohr magneton (c = velocity of light). Instead of β, the abbreviation BM is often used for the Bohr magneton.

So far we have neglected the spin of the electron. In the classical picture it would correspond to a rotation of the electron around its own axis. In wave mechanics it appears automatically as soon as the Schrödinger equation is made relativistically satisfactory, as Dirac showed 1926. In our context it suffices to realize that spin attributes a further angular momentum $\vec{s}$ (spin angular momentum) and a magnetic dipole moment μ_s to an electron in a particular orbital.

The magnitude of the spin angular momentum is:

$$|\vec{s}| = \sqrt{s(s+1)}\,\hbar \tag{2.26}$$

where $s = 1/2$ is the spin quantum number. Only two directions of the spin angular momentum in space are allowed; they are such that the projection of the vector on a reference axis is restricted to the values $m_s\hbar$, where:

$$m_s = \pm 1/2$$

The magnetic dipole moment of a single electron, due to its spin, is:

$$\mu_s = 2\,|\vec{s}|\,\beta = 2\,\sqrt{s(s+1)}\,\beta \tag{2.27}$$

where β is again the Bohr magneton. Whereas the magnetic moment due to orbital motion, Eq. (2.25), measured in Bohr magnetons, is equal to the magnitude of the orbital angular momentum, the magnetic moment due to spin is twice the magnitude of the relevant angular momentum. This fact is often referred to as the magneto-mechanical anomaly of the spinning electron.

To describe the spin in the wave function, a "spin-coordinate" σ has to be added to the three spatial coordinates r, ϑ, φ. The total wave function is then composed of radial function, spherical harmonics and spin function, the latter is given the symbol $\chi(\sigma)$.

Thus:

$$\Psi_{n,l,m_l,m} = \Psi(r, \vartheta, \varphi, \sigma) = R(r)\, Y(\vartheta, \varphi)\, \chi(\sigma) \qquad (2.28)$$

(Note that on the left hand side of Eq. (2.28) an index l has been added to the magnetic quantum number, originally introduced as m. This is usually done to distinguish it clearly from the quantum number m_s.)

The Schrödinger equation as written down in Eq. (2.10) neglects spin interactions. The Hamiltonian used there is said not to operate on spin. Therefore the wave functions $\Psi_{n,l,m_l,1/2}$ and $\Psi_{n,l,m_l,-1/2}$ are solutions giving the same energy (they are degenerate). This means that the orientation of spin in space is irrelevant to the energy. If, however, a particular interaction with the spin is to be considered (*e.g.* the interaction of an external magnetic field with the magnetic moment of spin, in electron spin resonance), a special "spin-Hamiltonian" has to be applied.

Reference

[1] R. E. Powell, J. Chem. Educ. *45*, 45 (1968).

Suggested Additional Reading

[2] M. C. Day and J. Selbin, *Theoretical Inorganic Chemistry*, Reinhold Publ. Corp., New York 1962.
[3] W. Kauzmann, *Quantum Chemistry*, Academic Press Inc., New York 1957. [4] A. Messiah, *Quantum Mechanics*, North Holland Publ. Comp., 1970, Vol. I. [5] P. W. Atkins, *Molecular Quantum Mechanics*, Clarendon Press, Oxford, 1970.

3. Transition Metal Ions

3.1. General Characteristics

The transition metals are generally defined as those which, as elements, have partly filled d or f shells. This definition comprises, for instance, the elements from scandium to nickel in the first transition series. A chemically more meaningful definition includes also elements which have partly filled d or f shells in any of their commonly occuring oxidation states. In this sense, copper is also a transition element, since the electronic configuration of Cu^{2+} is $3 d^9$. Scandium, on the other hand, is not a typical representative of its class since its chemistry is almost entirely restricted to that of the trivalent ion which has a closed shell electronic structure ($3 s^2 3 p^6$).

There is an important difference between the three "main transition series" which have partly filled d shells, and the rare earth's series with partly filled f shells. The d orbitals project well out to the periphery of the atoms or ions, so that electrons occupying them can interact strongly with the surroundings. Electrons in the 4 f orbitals, on the other hand, are much more shielded from the chemical environment by the outlying 5 s and 5 p shells.

The interaction of partly filled shells of transition metal ions with electrons of surrounding ions and molecules plays a predominant part in the type of catalysis which is the subject matter of this book. It is evident therefore that suitable catalysts are to be looked for in the d series much more than among the lanthanides. For this reason we shall concentrate on the transition metals with partly filled d orbitals.

3.2. The Electronic Structure of Free Ions

In this section we are concerned with the wave functions and energy levels of free transition metal ions. As already mentioned earlier, the Schrödinger equation cannot be solved exactly for many-electron systems, primarily because of mathematical complications introduced by the Coulombic interaction between the various electrons. But it has been found satisfactory to assume that the higher atoms are built up by feeding the proper number of electrons into a set of "hydrogen-like" orbitals, taking care of the Aufbau principle and of the Pauli exclusion principle. The Aufbau principle requires that the orbitals are filled in the order of increasing energy; the Pauli principle states that no two electrons in a many-electron system can have all four quantum numbers equal. As a consequence, two electrons can be "housed" into each space orbital Ψ_{n,l,m_l}, with the condition that their quantum numbers m_s are different, namely $m_s = +^1/_2$ and $m_s = -^1/_2$. In this situation it is said that the two electrons have their spins paired.

3.2.1. Electronic Configuration of Many-Electron Ions

The electronic configuration of an atom or ion is defined as the assignment of the available electrons to the given set of orbitals. The filling order for the hydrogen-like orbitals depends evidently on their energies. We anticipate that one important consequence of the inter-electronic repulsion is the removal of the orbital degeneracy observed in the solution of the hydrogen wave equation. Whereas all orbitals having the same principal quantum number are degenerate in the hydrogen atom [*cf.* Eq. (2.18)], the energy increases in the series:

$$1s < 2s < 2p < 3s < 3p < 4s < 3d < 4p \ldots$$

for higher atoms. (Note that $4s < 3d$.) When writing electronic configurations it is usual to omit the completely filled shells, because most properties (bonding, reactivity, spectra, magnetism, *etc.*) depend only on the electrons in the outermost shell (valence electrons). Thus, the electronic configuration of the first few atoms in the Periodic System is written:

$$H\,(1s^1);\ He\,(1s^2);\ Li\,(2s^1);\ Be\,(2s^2);\ B\,(2s^2 2p^1);\ C\,(2s^2 2p^2).$$

The reversed order in energy, $3p < 4s < 3d$, is the reason for the occurrence of the first series of transition elements after potassium and calcium. However, the ordering of the orbitals is not fixed once and for all, but depends on the number of electrons present. Thus, although the order of filling is:

$$K\,(4s^1);\ Ca\,(4s^2);\ Sc\,(4s^2 3d^1);\ Ti\,(4s^2 3d^2)$$

the $4s$ orbital is not always more stable than $3d$. As a transition metal looses electrons from its valence shell (*i.e.* it becomes an ion) the $3d$ energy level is lowered; in fact it can become lower than the $4s$ level. Thus a Ti^{3+} ion has its remaining valence electron predominantly in a $3d$ and not in the $4s$ orbital, as deduced from spectroscopic data.

3.2.2. Approximate Wave Functions

The angular part of the "hydrogen-like orbitals" is the same as for the hydrogen atom itself [Eqs. (2.15) and (2.16)]. However, the radial part of the wave functions for the higher atoms is different from those of the hydrogen atom for two major reasons: 1. The nuclear charge increases with increasing atomic number thus increasing the Coulombic attraction between the nucleus and the electrons, and 2. the electrons suffer Coulombic mutual repulsion forces. To a very crude approximation one may neglect the latter effect, and describe each electron by a hydrogen wave function in which only the greater nuclear charge Z is taken into account (see Table 2.3). Evidently, the Coulombic attraction between nucleus and electrons is overestimated in these so-called zero-order wave functions, because any particular electron is effectively shielded or "screened" from the nucleus by the other electrons. Therefore, the true orbitals are much more spread out, away from the nucleus, than would be expected from the zero-order wave functions. The shielding is not exactly the same for an electron in an ns, np, nd, ... orbital. This is in fact the reason why these orbitals have somewhat different energies (*cf.* preceding section).

A better approximation to the true orbitals has been achieved by replacing the nuclear charge Z with an effective nuclear charge $(Z - \sigma)$, where σ is known as the screening constant. Slater has given a number of simple rules for calculating screening constants, and for writing down approximate analytical functions (Slater orbitals) for the higher atoms. The Slater orbitals (including spherical harmonics and spin) are of the general form:

$$\Psi = N r^{n^*-1} \exp\left[-(Z-\sigma)r/n^*\right] Y(\vartheta, \varphi)\, \chi(s)$$

where n^* is equal to the main quantum number n for $n < 3$, and somewhat smaller than n for $n > 3$; N is a normalization constant. Slater orbitals are widely used in making quantum chemical calculations. The "Slater rules" for finding numerical values for n^* and σ are found in most relevant textbooks (e.g. [5, 6]).Without computing interelectronic effects explicitly, the "rules" have been selected such as to give the best possible fit to observed energies (spectra).

A somewaht different approach has been introduced by Hartree. The potential energy term in the differential equation of the hydrogen radial wave function [Eq. (2.14)] is replaced by a screening potential V_{scr}, which duly takes into account the screening effect of the other electrons. This potential could be calculated exactly, if the interelectronic repulsion could be handled correctly, i.e. if the accurate wave functions of all other electrons were known. As a first step, the screening potential for each orbital is calculated approximately from zero-order wave functions for the other electrons (or a reasonable estimate is made concerning V_{scr}, which is possible with some experience). With these approximate potentials, a set of wave functions is calculated, which is somewhat better than the zero-order orbitals. This set is used then to calculate better potentials, etc. The iterative procedure is repeated until the orbitals of the nth set differ negligibly from those of the $(n-1)^{th}$ set. The resulting orbitals are called self-consistent field (SCF) orbitals. They do not come out in form of analytical functions, but as numerical tables.

3.3. The Terms of the Free Ion

The terms are energy levels, arising from an electronic configuration with a partly filled outer shell, because of the effects of interelectronic repulsion. Their knowledge is important for the interpretation of electronic spectra, since these originate from transitions between terms.

We shall consider here particularly d shells. If a d shell is only partly filled, there are always several possibilities to distribute the available electrons among the five d orbitals. Each arrangement is called a microstate. The rule of combination tells us that there are $\dfrac{10!}{n!\,(10-n)!}$ microstates for n d electrons, if due care is taken of the Pauli exclusion principle. Take as an

Table 3.1. Some of the microstates of a d^2 configuration.

$m_l =$	2	1	0	−1	−2	M_L	M_S
	↑	↑				3	1
	↑		↑			2	1
	↑			↑		1	1
	↑				↑	0	1
	↑	↓				3	0
	↓	↑				3	0
	↑	↓				4	0
					↑↓	−4	0

example a d^2 ion (for instance V^{3+}). There is a total of 45 microstates; only a choice of them is given in Table 3.1. The five d orbitals are characterized by their quantum number m_l; for $m_s = +^1/_2$ we use the symbol ↑ ("spin up"), for $m_s = -^1/_2$ the symbol ↓ ("spin down"). The meaning of the symbols M_S and M_L will become clear later on.

The interelectronic repulsion is not the same for all of these distributions. As a consequence there are small differences in energy, small compared with the differences between shells and subshells. Certain microstates have, however, exactly or very nearly the same energy, together they form a term. In the ground state the atom will choose the energetically most favorable term (minimum interelectronic repulsion). The other terms correspond to excited states. We shall now first outline a suitable characterization and labeling of microstates and terms, and then proceed to discuss the terms arising from a certain d^n configuration.

3.3.1. Total Angular Momenta

Angular momenta are used to classify microstates and terms. In a many-electron system, the angular momenta of the single electrons interact with each other via the associated magnetic properties. Just as a little bar magnet is forced into a certain direction by an outer magnetic field, each electron is compelled to align itself in the field of the other electrons. The orbital and spin angular momenta of the individual electrons are said to be coupled together. There are two limiting cases of how the momenta may be assumed to couple. In the first case the interaction among all individual orbital angular momenta and among all individual spin angular momenta is strong compared with the interaction between spin and orbital angular momentum of each electron. The orbital angular momenta $\vec{l}$ of all electrons then couple to give a total orbital angular momentum $\vec{L}$; the spin angular momenta $\vec{s}$ also couple to give a total spin angular momentum $\vec{S}$. Quantum mechanics require $\vec{L}$ and $\vec{S}$ to be also quantized. This kind of coupling is chiefly applicable to the lighter elements (up to atomic number *ca.* 30; the evidence comes from spectroscopy). It is called *LS* coupling or Russell-Saunders coupling. The latter name is a tribute to the pioneering work of A. N. Russell and F. A. Saunders in early characterization of atomic spectra, in the mid-nineteen twenties.

The other limiting way to look at atoms is to assume that $\vec{l}$ and $\vec{s}$ of each electron couple together to give an angular momentum $\vec{j}$, and that all $\vec{j}$ couple to give the total angular momentum $\vec{J}$. This $j-j$ coupling appears to be a reasonable approximation for excited states of the heaviest atoms. For most elements the actual coupling of angular momenta is somewhere between the two extremes. But in most cases that will concern us here, the Russell-Saunders coupling scheme may safely be used to classify terms.

Given the vectorial character of angular momenta, the coupling may be handled by vector addition. The total orbital angular momentum is then:

$$\vec{L} = \sum_i \vec{l}_i$$

(It is customary to use capital letters for all many-electron, lower case letters for all one-electron quantities.)

There is a considerable formal analogy between the individual $\vec{l}$ and the resultant $\vec{L}$. The latter is related to a total orbital angular momentum quantum number L by [compare Eq. (2.23)]:

$$|\vec{L}| = \sqrt{L(L+1)}\,\hbar \tag{3.1}$$

L can be zero or a positive integer $\leqslant \sum l_i$. Only certain orientations in space are allowed for $\vec{L}$, which are such that the projection of the vector on a reference axis (*e.g.* the z axis) must be $M_L \hbar$, where

$$M_L = L, \ L-1, \ L-2, \ L-3, \ \ldots \ -L \tag{3.2}$$

Due to the rules of vector addition and the restrictions on m_l, it turns out that M_L can always be calculated as the algebraic sum of the $m_{l,i}$ of the individual electrons:

$$M_L = \sum_i m_{l,i} \tag{3.3}$$

This is exemplified in Fig. 3.1 for two d electrons ($l = 2$), where m_l can have the values 2, 1, 0, -1, -2 [*cf.* Eq. (2.24)]. Let us assume that vector $\vec{a}$ represents the orbital angular momentum $\vec{l}$ of electron number 1, vector $\vec{b}$ that of electron number 2. The ordinate in Fig. 3.1 marks the projection of the vectors on the z axis in units of $\hbar$, which is identical to the magnetic quantum number m_l. In the case drawn, electron 1 has $m_{l,1} = 1$ and electron 2 has $m_{l,2} = 2$. Vector $\vec{c}$ results from the vectorial addition of $\vec{a}$ and $\vec{b}$; it represents hence the total angular momentum $\vec{L}$. Its z component (in units of $\hbar$) gives M_L, showing that $M_L = m_{l,1} + m_{l,2} = 3$.

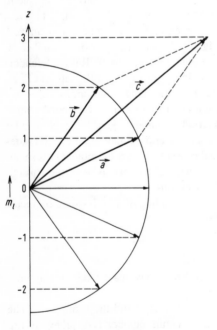

Fig. 3.1. Vectorial addition of the orbital angular momenta $\vec{l}$ of two d-electrons (vectors $\vec{a}$ and $\vec{b}$) to give the total orbital angular momentum $\vec{L}$ (vector $\vec{c}$). M_L (the z component of $\vec{L}$) results as the arithmetical sum of the individual $m_{l,i}$.

The total spin angular momentum is given by:

$$\vec{S} = \sum_i \vec{s}_i$$

with

$$|\vec{S}| = \sqrt{S(S+1)}\,\hbar \tag{3.4}$$

The total spin angular momentum quantum number S is zero, or a positive integer or half integer. The allowed orientations of $\vec{S}$ are such that the projections on the z axis are $M_S\,\hbar$ with:

$$M_S = S,\ S-1,\ S-2,\ S-3,\ \ldots\ -S \tag{3.5}$$

and

$$M_S = \sum_i m_{s,i} \tag{3.6}$$

When summing the individual orbital and spin angular momenta, completely filled shells and subshells can be neglected, since their momenta always compensate to total zero. Take as an example the completely filled $3p^6$ subshell of a transition metal ion. There is only one possibility to arrange the six electrons in the three $2p$ orbitals, two electrons with paired spin in each. The spins evidently compensate to total spin zero. There are two electrons with orbital angular momentum $m_l = 1$, two with $m_l = -1$, and two with $m_l = 0$, and therefore $M_L = 2 \times 1 + 2 \times (-1) + 0 = 0$. From Eq. (3.2) it follows that the number of allowed projections of L on the z axis is $2L+1$. Since for the $3p^6$ subshell only one projection ($M_L = 0$) is possible, $2L+1 = 1$, and hence $L = 0$; *i.e.* also the orbital angular momentum of the filled subshell is zero.

The angular momenta and the relevant quantum numbers are tabulated in Appendix 3.1 at the end of this Chapter. The formal analogy between the formulae for one-electron orbitals and many-electron terms will help to memorize them.

3.3.2. Russell-Saunders Term Symbols

The quantum numbers L and S, defined in the preceding Section, are used to label terms, in form of the "Russell-Saunders term symbol":

$$^{2S+1}L$$

As with single electrons, letters are used to designate a certain value of L[*)]:

L:	0	1	2	3	4	5	...
Symbol:	S	P	D	F	G	H	...

---

[*] The use of the same letter, S and S respectively, for a term and for a quantum number is an unfortunate but, for historical reasons, common practice. However, use of roman characters for term symbols and of italic characters for quantum numbers, as in this book, helps to keep things apart.

The superscript $(2S+1)$ in the term symbol is known as the multiplicity of the term; it is equal to the number of unpaired electrons plus one. Terms with $(2S+1)=1, 2, 3,\ldots$ are called singlet, doublet, triplet, ... terms. (These names come from spectroscopy; we return to this topic in Section 3.4). The symbol ^3P, for instance, stands for a triplet P term, ^1G for a singlet G term, *etc.*

A term with quantum numbers L and S is composed of $(2L+1)(2S+1)$ microstates, which, in first approximation*[)], have the same energy. This is because of Eqs. (3.2) and (3.5): any of the $(2L+1)$ allowed orientations of the total orbital angular momentum $\vec{L}$ may be combined with any of the $(2S+1)$ orientations of the total spin angular momentum $\vec{S}$, and the energy depends only on the absolute values of $\vec{L}$ and $\vec{S}$, but not on their orientation in space. The microstates are characterized by the quantum numbers M_L and M_S which are obtained with the aid of Eqs. (3.3) and (3.6). They are included in Table 3.1 for the few examples given there.

3.3.3. The Terms Arising from d^n Configurations

We come now to the problem of finding out which terms arise from a given d^n configuration. The procedure will be exemplified for a d^2 ion. It includes the following steps: 1) Write down all possible microstates which are compatible with Pauli's exclusion principle, and label them with their M_L and M_S values. This is done in Table 3.2 in a systematic way. Each electron is symbolized by an X. Since each electron may have $m_s = +^1/_2$ or $m_s = -^1/_2$, there are four possible arrangements of spin in each row, namely $\uparrow\uparrow$, $\uparrow\downarrow$, $\downarrow\uparrow$, and $\downarrow\downarrow$, giving rise to

Table 3.2. The 45 microstates of the configuration d^2.

$m_l =$	2	1	0	-1	-2	M_L	M_S
	×	×				3	$1, 0, 0, -1$
	×		×			2	$1, 0, 0, -1$
	×			×		1	$1, 0, 0, -1$
	×				×	0	$1, 0, 0, -1$
		×	×			1	$1, 0, 0, -1$
		×		×		0	$1, 0, 0, -1$
		×			×	-1	$1, 0, 0, -1$
			×	×		-1	$1, 0, 0, -1$
			×		×	-2	$1, 0, 0, -1$
				×	×	-3	$1, 0, 0, -1$
	× ×					4	0
		× ×				2	0
			× ×			0	0
				× ×		-2	0
					× ×	-4	0

---

* Small differences in energy which appear as a consequence of spin-orbit coupling will be discussed in Section 3.4.

$M_S = 1, 0, 0, -1$ respectively. This is, of course, different for the microstates with both electrons in the same orbital; they must have $M_S = 0$ (spins paired). Note that a certain combination of M_L and M_S may occur several times.

2) Select from the microstates with highest M_L value the one with highest M_S. For d^2 this is $M_L = 4$, $M_S = 0$. ($M_L = 4$ is incompatible with $M_S \neq 0$ because of the Pauli principle.) Eqs. (3.2) and (3.5) reveal that this microstate is indicative for the existence of a term with $L = M_{L,max} = 4$ and $S = M_{S,max} = 0$, *i.e.* a ^1G term. This term comprises also microstates with $M_L = 3, 2, 1, 0, -1, -2, -3, -4$, each with $M_S = 0$ (a total of 9 microstates). Substract microstates with these combinations from the total list of 45.

3) Select from the remaining set of microstates among those with highest M_L the one with highest M_S. For d^2 this happens to be $M_L = 3$, $M_S = 1$, indicative for a ^3F term. This term comprises microstates which are possible as combinations of $M_L = \pm 3, \pm 2, \pm 1, 0$ and $M_S = \pm 1, 0$; a total of $(2L+1)(2S+1) = 7 \times 3 = 21$. Substract these, and repeat until nothing is left.

The result of this computation is the following: a d^2 configuration is split by the interelectronic repulsion into the terms ^3F, ^3P, ^1G, ^1D and ^1S.

If there is only one d electron (d^1), it may be placed in ten possible situations, spin up or spin down, in either of the five d orbitals. The highest M_L, M_S combination is $M_L = m_l = 2$, $M_S = m_s = 1/2$, indicative for a ^2D term; all other microstates form part of this term. Thus, a d^1 configuration gives rise to just one ^2D term. The number of possible arrangements increases dramatically with the number of electrons. There are 120 microstates for 3 electrons, 210 for 4, and 256 for 5. (For a more than half filled shell, $n > 5$, the number of microstates goes down again.) The labor to be invested in working out the terms increases proportionally. Fortunately this has been done once and for all, and the terms of all possible configurations may be found tabulated. For the d^n configurations the terms are given in Table 3.3.

It may be seen from this table that a configuration d^{10-n} gives rise to the same terms as does d^n. This may be understood in terms of what is usually called the "hole-formalism". "Holes" in more than half filled shells may, from an electrostatic point of view, be treated like positrons. They repell each other in exactly the same way as electrons do. Evidently the same correspondence holds for p^n and p^{6-n} configurations, as well as for f^n and f^{14-n} configurations.

Table 3.3 The terms arising from d^n configurations.

Configuration			Terms				
d^1, d^9	^2D						
d^2, d^8	^3F	^3P					
	^1G	^1D	^1S				
d^3, d^7	^4F	^4P					
	^2H	^2G	^2F	$2 \times {}^2D$	^2P		
d^4, d^6	^5D						
	^3H	^3G	$2 \times {}^3F$	^3D	$2 \times {}^3P$		
	^1I	$2 \times {}^1G$	^1F	$2 \times {}^1D$	$2 \times {}^1S$		
d^5	^6S						
	^4G	^4F	^4D	^4P			
	^2I	^2H	$2 \times {}^2G$	$2 \times {}^2F$	$3 \times {}^2D$	^2P	^2S

3.3.4. The Ground Terms of the d^n Configurations

So far, we have seen what terms arise from a configuration, but not in which order they lie in energy. It is particularly important to know which term lies lowest, since this determines the ground state of the atom or ion under consideration. Hund's rules may be applied to find the ground term easily, as the one ensuring minimum repulsion between the electrons, having them separated from each other as far as possible. Hund's rules state:

1. of the terms of a configuration those with maximum spin multiplicity tend to lie lowest;
2. of the terms with maximum multiplicity that with largest orbital angular momentum lies lowest.

Based on these rules, the following procedure allows one to write down rapidly the ground term of a d^n configuration (or any other configuration). List horizontally the m_l values of the orbitals of the relevant incomplete shell (see Table 3.4). Fill these orbitals with the available electrons, starting from the left. Single electrons are added to each orbital first. If pairing is necessary, start from the right. This step takes care of both Hund's rules. The m_l values of unpaired electrons are then added algebraically. The sum is $M_{L,\mathrm{max}} = L$. The corresponding letter is assigned to the term. The number of unpaired electrons plus one gives the multiplicity of the term. (Note again the consequences of the hole formalism.)

Table 3.4. The ground terms for d^n configurations.

Configuration	m_l	2	1	0	−1	−2	L	Ground term
d^1		↑					2	^2D
d^2		↑	↑				3	^3F
d^3		↑	↑	↑			3	^4F
d^4		↑	↑	↑	↑		2	^5D
d^5		↑	↑	↑	↑	↑	0	^6S
d^6		↑	↑	↑	↑	↑↓	2	^5D
d^7		↑	↑	↑	↑↓	↑↓	3	^4F
d^8		↑	↑	↑↓	↑↓	↑↓	3	^3F
d^9		↑	↑↓	↑↓	↑↓	↑↓	2	^2D

3.3.5. Term Energies and Racah Parameters

The order of the other terms and their relative energies cannot be predicted in a simple manner. These data are available only as approximate solutions of the corresponding wave equations, introducing repulsion terms. The relevant Hamiltonian is:

$$\mathscr{H} = \sum_{i=1}^{n} (H_i^0 + {}^{1}/{}_2 \sum_{j=1}^{n} e^2/r_{ij}) \qquad i \neq j$$

where H_i^0 is a hydrogen-like Hamiltonian for each of the n electrons, and e^2/r_{ij} expresses the repulsion between electrons i and j. The approximate solution of the problem has been achieved by application of first order perturbation theory. The system is first treated as if the electrons do not interact. The influence of the interelectronic repulsion is then evaluated by means of a perturbation operator. This procedure is known in the literature as Slater's theory of atoms and ions. We shall discuss here only briefly the results of this somewhat lengthy mathematical treatment [6, 7]. The energy levels of a configuration (the terms) are given as functions of three parameters, A, B, and C. These parameters, which have been introduced by Racah, are linear combinations of radial functions which occur in the course of the perturbation treatment. As an example, the energy levels of the d^2 configuration are given in Eqs. (3.7); the corresponding data for all other d^n configurations may be found in the literature (e.g. [6]).

$$
\begin{aligned}
{}^3F &\,\triangleq A - 8B \\
{}^3P &\,\triangleq A + 7B \\
{}^1G &\,\triangleq A + 4B + 2C \\
{}^1D &\,\triangleq A - 3B + 2C \\
{}^1S &\,\triangleq A + 14B + 7C
\end{aligned}
\tag{3.7}
$$

The Racah parameters are different for each atom or ion. The two most important of them are B and C, since they determine the energy difference between the terms, and therefore the position of the spectral lines of the corresponding transitions. In fact they are accessible from the emission spectra of the free ions by fitting experimental term energies to Eqs. (3.7) by the method of least squares. In the first transition series, B varies from 560 to 1240 cm^{-1}, and C/B is generally *ca.* 4 [1].

3.4. Spin-Orbit Coupling

When introducing the Russell-Saunders coupling scheme in Section 3.3.2, we neglected that the total orbital angular momentum $\vec{L}$ and the total spin angular momentum $\vec{S}$ also couple, to give the total angular momentum $\vec{J}$.

$$\vec{J} = \vec{L} + \vec{S}$$

The energies involved are generally small ($\simeq 10^2$ cm^{-1}) compared with the energy differences between terms ($\simeq 10^4$ cm^{-1}). This is the reason why, in first approximation, we could forget about this coupling when presenting the Russell-Saunders term symbolism. Spin-orbit coupling plays, however, a considerable role in determining the detailed magnetic properties of many transition metal compounds (see Section 5.3).

Physically, spin-orbit coupling means that $\vec{L}$ can have only certain quantized directions in the magnetic field originating from $\vec{S}$, and *vice versa*. As in all previously discussed angular momenta $\vec{J}$ is also related to a quantum number J, by:

$$|\vec{J}| = \sqrt{J(J+1)}\,\hbar$$

where J can have all integer or all half-integer values between $(L + S)$ and $(L - S)$:

$$J = L + S, \; L + S - 1, \; L + S - 2, \; \ldots \; L - S \tag{3.8}$$

The projections in the z-direction are $M_J \hbar$, where

$$M_J = J, \ J-1, \ J-2, \ J-3, \ \ldots \ -J \tag{3.9}$$

and

$$M_J = M_L + M_S \tag{3.10}$$

From Eq. (3.8) it follows that a term with given quantum numbers L and S can have $(2S+1)$ different values of J, and hence of total angular momentum $\vec{J}$. Slightly different energies are associated with these different values of $\vec{J}$. This means a term is split into $(2S+1)$ relatively near sublevels (often named states*)). There is one exception: terms with $L=0$ (S terms) never split, the reason being that, with zero orbital angular momentum, there is no resulting magnetic field which would force the spin angular momentum to take certain quantized orientations.

Optical spectra arise from transitions between terms. If the relevant terms consist of $(2S+1)$ energetically close components, the spectra do not show just single lines, but multiplets. From this situation comes the designation "multiplicity" for $(2S+1)$ in the Russell-Saunders term symbols. (Only transitions between terms of equal spin multiplicity are allowed.)

The various states of a term are characterized by adding the quantum number J as a subscript to the Russell-Saunders term symbol. The ground term of the d^2 configuration, ^3F ($L=3$, $S=1$), consists of the states [cf. Eq. (3.8)]:

$$^3F_4, \ ^3F_3, \ ^3F_2$$

The state with lowest J lies lowest in energy for terms arising from configurations with less than half filled shells; the reverse is true for configurations with more than half filled shells (third Hund's rule). Half filled shells give rise to S terms, which do not split.

The energy gap between adjacent J levels is λJ_i, where J_i is the larger of the two quantum numbers involved (cf. Fig. 3.2). This statement is called the Landé interval rule. The parameter λ is known as the spin-orbit coupling constant. It is in fact a constant for any given term, but it may vary from one term to another for the same ion. It should be noted that the splitting of a term into states obeys a center of gravity rule; in other words, if ΔE_J is the energy difference between the mother term and the state with quantum number J, then

$$\sum_{J=L-S}^{L+S} \Delta E_J = 0$$

Obtained from the multiplet pattern of the emission spectra of the free ions, the spin-orbit coupling constant λ may be found tabulated in the literature [2]. It increases as the atomic number increases. For the higher transition metals (and particularly for the lanthanides) the energy related to spin-orbit coupling becomes comparable to that related to interelectronic

---

* There is some confusion in the literature concerning the denomination of terms, states and micro-states. The nomenclature used in this book appears to be the most common one.

repulsion (under these conditions the Russell-Saunders coupling scheme looses its significance).

Finally it should be noted that each state arising from a term is still $(2J + 1)$-fold degenerate [see Eq. (3.9)]. In other words, the $(2J + 1)$ allowed orientations of the total angular momentum $\vec{J}$ have all the same energy. This last degeneracy is lifted only by the application of an outer magnetic field. The separation between adjacent levels is then $g\beta H_0$, *i.e.* it is proportional to the magnetic field strength H_0. The parameter g is the Landé splitting factor, and β is the Bohr magneton. The final splitting of the components of a state by the magnetic field is of the order of 1 cm^{-1} for fields available in the laboratory. It is the basis for EPR and related measurements (see Section 6.4.2).

Fig. 3.2 is an attempt to summarize the splitting pattern for the case of the d^2 configuration. In the interest of clarity, the various splittings are not drawn to scale.

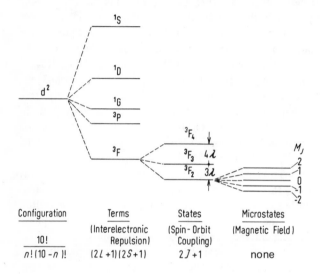

Configuration	Terms	States	Microstates
	(Interelectronic	(Spin-Orbit	(Magnetic Field)
$\dfrac{10!}{n!\,(10-n)!}$	Repulsion) $(2L+1)(2S+1)$	Coupling) $2J+1$	none

Fig. 3.2. The splitting of a configuration; d^2 is used as an example. The numbers in the last line indicate the degeneracy.

3.5. Transition Metal Complexes

In homogeneous catalysis we are generally concerned not with free transition metal ions but with transition metal complexes. We mean by this a central metal ion associated with a number of attached ions or neutral molecules, the whole forming a distinguishable entity in solution. Coordination compound or coordination cluster are used synonymously instead of the word complex. It is customary to call the ions and molecules surrounding the central ion "ligands". Typical examples are Cl$^-$, Br$^-$, CN$^-$, H$_2$O, NH$_3$, (C$_6$H$_5$)$_3$P, C$_2$H$_4$. The neutral molecules use either lone pairs of electrons or electrons in π orbitals for the bonding to the metal center. There are also double-ended chelating ligands, like ethylenediamine and the acetylacetonate anion which attach themselves to the metal ion in two places and are called bidentate ligands. Correspondingly, diethylenetriamine may act as a tridentate and triethylenetetramine as a quadridentate ligand.

$$H_2N \overset{\cdot\cdot}{\underset{CH_2-CH_2}{\diagdown}} \overset{\cdot\cdot}{NH_2} \qquad \left[\begin{array}{c} \overset{O}{\underset{H_3C}{\parallel}} \overset{}{C} \overset{O}{\underset{CH}{\parallel}} \overset{}{C} \overset{}{CH_3} \end{array} \right]$$

The number of ligands a metal ion can accomodate is called the coordination number. Although some metal ions have characteristic coordination numbers, one and the same central ion may also exhibit different coordination numbers toward different ligands, a typical example being cobalt in the complexes $[CoCl_4]^{2-}$ and $[Co(H_2O)_6]^{2+}$.

The coordination compound may or may not have a net electrical charge. This question is intimately related to that of the valency of the central metal atom. We define the valency as the formal positive charge on the metal, which is determined by assigning the usual formal charges to the ligands, and having regard to the charge balance. Thus in $[CoCl_4]^{2-}$ we assign one negative charge to each chlorine and hence say that the metal is divalent. Frequently, the formal charge is noted by a Roman number in parentheses, particularly if the name of the compound is written down explicitly, e.g. hexaminecobalt(III) chloride, $[Co(NH_3)_6]Cl_3$, or pentacarbonyltriphenylphosphinechromium(0), $[Cr(0) (CO)_5P(C_6H_5)_3]$. The formal valency of a metal in a complex is correlated sometimes only very roughly with the true electronic distribution within the compound. Usually the charge is dissipated over the ligands. The classical example is the vanadyl ion, $[V(IV)O(H_2O)_5]^{2+}$, where molecular orbital calculations have indicated that roughly only one positive charge is located at the metal; the calculated charge distribution is $V^{+0.97}O^{-0.60}(5\,H_2O)^{+1.63}$ [3].

The structure of a great number of transition metal complexes has been determined by X-ray crystallographic analysis. It turns out that the most common arrangement of six ligands about the metal is that of a more or less distorted octahedron, with the metal in the center. Four ligands are generally arranged at the corners of a tetrahedron or of a planar square, again with the metal in the center; five ligands may form a trigonal bypyramid or a tetragonal pyramid. It has been proved by optical spectroscopy, EPR and magnetic susceptibility data that this approximate symmetry of the complexes is also maintained in solution. The term "approximate symmetry" needs further comment. We use the nomenclature octahedral, tetrahedral, etc. whether or not all ligands are identical. Moreover only the atoms adjacent to the metal center are taken into account in the assessment of symmetry. Thus $[Ti(H_2O)_6]^{3+}$, $[Co(NH_3)_4Br_2]^+$ or $[Co(en)_3]^{3+}$ are all considered as octahedral complexes. (The abbreviation "en" is commonly used for the bidentate ligand ethylenediamine).

The ligands affect the electrons of the metal in two distinct ways. One is through the electrostatic field of their negative charges, the other is by more or less covalent bonding. Two important approaches to the description of the electronic structure of transition metal complexes have been advanced, each of them taking into account predominantly one of the two influences of the ligands.

The ligand field theory considers the electrostatic field of the ligands to provide an additional potential perturbing the electrons of the free metal ion. The detailed changes in the electronic structure of the metal depend on the symmetry of the perturbing potential, which is evidently given by the distribution of the ligands around the central ion. This approach clearly puts the emphasis on the central atom, and therefore it is particularly suited to describe and predict those properties of transition metal complexes which depend primarily on the electronic structure of the metal: absorption spectra, magnetic susceptibility and, to a certain extent, thermodynamic properties.

The molecular orbital (MO) theory, on the other hand, considers the overlap between metal and ligand orbitals and takes care of electron sharing between them. In other words, it is concerned primarily with the bonding within the coordination compound, taking into account any degree of covalency and accomodating σ as well as π bonding. Molecular orbital calculations in general, and particularly for transition metal complexes, are greatly simplified by symmetry arguments.

These brief remarks may indicate the great importance of the symmetry of the ligand arrangement for most physical and chemical properties of the transition metal complexes. The mathematical formalism in which symmetry arguments are couched most lucidly is group theory. Therefore, this theory has a central position even in the qualitative treatment of coordination compounds. For this reason, and for the benefit of readers not aquainted with group theory, a brief account of the most important principles and rules is given in the next chapter, before ligand field theory and MO theory are discussed in more detail in Chapters 5 and 6.

Appendix 3.1. Angular Momenta and Quantum Numbers.

One-electron orbitals	Many-electron terms
Orbital angular momentum: $\vec{l}$	$\vec{L} = \sum_i \vec{l_i}$
$\lvert \vec{l} \rvert = \sqrt{l(l+1)}\,\hbar$	$\lvert \vec{L} \rvert = \sqrt{L(L+1)}\,\hbar$
$l = (n-1),\ (n-2) \ldots 0$	$L =$ positive integer $\leqslant \sum_i l_i$, or zero
$m_l = l,\ l-1,\ l-2 \ldots -l$	$M_L = L,\ L-1,\ L-2 \ldots -L$
	$M_L = \sum_i m_{l,i}$
Spin angular momentum: $\vec{s}$	$\vec{S} = \sum_i \vec{s_i}$
$\lvert \vec{s} \rvert = \sqrt{s(s+1)}\,\hbar$	$\lvert \vec{S} \rvert = \sqrt{S(S+1)}\,\hbar$
$s = {}^1/_2$	$S =$ positive integer or halfinteger $\leqslant \sum_i s_i$, or zero
$m_s = \pm {}^1/_2$	$M_S = S,\ S-1,\ S-2 \ldots -S$
	$M_S = \sum_i m_{s,i}$

References

[1] B. N. Figgis, *Introduction to Ligand Fields*, Interscience Publ., New York, 1966, Ch. 3.   [2] T. M. Dunn, Trans. Faraday Soc., *57*, 1441 (1961).   [3] C. J. Ballhausen and H. B. Gray, Inorg. Chem., *1*, 111 (1962).

Suggested Additional Reading

[4] F. A. Cotton and G. Wilkinson, *Advanced Inorganic Chemistry*, Interscience Publ., New York, 3. ed. 1972.   [5] W. Kauzmann, *Quantum Chemistry*, Academic Press Inc., New York, 1957.   [6] J. S. Griffith, *The Theory of Transition-Metal Ions*, Cambridge University Press, 1961.   [7] J. C. Slater, *Quantum Theory of Atomic Structure*, Vol. I and II, McGraw-Hill Book Comp., New York, 1960.

4. Symmetry and Group Theory

4.1. The Concept of a Group

Group theory is a branch of mathematics. For the nonmathematician it appears to be a highly sophisticated device for handling sets of abstract "elements" which are interconnected by certain rules. In the present context, we shall be concerned only with limited aspects of group theory presented in a pragmatic way, leaving all derivations and proofs to the special literature cited at the end of the Chapter. In other words, we shall use group theory as a tool, and not as a science. We must, of course, make use of the terminology of group theory, and we shall try to provide a minimum of background meaning for this terminology.

A mathematical group is composed of a set of elements A, B, C, ..., and a rule for combining any two of the elements to form their "product", such that:

1. every "product" of two elements and the "square" of each element is a member of the set;
2. the associative law of algebra holds, i.e., $A(BC) = (AB)C$;
3. the set contains an identity element E such that $EX = XE = X$ for every element X of the set;
4. each element X has an inverse X^{-1} such that $XX^{-1} = X^{-1}X = E$.

In these four conditions, defining a set of elements A, B, C, ... as a group, we have set "product" and "square" between quotation marks. The reason is that in group theory the operation of forming a "product" may or may not be a multiplication in the ordinary arithmetic sense. This is best illustrated by some simple examples.

The four numbers 1, -1, i and $-$i form a group, if the law of combining (of forming the "product") is multiplication in the ordinary sense, and the identity element is 1. It is easy to see that all four defining conditions of a group are fulfilled: the product of any two elements (*e.g.* $1 \times -1 = -1$; $i \times -i = 1$) and the squares (*e.g.* $-i \times -i = -1$) are members of the set; the associative law holds: $1 \times (i \times -1) = (1 \times i) \times -1 = -i$; furthermore $1 \times -i = -i \times 1 = -i$; and finally the inverse of i is $-$i since $i \times -i = 1$; and so on.

The set of all integers, positive, negative and zero, form a group, if the law of combination is addition and the identity element is zero. Let us check the defining properties (1) through (4) with any members of the group, say $+320$, -12 and $+5$. The "product" $+320 + (-12) = +308$, as well as the "square" $(-12) + (-12) = -24$ are members of the set. Furthermore: $5 + (320 - 12) = (5 + 320) - 12$; $0 + 320 = 320 + 0 = 320$; $5 + (-5) = 0$.

Our third example is a "rotation group". Take a plane lamina with the shape of a regular hexagon; consider those rotations which leave the aspect of the lamina unchanged (but do not turn the lamina over). A rotation of $2\pi/6 = 60°$ satisfies this condition; so also does a rotation by 120°, 180°, etc. Each movement may be expressed as a multiple of the first rotation by $2\pi/6 = 60°$. If we denote rotation by 60° as C_6, the rotation by 120° may be expressed as C_6^2, which means that the effect is the same as if we had moved the hexagon twice by 60°. A total of six movements is possible: C_6^1, C_6^2, C_6^3, C_6^4, C_6^5, and $C_6^6 = E$. The last has the same effect as if we had not moved the hexagon at all. The law of combination is now "carry out two rotations, one after the other". The identity element is "leave the hexagon unmoved"; the inverse of each movement is that which undoes the change obtained with the first, *i.e.* rotation in the opposite direction. The set of these movements satisfies all requirements of a group. The inherent property of our lamina engendering this group is evidently its hexagonal symmetry.

This last example brings us near to our present problem: that of the symmetry of molecules and coordination compounds.

4.2. Symmetry Groups

It will be shown in this section that the symmetry properties of any compound having some sort of symmetry may be expressed in such a way that they form the elements of a group. Groups based entirely on symmetry arguments are called symmetry groups. (In fact the rotation group of the hexagon in the last section represents one of the symmetry groups.) The importance of showing that molecular symmetry forms a base for mathematical groups lies in the fact that some of the powerful methods and lucid results of group theory can be used to study symmetric molecules.

4.2.1. Symmetry Operations and Symmetry Elements

Consider a transition metal complex having six identical ligands arranged at the vertices of a regular octahedron, say MX_6. By intuition we consider this molecule as a highly symmetric entity. For defining its symmetry in a mathematical sense we place it in a system of Cartesian coordinates, with the origin at the nucleus of the central metal ion (Fig. 4.1).

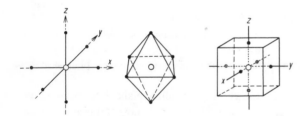

Fig. 4.1. Different representations of an octahedral complex; ○: central ion; ●: ligands.

It will be important to distinguish clearly between symmetry operations and symmetry elements. A symmetry operation is a movement of our octahedral complex such that after the movement the compound is in an equivalent position. Rotation about the z axis by $2\pi/4$ is such an operation; rotation by $2\pi/2$ is another. Symmetry operations are "generated" by a symmetry element which in this case is a fourfold rotation axis (colinear with the z axis). Rotation axes are designed with the symbol C_n, the subscript n giving the order of the axis. A C_n generates n operations; for C_4 they are:

$$C_4^1 \qquad\qquad C_4^2 = C_2 \qquad\qquad C_4^3 \qquad\qquad C_4^4 = E$$

Rotation by: $2\pi/4$ $\qquad\quad$ $2\pi/2$ $\qquad\quad$ $3\times 2\pi/4$ $\qquad\quad$ 2π

Note that the operations C_4^1 and C_4^3 are similar in a certain sense; both can only be produced by a fourfold axis. The situation obtained by C_4^3 could also have been reached by C_4^{-1} (rotating 90° in the opposite direction). Therefore these two operations are grouped together and written as $2\,C_4$. They are said to form a class*). Operation C_4^2, on the other hand, has the same result as if the axis would have been only of order 2. This operation is different in type, it forms a class by itself. Operation C_4^4 comes to the same result as if we had done nothing; it also forms a class by itself. Summarizing: The symmetry element C_4 generates four symmetry operations, grouped in three classes: $2\,C_4$, C_2 and E. There are three fourfold axes in a regular octahedron, coincident with the three Cartesian coordinates.

There are also four threefold axes, each passing through the center of two opposite triangular faces of the octahedron. One of them is represented in Fig. 4.2a. Each C_3 generates C_3^1, C_3^2, and E. Evidently C_3^1 and C_3^2 belong to the same class. Thus the three operations generated by each threefold axis are $2\,C_3$ and E. However E has already been generated by the fourfould axes, only $2\,C_3$ is new. Moreover there are six twofold axes which bisect opposite edges; two are represented in Fig. 4.2b. Each C_2 generates C_2^1 and E; C_2^1 is new.

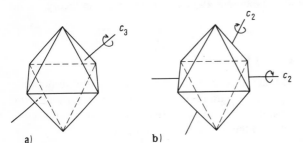

a)

b)

Fig. 4.2. Threefold and twofold axes in an octahedral complex.

Proceeding from the rotation axes we have already 24 symmetry operations: E, $8\,C_3$, $6\,C_2$, $6\,C_4$, $3\,C_2'\,(=3\,C_4^2)$. They will be found in Table 4.6, called a character table, which is given in Appendix 4.2 at the end of this chapter, and which will be discussed in detail later on.

However, we have still to discuss further symmetry elements, namely mirror planes, inversion center and improper axes.

By inspection of Fig. 4.1 it is easy to see that the xy plane is a mirror plane of the octahedron. Move any point situated above this plane to the equivalent point below, and *vice versa*, and the same octahedron will result. Symmetry planes are designated σ; each σ generates just one symmetry operation, denoted by the same symbol σ. There are 9 such planes in the octahedron; three of them are perpendicular to the main axes, in the xy, xz and yz planes. The operations generated by them together form a class: $3\,\sigma_h$**). The other six mirror planes are called diagonal (with subscript d) and form also a class: $6\,\sigma_d$. They are best recognized if the octahedron is placed into a cube (right hand side of Fig. 4.1). Take a diagonal of any face of the cube, imagine the cube cut along this diagonal, and through to the opposite diagonal. The plane defined by this "cut" is a mirror plane of the octahedron; it bisects two axes and includes the third.

An inversion center requires that any point of coordinates x, y, z has an equivalent at $-x$, $-y$, $-z$. Evidently this is true for the octahedron. The symbol for the element as well as for the operation of inversion is i.

---

*   For a more rigorous definition of classes see [1].
** The subscript h comes from horizontal, as opposed to vertical (σ_v). This distinction plays a role with compounds having one main axis, see Section 4.2.2.

The last type of symmetry elements is the improper axis, with the symbol S_n. The operations generated are somewhat complicated, and best thought of as taking place in two steps: first a rotation by $2\pi/n$ and then a reflection in a plane perpendicular to the axis of rotation. Such axes are present in the octahedral complexes, but it is easier to recognize a S_n in another type of molecule, ferrocene (Fig. 4.3). The main axis is a C_5, but this is at the same time a S_{10}: rotation by $2\pi/10$ and reflection through a plane parallel to the rings and including the central ion brings any point of the upper five membered ring to an equivalent point in the lower ring. A further rotation by $2\pi/10$ and subsequent reflection brings the molecule into a situation which could also have been the result of a mere rotation by $2\pi/5$ from the starting point, *i.e.* $S_{10}^2 = C_5^1$.

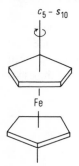

Fig. 4.3. Ferrocene.

In the octahedron, there are four S_6 axes, colinear with the C_3 axes (*cf.* Fig. 4.2). Each generates S_6^1, $S_6^2 = C_3^1$, $S_6^3 = i$, $S_6^4 = C_3^2$, S_6^5 and $S_6^6 = E$. However, the 2 C_3 have already been generated by the threefold axes, and also i and E are already accounted for. Thus each S_6 generates 2 S_6 new operations. Moreover, the octahedron possesses three S_4 axes, colinear with the C_4 axes, each generating S_4^1, $S_4^2 = C_2$, S_4^3 and E, but only 2 S_4 are new.

Arising from the symmetry elements inversion center, improper axes and mirror planes, we have 24 more symmetry operations, namely i, 6 S_4, 8 S_6, 3 σ_h and 6 σ_d; so we find a total of 48 operations for a regular octahedron (*cf.* Table 4.6, Appendix 4.2).

The highly symmetric octahedral complex has helped us to become acquainted with all symmetry elements and operations relevant to molecular symmetry. Of course, many transition metal complexes and most organic and inorganic molecules are less symmetric, and will be subject to only a selection of symmetry operations.

The water molecule, for instance, has only one twofold axis, and two mirror planes perpendicular to each other. One mirror plane lies in the yz plane of the coordinate system (molecular plane); the other lies in the xz plane. These symmetry elements generate the operations C_2, $\sigma_v(xz)$, $\sigma_v(yz)$, E.

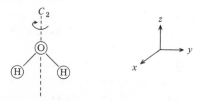

With molecules of less symmetry, certain conventions exist related to the placement of the molecules in a coordinate system. The origin is located at the center of gravity of the molecule; the axis of highest order is taken as the z direction; the other directions are fixed taking into account the "right-hand rule" (*cf*. Fig. 2.2).

4.2.2. Point Groups and the Classification of Molecules

The complete set of symmetry operations that can be performed on a molecule forms a group. Using the relatively simple case of the water molecule, we shall check that the four operations, C_2, $\sigma_v(xz)$, $\sigma_v(yz)$, E, in fact satisfy the mathematical requirements of a group (*cf*. Section 4.1).

The rule for combining any two operations to form their "product" is now: carry out the two operations one after the other. We shall work out the product of C_2 and $\sigma_v(yz)$. Consider any point of the molecule, having the coordinates x, y, z. (Refer to the coordinate system given with the water molecule in the preceding section.) Rotation through 180° around the z axis brings this point to:

$$x' = -x; \quad y' = -y; \quad z' = z.$$

Subsequent reflection at $\sigma_v(yz)$ results in:

$$x'' = x; \quad y'' = -y; \quad z'' = z.$$

This last result could have been obtained in one step reflecting the original point at $\sigma_v(xz)$. We may write this as

$$\sigma_v(yz)\,C_2 = \sigma_v(xz) \tag{4.1}$$

i.e. the product of two symmetry operations is another operation included in the set, which is one of the requirements of a group. The same is true for all other combinations within the set.

In connection with Eq. (4.1) note that the order in which symmetry operations are applied is, by convention, the order in which they are written from right to left. In general the order makes a difference although there are cases (such as the present one) where it does not.

The associative law is obviously valid for the successive application of symmetry operations. The identity operation E is: performing no operation at all; we have been using it in this sense all along. Finally, the inverse of each operation is a second operation which will exactly undo what the given operation does. For a reflection σ, the inverse is evidently σ itself: $E = \sigma\sigma$. For a rotation, it is rotation in the opposite direction. These reverse rotations have not been explicitly quoted in the catalogue of symmetry operations for the octahedral complex and for the water molecule in the preceding section, but they are implied: remember that C_4^3 has the same effect as C_4^{-1}, C_3^1 the same as C_3^{-2}, etc.

The symmetry groups we are dealing with are particularly suited to study the symmetry of isolated molecules or complexes. They are defined in such a way as to leave the center of gravity invariant to all symmetry operations. They are called point groups. There exist other symmetry groups which contain elements of translation. They are useful in crystallography, and are called space groups.

We shall now introduce symbols and descriptions for all possible point groups. After some practice any molecule may be easily assigned to one or the other of the groups. Although the groups are made up from the symmetry operations, it will be found more convenient to describe them enumerating those symmetry elements which are necessary to recognize the group.

C_1:	no symmetry at all
C_s:	only one mirror plane
C_i:	only an inversion center
C_n:	only one n-fold axis
D_n:	n-fold axis; n perpendicular*) twofold axes
C_{nv}:	n-fold axis; n vertical*) mirror planes
C_{nh}:	n-fold axis; a horizontal*) mirror plane
D_{nh}:	n-fold axis; horizontal mirror plane; n perpendicular twofold axes
D_{nd}:	n-fold axis; n perpendicular twofold axes; vertical mirror planes intersecting the twofold axes
S_{2n}:	only one $2n$-fold improper axis
T_d:	symmetry of a regular tetrahedron
O_h:	symmetry of a regular octahedron
$D_{\infty h}$:	linear molecules with a mirror plane perpendicular to the molecular axis
$C_{\infty v}$:	linear molecules without a mirror plane perpendicular to the molecular axis.

Evidently all octahedral complexes belong to the group O_h, tetrahedral complexes to T_d. The water molecule will be easily recognized as C_{2v}, the ethylene molecule as D_{2h}, acetylene as $D_{\infty h}$. A square planar transition metal complex, say $[PtCl_4]^{2-}$, is D_{4h}; a trigonal bipyramid is D_{3h}, a tetragonal pyramid C_{4v}, etc.

4.3. Representations of Groups

4.3.1. Transformation Matrices

We examined earlier the behaviour of a general point with the coordinates x, y, z under the symmetry operations of the group of the water molecule, which by now is defined as the group C_{2v}. In group theory, this sort of geometric transformation is handled by matrix algebra. (A brief account on the relevant rules of matrix algebra is given in Appendix 4.1, at the end of this Chapter.)

Consider once more the symmetry operations C_2, $\sigma_v(xz)$, $\sigma_v(yz)$ and E of the group C_{2v}, and study separately their effect on a general point xyz. The operation C_2 means rotate by 180° about the z axis. We may write the coordinates of the new point, $x'\,y'\,z'$, explicitly:

---

* The expressions perpendicular, vertical, horizontal refer to the main axis.

$$x' = -1\,x + 0\,y + 0\,z = -x$$
$$y' = 0\,x - 1\,y + 0\,z = -y \qquad\qquad (4.2)$$
$$z' = 0\,x + 0\,y - 1\,z = +z$$

In matrix notation, this is written:

$$\begin{bmatrix} -1 & 0 & 0 \\ 0 & -1 & 0 \\ 0 & 0 & 1 \end{bmatrix} \begin{bmatrix} x \\ y \\ z \end{bmatrix} = \begin{bmatrix} -x \\ -y \\ +z \end{bmatrix}$$

Application of the rules of matrix multiplications gives Eqs. (4.2). The square matrix is called a transformation matrix; it transforms the coordinates of the original point to that of the new one. In the same way the transformation matrices for the other operations of C_{2v} may be worked out. They are:

$$E:\quad \begin{bmatrix} 1 & 0 & 0 \\ 0 & 1 & 0 \\ 0 & 0 & 1 \end{bmatrix} \qquad\qquad \sigma_v(xz):\quad \begin{bmatrix} 1 & 0 & 0 \\ 0 & -1 & 0 \\ 0 & 0 & 1 \end{bmatrix}$$

$$C_2:\quad \begin{bmatrix} -1 & 0 & 0 \\ 0 & -1 & 0 \\ 0 & 0 & 1 \end{bmatrix} \qquad\qquad \sigma_v(yz):\quad \begin{bmatrix} -1 & 0 & 0 \\ 0 & 1 & 0 \\ 0 & 0 & 1 \end{bmatrix}$$

Such a set of transformation matrices possesses also the properties required by a mathematical group; e.g. (cf. Eq. 4.1):

$$\begin{bmatrix} -1 & 0 & 0 \\ 0 & 1 & 0 \\ 0 & 0 & 1 \end{bmatrix} \begin{bmatrix} -1 & 0 & 0 \\ 0 & -1 & 0 \\ 0 & 0 & 1 \end{bmatrix} = \begin{bmatrix} 1 & 0 & 0 \\ 0 & -1 & 0 \\ 0 & 0 & 1 \end{bmatrix}$$

$$\sigma_v(yz)\,C_2 = \sigma_v(xz)$$

Therefore the set of transformation matrices represents equally well the group C_{2v}, as the set of symmetry operations does. The set of matrices is said to form a "representation" of the group.

A very important feature with regard to a matrix is its "character" (or trace). It is the sum of the elements in the principal diagonal (left hand top to right hand bottom). For most purposes the knowledge of the set of the characters of the transformation matrices is sufficient. The characters are denoted χ; thus for the representation of C_{2v} given above we have $\chi(E) = 3$, $\chi(C_2) = -1$, $\chi(\sigma_{xz}) = 1$, $\chi(\sigma_{yz}) = 1$. The symbol Γ is used for the total of characters of one representation. We may summarize the representation of C_{2v}, based on the general point xyz as follows:

C_{2v}	E	C_2	$\sigma_v(xz)$	$\sigma_v(yz)$
$\Gamma(xyz)$	3	-1	1	1

Since the transformation matrix of E is always a unit matrix (all diagonal elements unity), the character of E indicates the dimension of the representation, *i.e.* the dimension of the matrices forming this particular representation.

4.3.2. Wave Functions as Bases for Representations

Not only the general point xyz, but any algebraic function, or even a set of functions may be used as a basis for a representation of a group. In other words, there is an unlimited number of possible representations. Of particular interest are, of course, wave functions as a base for a representation.

Let us consider the following problem. The transition metal complex bis(cyclopentadienyl)-titanium dichloride (Fig. 4.4) belongs to the same symmetry group as the water molecule*). A twofold axis bisects the $Cl-Ti-Cl$ and the $Cp-Ti-Cp$ angles; there are also two vertical mirror planes, the one including Ti and both Cl, the other including Ti and the centers of the cyclopentadienyl rings. We shall now use the p_z orbital of the titanium ion as a base for a representation of the group C_{2v}, to which the titanium complex belongs.

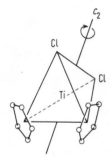

Fig. 4.4. Bis(η^5-cyclopentadienyl)titanium dichloride.

The algebraic functions of atomic orbitals have been given in Table 2.4, and their graphical descriptions in Section 2.2.3. The usual "dual-purpose" sketches of the orbitals (Fig. 2.4) actually represent the squares of the wave functions, but they have the same symmetry behavior as the functions themselves, provided the changes in sign of the wave functions are duly taken into account. Thus, if we want to find out how the p_z orbital varies under the symmetry operations of the group C_{2v}, we just have to apply the operations to the dumbbell-like picture of the orbital. We begin with the operation C_2, that is rotation by $2\pi/2$ about the twofold axis (the z axis).

---

* For many purposes it is satisfactory to describe this complex as "approximately tetrahedral" which means that the central titanium ion is surrounded by four ligands located approximately at the corners of a tetrahedron. Taking into account, however, that the ligands are not equal, the exact symmetry is C_{2v}.

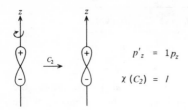

$$p'_z = 1p_z$$

$$\chi(C_2) = 1$$

Evidently, the transformation matrix is just unity. In this case the matrix itself and its character are identical. Likewise, reflection in the two vertical mirror planes leaves the orbital invariant and, of course, also E. Therefore, using the p_z orbital as a base, we obtain the following unidimensional representation:

C_{2v}	E	C_2	$\sigma_v(xz)$	$\sigma_v(yz)$
$\Gamma(p_z)$	1	1	1	1

Based upon the p_y orbital, a different representation results. C_2 transforms the orbital to its negative:

$$p'_y = -1p_y$$

$$\chi(C_2) = -1$$

The same is true for reflection in σ_{xz}, whereas σ_{yz} and E leave the orbital unchanged. The resulting representation is also unidimensional:

C_{2v}	E	C_2	$\sigma_v(xz)$	$\sigma_v(yz)$
$\Gamma(p_y)$	1	−1	−1	1

4.3.3. Character Tables and Irreducible Representations

Certain representations called irreducible representations are of particular interest. Their number in each group is equal to the number of classes in the group. The word irreducible indicates that all other representations are reducible, *i.e.* they may be reduced (or divided) to irreducible ones.

The irreducible representations of each group have been determined once and for all and their most important feature, their characters, are available in form of tables. These are called character tables. They more or less summarize the most important properties of a group, and contain all information we need for the discussion of molecular symmetry.

The character table of the group C_{2v} is reproduced below:

C_{2v}	E	C_2	$\sigma_v(xz)$	$\sigma_v(yz)$
A_1	1	1	1	1
A_2	1	1	-1	-1
B_1	1	-1	1	-1
B_2	1	-1	-1	1

The group C_{2v} is of relatively low symmetry. There are only four classes of symmetry operations, and therefore only four irreducible representations. All are of dimension one. This is not so for groups of higher symmetry, as may be seen in Appendix 4.2, where the character tables of some of the groups more commonly found in coordination compounds are given.

Every group has one representation for which all characters are 1; it is called the "totally symmetric representation". Whatever might be the basis of this representation, it remains unchanged under all symmetry operations. As an example we saw previously that this is the case with the p_z orbital in C_{2v} symmetry. We say that the p_z orbital, in this symmetry, transforms as the totally symmetric representation.

Whereas in general a representation is given the symbol Γ, the irreducible ones are characterized by shorthand symbols which indicate their symmetry properties. These symbols have been introduced by R. S. Mulliken, and have the following meaning:

One-dimensional representations are designated either A or B, two-dimensional E, three-dimensional T.

A denotes representations which are "symmetric" with respect to rotation by $2\pi/n$ about the principal axis, and that means that $\chi(C_n) = 1$; B denotes those which are "antisymmetric", i.e. $\chi(C_n) = -1$.

A subscript 1 (or 2) attached to A's and B's means that the representation is symmetric (antisymmetric) with respect to a C_2 perpendicular to the main axis or to a σ_v; primes and double primes indicate symmetry and antisymmetry with respect to a σ_h. Subscripts 1 or 2 are also used for E's and T's when appropriate, to differentiate between several representations of the same dimension.

In groups with a center of inversion, subscripts g and u differentiate between representations which are symmetric or antisymmetric with respect to the inversion.

4.3.4. Reduction Formula for General Representations

As already mentioned, the p_z orbital transforms as A_1 in C_{2v}. Comparing now the representation $\Gamma(p_y)$ resulting from the p_y orbital with the character table of C_{2v}, we see that p_y transforms as B_2.

The representation $\Gamma(xyz)$, based on the general point xyz, and shown in Section 4.3.1, does not occur in the character table; it is a reducible representation. We shall now present the procedure for reducing such a "general representation".

Our problem is to determine the number of times (a_i), the ith irreducible representation is contained in the reducible one. This is done with the following formula:

$$a_i = \frac{1}{h} \sum \chi(R)\,\chi_i(R) \tag{4.3}$$

Herein h is the number of symmetry operations of the group (also named the order of the group); $\chi(R)$ is the character corresponding to symmetry operation R in the reducible representation; $\chi_i(R)$ is the character of the same symmetry operation in the ith irreducible representation. The sum has to be taken over all operations.

The application of Eq. (4.3) is straightforward, making use of the character tables. Remember, however, that in most groups (although not in C_{2v}) several symmetry operations are collected together in classes[*]; this has to be taken into account for the determination of h as well as for the sum.

To illustrate the procedure, we shall reduce our representation $\Gamma(xyz)$ of the group C_{2v}. For convenience, we repeat the character table and the characters of the representation in question.

C_{2v}	E	C_2	$\sigma_v(xz)$	$\sigma_v(yz)$
A_1	1	1	1	1
A_2	1	1	-1	-1
B_1	1	-1	1	-1
B_2	1	-1	-1	1
$\Gamma(xyz)$	3	-1	1	1

Using Eq. (4.3), we find:

$$a\,(A_1) = \frac{1}{4}\,[(3)\,(1) + (-1)\,(1) \quad + (1)\,(1) \quad + (1)\,(1) \quad] = 1$$

$$a\,(A_2) = \frac{1}{4}\,[(3)\,(1) + (-1)\,(1) \quad + (1)\,(-1) + (1)\,(-1)] = 0$$

$$a\,(B_1) = \frac{1}{4}\,[(3)\,(1) + (-1)\,(-1) + (1)\,(1) \quad + (1)\,(-1)] = 1$$

$$a\,(B_2) = \frac{1}{4}\,[(3)\,(1) + (-1)\,(-1) + (1)\,(-1) + (1)\,(1) \quad] = 1$$

The result: the representation $\Gamma(xyz)$ is reducible to $A_1 + B_1 + B_2$.

To familiarize somewhat more with reducible representations and their reduction, we shall now use the three p-orbitals as a base for a representation of C_{4v}, the group which includes tetragonal pyramidal complexes.

The symmetry elements of C_{4v} are: one fourfold axis and four vertical mirror planes. Two of the planes lie in the xz and yz planes, the other two bisect these planes (see Fig. 4.5). These elements generate the operations $2\,C_4$, C_2, $2\,\sigma_v$, $2\,\sigma_d$, E.

---

[*] Symmetry operations belonging to the same class have the same character; this is in fact the reason why they are grouped together.

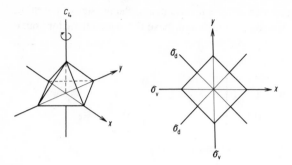

Fig. 4.5. C_{4v}: the symmetry group of the tetragonal pyramid.

We write the three p-orbitals as a column matrix, and find the transformation matrices for the symmetry operations. The class $2\,C_4$ includes C_4^1 and C_4^3, *i.e.* rotation by $2\pi/4$ and $3 \times 2\pi/4$. Rotation by $2\pi/4$ about the z axis (C_4) does not alter the p_z orbital; but p_x and p_y are interchanged, and one of them changes sign:

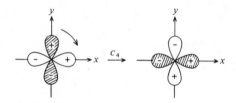

We may write this:

$$
\begin{aligned}
p'_x &= 0\,p_x + 1\,p_y + 0\,p_z \\
p'_y &= -1\,p_x + 0\,p_y + 0\,p_z \\
p'_z &= 0\,p_x + 0\,p_y + 1\,p_z
\end{aligned}
$$

In matrix notation this becomes:

$$
C_4: \quad
\begin{bmatrix} 0 & 1 & 0 \\ -1 & 0 & 0 \\ 0 & 0 & 1 \end{bmatrix}
\begin{bmatrix} p_x \\ p_y \\ p_z \end{bmatrix}
=
\begin{bmatrix} p'_x \\ p'_y \\ p'_z \end{bmatrix}
\qquad \chi = 1
$$

The character of the transformation matrix is 1. Rotation by $3 \times 2\pi/4$, belonging to the same class, must have the same character.

C_2 ($= C_4^2$ = rotation by $2\pi/2$) does not change the p_z orbital, but changes sign of p_x and of p_y:

$$
C_2: \quad
\begin{bmatrix} -1 & 0 & 0 \\ 0 & -1 & 0 \\ 0 & 0 & 1 \end{bmatrix}
\begin{bmatrix} p_x \\ p_y \\ p_z \end{bmatrix}
=
\begin{bmatrix} p'_x \\ p'_y \\ p'_z \end{bmatrix}
\qquad \chi = -1
$$

The σ_v's leave the p_z and one of the other two orbitals invariant, and change sign of the third. The transformation matrices, although somewhat different, have the same character (same class):

$$\sigma_v: \begin{bmatrix} 1 & 0 & 0 \\ 0 & -1 & 0 \\ 0 & 0 & 1 \end{bmatrix} \text{ or } \begin{bmatrix} -1 & 0 & 0 \\ 0 & 1 & 0 \\ 0 & 0 & 1 \end{bmatrix} \quad \chi = 1$$

The σ_d's have much the same effect as C_4, interchanging p_x and p_y:

$$\sigma_d: \begin{bmatrix} 0 & 1 & 0 \\ 1 & 0 & 0 \\ 0 & 0 & 1 \end{bmatrix} \text{ or } \begin{bmatrix} 0 & -1 & 0 \\ -1 & 0 & 0 \\ 0 & 0 & 1 \end{bmatrix} \quad \chi = 1$$

E does not change anything, hence:

$$E: \begin{bmatrix} 1 & 0 & 0 \\ 0 & 1 & 0 \\ 0 & 0 & 1 \end{bmatrix} \quad \chi = 3$$

The character table*) of C_{4v} as well as the representation obtained on the base of the three p orbitals is given below:

C_{4v}	E	$2\,C_4$	C_2	$2\,\sigma_v$	$2\,\sigma_d$
A_1	1	1	1	1	1
A_2	1	1	1	-1	-1
B_1	1	-1	1	1	-1
B_2	1	-1	1	-1	1
E	2	0	-2	0	0
$\Gamma(p_x p_y p_z)$	3	1	-1	1	1

Application of Eq. (4.3) gives:

$$a(A_1) = \frac{1}{8}[(3)(1) + 2(1)(1) + (-1)(1) + 2(1)(1) + 2(1)(1)] = 1$$

$$a(A_2) = \frac{1}{8}[(3)(1) + 2(1)(1) + (-1)(1) + 2(1)(-1) + 2(1)(-1)] = 0$$

---

* Note that the same letter E is used for the identity operation and for the two-dimensional irreducible representations in the character tables. This is unfortunate, but common practice.

$$a\,(B_1) = \frac{1}{8}\,[(3)\,(1) + 2(1)\,(-1) + (-1)\,(1) \quad + 2(1)\,(1) \quad + 2(1)\,(-1)] = 0$$

$$a\,(B_2) = \frac{1}{8}\,[(3)\,(1) + 2(1)\,(-1) + (-1)\,(1) \quad + 2(1)\,(-1) + 2(1)\,(1) \quad] = 0$$

$$a\,(E) \ = \frac{1}{8}\,[(3)\,(2) + 2(1)\,(0) \quad + (-1)\,(-2) + 2(1)\,(0) \quad + 2(1)\,(0) \quad] = 1$$

$$\Gamma(p_x p_y p_z) = A_1 + E$$

We find only two irreducible representations for the three p orbitals. One of them however, is of dimension 2. We could have used the p_z orbital alone as a basis for a representation; since all operations let this orbital invariant, it clearly transforms as the totally symmetric representation A_1. The orbital p_x could not have been used on its own as a base, since some of the operations of the group C_{4v} transform it to p_y, and *vice versa*. Evidently these two orbitals are coupled together in the symmetry group C_{4v}. We say, they together form a base for a representation. Working it out as done above, it turns out that the representation based on p_x and p_y only, is of dimension 2, *i.e.* it is an E representation. We may generalize this result: a set of two orbitals which "transform together" always belongs to an E representation, a set of three such orbitals to a T representation.

4.4. The Transformation Properties of Orbitals and Terms

It was relatively easy to determine the transformation properties of the three p orbitals in C_{4v}. Evidently, the task will become more complicated with the groups containing more symmetry operations. But fortunately this kind of information has been elaborated once and for all, and is available from the character tables. The complete tables (Appendix 4.2) have, at the right hand side, two additional columns not yet discussed. One contains always the six symbols x, y, z, R_x, R_y, and R_z; the first three represent the coordinates x, y, z, whereas the R's stand for rotations. The other column lists squares and binary products of the coordinates. The position of these functions within the table indicates the transformation properties of the function in question. These data are not at all restricted to their use in connection with orbitals; in fact the mathematical device of group theory and the character tables are much older than the concept of orbitals. But the orbitals, in their real forms (*cf.* Table 2.4) are functions of x, y, z or squares or binary products thereof. Consequently their transformation properties can directly be read from the tables. (We are not concerned with the R's at present.)

Thus, we could have seen directly from the C_{4v} character table (Table 4.2, Appendix 4.2) that in this group the p_z orbital "belongs" to the totally symmetric representation A_1, which means that it forms a basis for this representation, and hence has the same transformation properties under all symmetry operations of the group. The orbitals p_x and p_y together form a basis for the two dimensional representation E in the C_{4v} group. The character table of the O_h group (Table 4.6) shows that in octahedral symmetry the three p orbitals together transform as T_{1u}. Of the d orbitals, d_{xy}, d_{xz} and d_{yz} together transform as T_{2g}, whereas $d_{x^2-y^2}$ and d_{z^2}[*)] together form a basis for E_g.

Irreducible representations of dimensions greater than 1 are said to be degenerate. The important corollary is: orbitals which belong to degenerate representations are themselves degenerate. Thus, the set of five d orbitals which we learned were degenerate in a free transition metal ion, is split, in octahedral symmetry, into a pair of degenerate orbitals transforming as E_g, and a group of three degenerate orbitals transforming as T_{2g}. This and similar statements will gain in significance in the following Chapter.

The results we have obtained so far for orbitals also apply to the behavior of terms arising from groups of orbitals. This is not surprising if one considers the near formal equivalence of the relevant relationships between angular momenta and quantum numbers for one electron orbitals and for terms (*cf.* Appendix 3.1). Whereas the angular part of the wave function of an electron contains a factor $\Phi(\varphi) = e^{\pm im\varphi}$ [*cf.* Eq. (2.15)], the corresponding factor in the wave function of a term is just $\Phi(\varphi) = e^{\pm iM_L\varphi}$, *etc.* Thus, a D term splits, in octahedral symmetry, into a threefold degenerate T_{2g} term and a twofold degenerate E_g term.

Knowledge of the splitting of terms in different symmetries is indispensable for a meaningful interpretation of optical spectra (Section 5.2). Contrarily to the one electron orbitals, where our interest is limited to s, p, and d species, for terms also the higher values of the quantum number L are important (F, G, H ... terms; see *e.g.* Table 3.3). The splittings of these higher terms cannot be read directly from the character tables. They are, however, also available in form of tables (*e.g.* [1]). For some of the more important symmetries this splitting is given in Appendix 4.3. (Note that for terms the g or u character is determined by the nature of the orbitals of the individual electrons making up the configuration from which the term is derived. We shall be interested only in terms derived from d^n configurations. Since d orbitals are inherently g, all subscripts in Table 4.3 are g.)

Appendix 4.1. Matrices

A matrix is a rectangular array of numbers or symbols of numbers, such as A. In the general formulation each element of the matrix has two indices, the first indicating the row, the second the column.[**]

$$
A = \begin{bmatrix}
a_{11} & a_{12} & a_{13} & \cdots & a_{1n} \\
a_{21} & a_{22} & a_{23} & \cdots & a_{2n} \\
a_{31} & a_{32} & a_{33} & \cdots & a_{3n} \\
\cdot & & & & \\
\cdot & & & & \\
\cdot & & & & \\
a_{m1} & a_{m2} & a_{m3} & \cdots & a_{mn}
\end{bmatrix}
$$

---

* Remember that d_{z^2} may be considered as a linear combination of $d_{z^2-x^2}$ and $d_{z^2-y^2}$ (see Section 2.2.3).
** Matrices are often printed in bold type characters. This is not a definite rule and is not followed here.

Square brackets are used to enclose the array*). (This distinguishes it visually from a determinant which is written between vertical lines.)

Matrices may or may not have the same number of rows and columns. We are only concerned with square matrices ($m = n$) and with column matrices (consisting of only one column).

Square matrices have some special properties. A set of square matrices of the same order (same number of rows and columns) behaves, in certain aspects, very much like a set of ordinary numbers. They are said to form an algebra, giving sums and products. Let a_{ij} be the elements of matrix A and b_{ij} those of matrix B, then $A \pm B = C$, with $c_{ij} = a_{ij} \pm b_{ij}$. For the product $AB = C$, the elements are given by

$$c_{ij} = \sum_{k=1}^{n} a_{ik} b_{kj}$$

Since multiplication is of greatest interest in the present context, an example calculation is given explicitly.

$$\begin{bmatrix} a_{11} & a_{12} \\ a_{21} & a_{22} \end{bmatrix} \begin{bmatrix} b_{11} & b_{12} \\ b_{21} & b_{22} \end{bmatrix} = \begin{bmatrix} c_{11} & c_{12} \\ c_{21} & c_{22} \end{bmatrix}$$

$$c_{11} = a_{11} b_{11} + a_{12} b_{21}$$
$$c_{12} = a_{11} b_{12} + a_{12} b_{22}$$
$$c_{21} = a_{21} b_{11} + a_{22} b_{21}$$
$$c_{22} = a_{21} b_{12} + a_{22} b_{22}$$

(Note that the rows of matrix A are combined with columns of matrix B. As a consequence, the product AB is generally not equal to BA; this is an important difference to the algebra of ordinary numbers.)

The role of unity is played by a matrix denoted by E, for which the elements of the principal diagonal**) are unity, and all other elements are zero, e.g.:

$$E = \begin{bmatrix} 1 & 0 \\ 0 & 1 \end{bmatrix}$$

A matrix of this type is called a unit matrix or identity matrix. It is easily verified that for any other matrix A of the same order

$$EA = AE = A$$

A column matrix may represent a vector, *e.g.*

---

* Some authors use curved, bold type brackets.
** Principal diagonal: top left to bottom right.

$$P = \begin{bmatrix} x \\ y \\ z \end{bmatrix}$$

This statement requires perhaps illustration. Consider a vector in a Cartesian coordinate system. The length and direction of a vector are defined by the coordinates x, y, z of one end of the vector, provided the other end is located at the origin. The three coordinates may be written as the column matrix P which then is said to represent the vector. Square matrices and column matrices can be multiplied together if they are of the same order. The result is again a vector.

For instance:

$$\begin{bmatrix} a_{11} & a_{12} & a_{13} \\ a_{21} & a_{22} & a_{23} \\ a_{31} & a_{32} & a_{33} \end{bmatrix} \begin{bmatrix} x_1 \\ y_1 \\ z_1 \end{bmatrix} = \begin{bmatrix} x_2 \\ y_2 \\ z_2 \end{bmatrix}$$

$$x_2 = a_{11} x_1 + a_{12} y_1 + a_{13} z_1$$
$$y_2 = a_{21} x_1 + a_{22} y_1 + a_{23} z_1$$
$$z_2 = a_{31} x_1 + a_{32} y_1 + a_{33} z_1$$

The importance of this representation of a vector lies in the matrix notation for geometric tranformations. Consider a vector r_1 in the xy plane, defined by the point with coordinates x_1, y_1 (cf. Fig. 4.6). Rotation by the angle ϑ in the xy plane transforms the vector into r_2, defined by x_2, y_2. According to the rules of analytical geometry:

$$x_2 = x_1 \cos \vartheta - y_1 \sin \vartheta$$
$$y_2 = x_1 \sin \vartheta + y_1 \cos \vartheta$$

In matrix formulation this reads

$$\begin{bmatrix} \cos \vartheta & -\sin \vartheta \\ \sin \vartheta & \cos \vartheta \end{bmatrix} \begin{bmatrix} x_1 \\ y_1 \end{bmatrix} = \begin{bmatrix} x_2 \\ y_2 \end{bmatrix}$$

Matrix multiplication results in the two above equations.

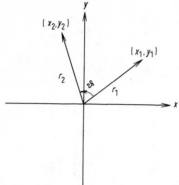

Fig. 4.6. Rotation of a vector in the *xy* plane.

Appendix 4.2. Character Tables of Some Relevant Groups

Table 4.1.

C_{2v}	E	C_2	$\sigma_v(xz)$	$\sigma_v(yz)$		
A_1	1	1	1	1	z	x^2, y^2, z^2
A_2	1	1	-1	-1	R_z	xy
B_1	1	-1	1	-1	x, R_y	xz
B_2	1	-1	-1	1	y, R_x	yz

Table 4.2.

C_{4v}	E	$2C_4$	C_2	$2\sigma_v$	$2\sigma_d$		
A_1	1	1	1	1	1	z	$x^2 + y^2, z^2$
A_2	1	1	1	-1	-1	R_z	
B_1	1	-1	1	1	-1		$x^2 - y^2$
B_2	1	-1	1	-1	1		xy
E	2	0	-2	0	0	$(x,y)\,(R_x, R_y)$	(xz, yz)

Table 4.3.

D_{3h}	E	$2C_3$	$3C_2$	σ_h	$2S_3$	$3\sigma_v$		
A_1'	1	1	1	1	1	1		$x^2 + y^2, z^2$
A_2'	1	1	-1	1	1	-1	R_z	
E'	2	-1	0	2	-1	0	(x,y)	$(x^2 - y^2, xy)$
A_1''	1	1	1	-1	-1	-1		
A_2''	1	1	-1	-1	-1	1	z	
E''	2	-1	0	-2	1	0	(R_x, R_y)	(xz, yz)

Table 4.4.

D_{4h}	E	$2C_4$	C_2	$2C_2'$	$2C_2''$	i	$2S_4$	σ_h	$2\sigma_v$	$2\sigma_d$		
A_{1g}	1	1	1	1	1	1	1	1	1	1		x^2+y^2, z^2
A_{2g}	1	1	1	-1	-1	1	1	1	-1	-1	R_z	
B_{1g}	1	-1	1	1	-1	1	-1	1	1	-1		x^2-y^2
B_{2g}	1	-1	1	-1	1	1	-1	1	-1	1		xy
E_g	2	0	-2	0	0	2	0	-2	0	0	(R_x, R_y)	(xz, yz)
A_{1u}	1	1	1	1	1	-1	-1	-1	-1	-1		
A_{2u}	1	1	1	-1	-1	-1	-1	-1	1	1	z	
B_{1u}	1	-1	1	1	-1	-1	1	-1	-1	1		
B_{2u}	1	-1	1	-1	1	-1	1	-1	1	-1		
E_u	2	0	-2	0	0	-2	0	2	0	0	(x, y)	

Table 4.5.

T_d	E	$8C_3$	$3C_2$	$6S_4$	$6\sigma_d$		
A_1	1	1	1	1	1		$x^2+y^2+z^2$
A_2	1	1	1	-1	-1		
E	2	-1	2	0	0		$(2z^2-x^2-y^2, x^2-y^2)$
T_1	3	0	-1	1	-1	(R_x, R_y, R_z)	
T_2	3	0	-1	-1	1	(x, y, z)	(xy, xz, yz)

Table 4.6.

O_h	E	$8C_3$	$6C_2$	$6C_4$	$3C_2'$ $(=C_4^2)$	i	$6S_4$	$8S_6$	$3\sigma_h$	$6\sigma_d$		
A_{1g}	1	1	1	1	1	1	1	1	1	1		$x^2+y^2+z^2$
A_{2g}	1	1	-1	-1	1	1	-1	1	1	-1		
E_g	2	-1	0	0	2	2	0	-1	2	0		$(2z^2-x^2-y^2, x^2-y^2)$
T_{1g}	3	0	-1	1	-1	3	1	0	-1	-1	(R_x, R_y, R_z)	
T_{2g}	3	0	1	-1	-1	3	-1	0	-1	1		(xz, yz, xy)
A_{1u}	1	1	1	1	1	-1	-1	-1	-1	-1		
A_{2u}	1	1	-1	-1	1	-1	1	-1	-1	1		
E_u	2	-1	0	0	2	-2	0	1	-2	0		
T_{1u}	3	0	-1	1	-1	-3	-1	0	1	1	(x, y, z)	
T_{2u}	3	0	1	-1	-1	-3	1	0	1	-1		

Appendix 4.3. Splitting of Terms, arising from d^n Configurations

Table 4.7.

Type of Term	Symmetry Environment	
	O_h	T_d*)
S	A_{1g}	A_1
P	T_{1g}	T_2
D	$E_g + T_{2g}$	$E + T_2$
F	$A_{2g} + T_{1g} + T_{2g}$	$A_2 + T_1 + T_2$
G	$A_{1g} + E_g + T_{1g} + T_{2g}$	$A_1 + E + T_1 + T_2$
H	$E_g + 2T_{1g} + T_{2g}$	$E + T_1 + 2T_2$
I	$A_{1g} + A_{2g} + E_g + T_{1g} + 2T_{2g}$	$A_1 + A_2 + E + T_1 + 2T_2$

Type of Term	Symmetry Environment	
	D_{4h}	D_3*)
S	A_{1g}	A_1
P	$A_{2g} + E_g$	$A_2 + E$
D	$A_{1g} + B_{1g} + B_{2g} + E_g$	$A_1 + 2E$
F	$A_{2g} + B_{1g} + B_{2g} + 2E_g$	$A_1 + 2A_2 + 2E$
G	$2A_{1g} + A_{2g} + B_{1g} + B_{2g} + 2E_g$	$2A_1 + A_2 + 3E$
H	$A_{1g} + 2A_{2g} + B_{1g} + B_{2g} + 3E_g$	$A_1 + 2A_2 + 4E$
I	$2A_{1g} + A_{2g} + 2B_{1g} + 2B_{2g} + 3E_g$	$3A_1 + 2A_2 + 4E$

Suggested Additional Reading

[1] F. A. Cotton, *Chemical Applications of Group Theory*, John Wiley and Sons, New York, 1963.
[2] M. Orchin and H. H. Jaffé, *Symmetry, Orbitals and Spectra*, Wiley Interscience, New York, 1971.

---

* No center of inversion; subscript g inapplicable.

5. Ligand Field Theory (LFT)

Ligand field theory is one of the two leading theories for interpreting the electronic structure of transition metal complexes (the other being molecular orbital theory, Chapter 6). A ligand field is an electrostatic field set up by the atoms immediately adjacent to the central transition metal ion. This field provides an additional potential to perturb the wave functions of the metal ion. LFT studies the correlation of physical properties – particularly spectral and magnetic properties – of transition metal complexes with the effects of the ligand field on the electronic structure of the central ion.

To a first approximation, the ligand atoms may be represented as point charges (or point dipoles in the case of neutral ligand molecules). A quantum mechanical treatment of this conceptionally simple model was given in 1929 by Bethe [1], but for crystals (NaCl) rather than for soluble transition metal complexes. As a consequence, ligand field theory in this approximation is often referred to as "crystal field theory". For soluble transition metal complexes, this name is a somewhat unfortunate one. The expression "crystal field" is normally used for the Madelung lattice potential which concerns the electrostatic effects not only of nearest neighbors, but of the whole crystal lattice, on a certain lattice element.

To a second approximation, LFT permits one also to take into account, to a certain extent, the electronic structure of the ligands. This is done, within the framework of the formalism of the purely electrostatic model, by adjusting empirically certain parameters which take care of interelectronic effects. Some authors apply the expression ligand field theory only to this second approach. In general, however, the term LFT appears now to be accepted to cover all aspects of the manner in which the electronic structure of the central transition metal ion is influenced by the nearest neighbors of the latter [16–20].

Particular questions of bonding between transition metal and ligands, on the other hand, are better understood by making use of the ideas and concepts of molecular orbital (MO) theory.

5.1. The Splitting of d Orbitals in Electrostatic Fields

In this section we shall treat qualitatively the effect of the electrostatic field of the ligands on the electronic structure of the central ion. Since we are dealing with transition metal complexes, we can limit ourselves to electrons in d orbitals.

5.1.1. The Octahedral Field

Consider first a transition metal ion possessing just one d electron, *e.g.* Ti^{3+}. In the free ion, the electron has the choice among the five d orbitals, because they are degenerate, and the total energy of the free ion will not depend upon which of the five d wave functions describes the electron.

Let us now place the ion in a ligand environment of, say, six water molecules arranged at the corners of a regular octahedron (*cf.* Fig. 4.1). The neutral water molecules are dipoles, they will arrange themselves with their negative ends directed towards the metal cation. We may represent the whole arrangement as in Fig. 5.1, in a Cartesian coordinate system, with the ligands on the axes and the metal ion at the origin.

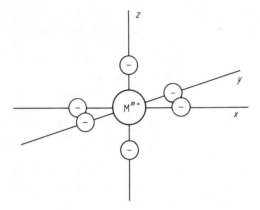

Fig. 5.1. The electrostatic environment of a transition metal cation, M^{m+}, surrounded by an octahedral array of negative charges.

Evidently the d electron of the Ti^{3+} ion will suffer a repulsion from the negative charges, increasing its energy. The repulsion will be the greater the closer the electron comes into the vicinity of the charges. Two of the five d orbitals point exactly in the direction of the ligands, namely d_{z^2} and $d_{x^2-y^2}$, whereas the other three, d_{xz}, d_{yz} and d_{xy}, have their lobes of greatest electron density located in the regions between the Cartesian axes (see Fig. 2.4). Hence it now becomes energetically more favorable for the system, if the electron remains in one of the latter orbitals where the repulsion is less. Since the spatial orientation of these three orbitals relative to the negative charges of the six ligands is identical, they remain degenerate; the same is true for d_{z^2} and $d_{x^2-y^2}$.

Fig. 5.2. gives schematically the effect of the ligand field on the energy levels of the five d orbitals of the central ion. It should be noted that the d orbitals are all affected by the presence of the ligand atoms, but d_{z^2} and $d_{x^2-y^2}$ more so than the other three. The levels are not drawn to scale. The elevation of the set of d orbitals as a whole is of the order of twenty to forty electron volts, whereas the energy gap between the two partial sets of d orbitals is generally of the order of only one to three electron volts [3]. But our main interest in this

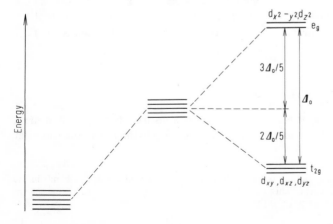

Fig. 5.2. Energy level diagram illustrating the splitting of the d orbitals in an octahedral ligand field.

chapter resides in the electronic population of the different d levels and in transitions between them. Therefore we shall focus our attention exclusively on the right hand part of Fig. 5.2, disregarding the shifts in the average energy of the d orbitals.

The energy difference between the two sets of d orbitals is designated as Δ_0, where the subscript o stands for octahedral. (Sometimes the older designation 10 Dq is used instead of Δ_0.) The energy gap Δ_0 depends on the strength of the ligand field. Ligands such as CO or CN^- produce a strong ligand field; they cause large splittings, in the range of 30,000 cm^{-1}, whereas Cl^- or Br^- are weak ligands, causing small Δ_0 of about 10,000 $cm^{-1*)}$; ammonia and water have intermediate Δ_0 values. The observation that neutral molecules may produce a stronger ligand field than negative ions can barely be understood on the basis of simple electrostatic theory alone; it indicates the necessity of taking also covalency into account. Nevertheless it is convenient to carry the simple electrostatic model of ligand influences as far as possible, and correct for covalent effects later on.

A theorem of quantum mechanics requires that, if a set of degenerate orbitals is split by a perturbation which is purely electrostatic in nature, the average energy of the perturbed levels remain unchanged ("center of gravity rule"). Hence the decrease in energy of the threefold degenerate level must equal the increase of the twofold (refer to Fig. 5.2). This is expressed by placing d_{xz}, d_{yz}, d_{xy} 2 $\Delta_0/5$ below, and d_{z^2}, $d_{x^2-y^2}$ 3 $\Delta_0/5$ above the level of the unperturbed d orbitals.

We could have used group theory to find out quickly that the fivefold degenerate d level is split into two levels in an octahedral field. Bethe [1] showed that this powerful tool can be used to solve problems of this type. Let us remember briefly how group theory is related to the problem under consideration. We have seen in Chapter 4 that if a molecule belongs to a certain symmetry group, then the wave functions describing the electrons of the molecule possess the same transformation properties under the symmetry operations of the group as do the irreducible representations of the group. For the d orbitals, this information is included in the character tables, and we can read directly from Table 4.6 that in the octahedral symmetry d_{z^2} and $d_{x^2-y^2}$ together form a basis for the twofold degenerate representation E_g, and d_{xy}, d_{xz}, d_{yz} together form a basis for the threefold degenerate representation T_{2g}. It is customary to label the orbitals with the group theoretical symbols; this has been done in Fig. 5.2. All through Section 5.1 we neglect interelectronic repulsion, *i.e.* we deal with "hydrogen-like" or "one-electron" orbitals. In Section 5.2 we shall consider interelectronic repulsion, *i.e.* deal with terms. Following the rules introduced in Chapters 2 and 3, we use lower case letters for the hydrogen-like orbitals and also for the group theoretical (or symmetry) labels associated with them, whereas capital letters are used for terms and their symmetry labels.

5.1.2. High-Spin and Low-Spin Complexes

We shall now discuss what happens if two or more electrons have to be placed in the d orbitals split by an octahedral field into the t_{2g} and e_g levels. Up to three electrons will enter

---

* Since the magnitude of orbital splitting is accessible from the position of absorption bands in the optical spectra (Section 5.2) it is convenient to use the same unit, cm^{-1}, for the frequency in the spectra and for Δ. Note that 10,000 $cm^{-1} = 1.24$ eV $\simeq 120$ kJ/mol.

the t_{2g} level, each occupying one of the three degenerate orbitals, with their spins all parallel (Hund's first rule, the rule of maximum multiplicity, *cf.* Section 3.3.4). According to the Aufbau principle, the next three electrons should also enter the t_{2g} level before the higher e_g level is started to be filled. The Pauli principle requires spin pairing to occur whenever two electrons have to occupy the same orbital. Spin pairing, however, is an unfavorable, energy consuming process (pairing energy P). So, if the energy gap Δ_0 is relatively small – more precisely, if $\Delta_0 < P$ – the fourth and the fifth electron will occupy the e_g level, and only the sixth electron will be obliged to enter a t_{2g} orbital with spin pairing. These considerations show that the occupancy of the two levels, and hence the number of unpaired electrons depend on the ligand field strength. This is shown in Fig. 5.3 for the case of a d^6 ion.

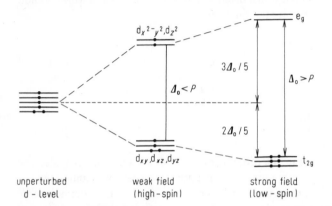

Fig. 5.3. Occupancy of the t_{2g} and e_g levels in an octahedral d^6 complex.

In a weak ligand field ($\Delta_0 < P$), the distribution of the electrons among the d orbitals is the same as it would be in the unperturbed, fivefold degenerate d level; this is called the "high-spin" case. An example is the Co^{3+} complex $[CoF_6]^{3-}$ which has four unpaired electrons, and hence is paramagnetic. With strong field ($\Delta_0 > P$), spin pairing causes "low-spin" complexes; $[Co(NH_3)_6]^{3+}$, *e.g.*, is diamagnetic.

For d^1–d^3 and d^8–d^{10}, there is only one possibility in each case to arrange the electrons in the scheme. For d^4–d^7, on the other hand, the configuration depends on the ligand field strength. This is shown in Table 5.1; the numbers of unpaired electrons are given in parentheses.

Table 5.1. Distribution of d electrons in octahedral ligand fields. (Number of unpaired electrons in parentheses.)

Configuration	Low field		High field	
d^1		t_{2g}^{1} (1)		
d^2		t_{2g}^{2} (2)		
d^3		t_{2g}^{3} (3)		
d^4	$t_{2g}^{3}e_{g}^{1}$ (4)		t_{2g}^{4}	(2)
d^5	$t_{2g}^{3}e_{g}^{2}$ (5)		t_{2g}^{5}	(1)
d^6	$t_{2g}^{4}e_{g}^{2}$ (4)		t_{2g}^{6}	(0)
d^7	$t_{2g}^{5}e_{g}^{2}$ (3)		$t_{2g}^{6}e_{g}^{1}$	(1)
d^8		$t_{2g}^{6}e_{g}^{2}$ (2)		
d^9		$t_{2g}^{6}e_{g}^{3}$ (1)		

Evidently, the distribution of the electrons among the two levels will have a bearing on the spectra and the magnetic properties of complexes with $d^4 - d^7$ configurations (Sections 5.2 and 5.3).

5.1.3. Fields other than Octahedral

So far we have considered only octahedral complexes. In a similar manner, we can determine the splitting of the d orbitals in environments of other symmetries which we may encounter in transition metal complexes, such as tetrahedral, square-planar, trigonal bipyramidal, *etc.* The most important point is to find out which orbitals remain degenerate and which do not. Group theory gives us the exact answer to this question, *e.g.*:

tetrahedral	d_{xy}, d_{xz}, d_{yz}:	t_2
(T_d, Table 4.5)	$d_{z^2}, d_{x^2-y^2}$:	e
square planar	d_{z^2}:	a_{1g}
(D_{4h}; Table 4.4)	$d_{x^2-y^2}$:	b_{1g}
	d_{xy}:	b_{2g}
	d_{xz}, d_{yz}:	e_g

But group theory is not able to tell us the ordering of the orbitals in the energy scale. The numerical evaluation of the energies is a matter of quantum chemical computation, and as such beyond the scope of this book. In the more simple cases, however, a common sense consideration as that carried out in Section 5.1.1. for the octahedral ligand environment will permit at least a qualitative ordering.

Let us consider what happens to the d levels if starting from an octahedral complex, we increase the distance between the central ion and the two ligands on the z axis (the axial ligands), maintaining the position of the four ligands in the *x-y* plane (the equatorial ligands) unchanged. From the point of view of geometry we go from an octahedron to a tetragonal bipyramid (tetragonally distorted octahedron) and finally, if the two ligands are removed completely, we have a square-planar arrangement of the four remaining ligands (Fig. 5.4; both types of complexes belong to the symmetry group D_{4h}).

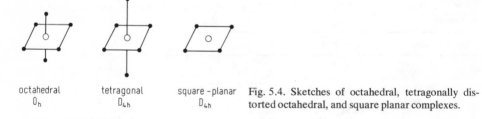

octahedral   tetragonal   square-planar   Fig. 5.4. Sketches of octahedral, tetragonally dis-
O_h D_{4h} D_{4h} torted octahedral, and square planar complexes.

Concerning the orbital energies, the repulsive effect between the two axial ligands and the d_{z^2} orbital will diminish as the ligands move away. In other words, the degeneracy of the e_g level is lifted, d_{z^2} becoming more stable than $d_{x^2-y^2}$ (see Fig. 5.5). A similar splitting occurs in the t_{2g} level. The energy of d_{xz} and d_{yz} is lowered, but these two orbitals remain degenerate, because their spatial distribution with respect to the axial as well as to the equatorial ligands is analogous.

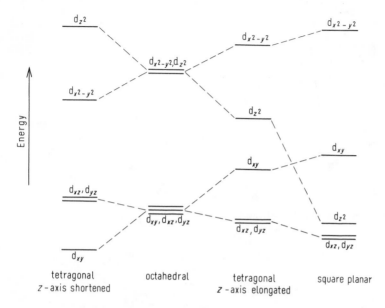

Fig. 5.5. Splitting of the d orbitals in ligand fields of several different symmetries.

Qualitatively, the same effect of orbital splitting would also be observed if, in an octahedral complex, the two axial positions were occupied by ligands which make a smaller contribution to the ligand field than do the equatorial ones.

For a square planar complex, *i.e.* in the complete absence of axial ligands, the separation of the originally degenerate orbitals is even more pronounced, as shown in Fig. 5.5. The orbital d_{z^2} has dropped below the d_{xy} level. (It may even become the most stable of all d orbitals, in certain square-planar ligand fields.)

It will be recognized easily that tetragonal distortion of the octahedron caused by bringing the two axial ligands nearer to the central ion splits the t_{2g} and e_g orbitals into exactly the same number of levels, but with the energy ordering reversed (repulsion in $d_{z^2} > d_{x^2-y^2}$, in $d_{xz}, d_{yz} > d_{xy}$).

For a tetrahedral complex, the d orbital splitting can be derived by an analogous line of reasoning. From the character table of the T_d group we know that one twofold degenerate set (e: d_{z^2}, $d_{x^2-y^2}$) and one threefold degenerate set (t_2: d_{xy}, d_{xz}, d_{yz}) result. In order to find out the relative energies, we place the tetrahedron within a cube (the regular tetrahedron is obtained by taking the alternate corners of the cube). The axes of a Cartesian coordinate system are chosen to pass through the cube face centers, as shown in Fig. 5.6.

The orbitals of the e set are directed through the cube face centers, along the axes; the t_2 orbitals are directed through the middle of the cube edges. An inspection of Fig. 5.6 indicates that the t_2 orbitals come nearer to the position of the ligand atoms, than do the e orbitals. Consequently it is the t_2 set which suffers a stronger repulsion in a tetrahedral ligand field. The relative energies of the two sets are inverted with respect to the splitting in an

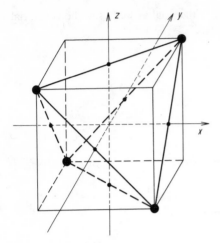

Fig. 5.6. Tetrahedron with coordinate system fixing the spatial orientation of the d orbitals.

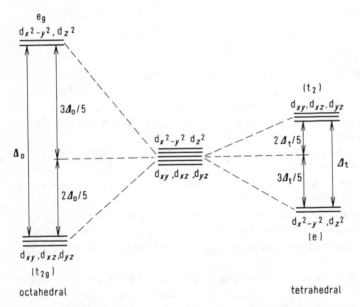

Fig. 5.7. Comparison of d orbital splitting in octahedral and tetrahedral ligand fields.

octahedral field (Fig. 5.7). It is also evident from Fig. 5.6 that the energy difference between the two sets is not as pronounced as in the case of an octahedral complex. It has been calculated that the energy gap in a tetrahedral complex, Δ_t, is roughly 4/9 of Δ_0 for comparable complexes (the same central metal ion and the same type of ligands). This has also been taken into account in the energy level diagram given in Fig. 5.7.

5.1.4. Ligand Field Stabilization Energy

We have recognized that the spatial orientation of the d orbitals is responsible for their energy separation in ligand fields. The preferential filling of the lower orbitals brings the metal ion into an energetically more favorable and hence more stable situation, as compared with the situation in which all electrons would be at the same (average) level. The energy difference is called the ligand field stabilization energy (LFSE). Electrons in the higher orbitals of course, offset this effect. The total LFSE may be expressed in terms of the energy gap Δ. For octahedral fields, each electron in the t_{2g} level brings a stabilization by $2\,\Delta_0/5$ (*cf.* Fig. 5.7); electrons in the e_g level destabilize by $3\,\Delta_0/5$ each. Correspondingly, for tetrahedral fields the stabilization per electron in the e level is $3\,\Delta_t/5$, whereas electrons in the t_2 level destabilize by $2\,\Delta_t/5$. The LFSE's for octahedral and tetrahedral fields and for all d^n configurations, are shown in Table 5.2.

Table 5.2. Ligand field stabilization energy (LFSE) for octahedral and tetrahedral complexes*).

Configuration	Octahedral		Tetrahedral	
	low field	high field	low field	high field
d^1	$-2\,\Delta_0/5$		$-3\,\Delta_t/5$	
d^2	$-4\,\Delta_0/5$		$-6\,\Delta_t/5$	
d^3	$-6\,\Delta_0/5$		$-4\,\Delta_t/5$	$-9\;\Delta_t/5 + \Pi$
d^4	$-3\,\Delta_0/5$	$-8\;\Delta_0/5 + \Pi$	$-2\,\Delta_t/5$	$-12\,\Delta_t/5 + \Pi$
d^5	0	$-10\,\Delta_0/5 + \Pi$	0	$-10\,\Delta_t/5 + \Pi$
d^6	$-2\,\Delta_0/5$	$-12\,\Delta_0/5 + \Pi$	$-3\,\Delta_t/5$	$-8\;\Delta_t/5 + \Pi$
d^7	$-4\,\Delta_0/5$	$-9\;\Delta_0/5 + \Pi$	$-6\,\Delta_t/5$	
d^8	$-6\,\Delta_0/5$		$-4\,\Delta_t/5$	
d^9	$-3\,\Delta_0/5$		$-2\,\Delta_t/5$	

From a first inspection of Table 5.2, and taking into account that $\Delta_t \simeq 4\,\Delta_0/9$ for comparable complexes, it appears that in all cases (except d^5 in a low field) the octahedral arrangement should be the more favorable. In fact, octahedral (or approximately octahedral) complexes are by far more frequent than tetrahedral ones. Nevertheless quite a number of tetrahedral complexes exist and are even relatively stable. This indicates that LFSE cannot be the only magnitude to decide the symmetry of a complex. Further influences are expected to arise from the bulkiness of the central ion and the ligands and from the metal-ligand distances.

5.2. Electronic Spectra of Transition Metal Complexes

Most of the transition metal complexes are colored, which means they are capable of absorbing energy in the visible region of the spectrum. We shall see in this section that the observed color is caused by electronic transitions among the various d levels established by the symmetry of the ligand field, and that the experimental spectra may be understood and interpreted very well in terms of ligand field theory.

---

* Π stands for the total pairing energy of the electrons involved which has to be taken into account when judging the energy gain effected by going from high spin to low spin complexes.

The electronic spectra may be utilized in a number of ways in the field of homogeneous catalysis. Spectral measurements may not only serve to characterize a new catalyst and to elucidate its stereochemistry, but they can also be employed to follow reaction paths and rates.

5.2.1. Band Intensities

The general appearance of the absorption spectrum of a transition metal complex is shown schematically in Fig. 5.8.

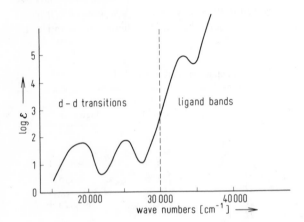

Fig. 5.8. General aspect of an absorption spectrum of a transition metal complex.

In the range from 10,000 to 30,000 cm^{-1}, *i.e.* in the visible and near infrared region, one usually observes one or several bands of relatively low intensity (molar extinction coefficient ε between 1 and 1000). These bands are ascribed to the d→d transitions. In the ultraviolet region the spectrum contains mostly several very intense bands, with ε values from 10,000 to 100,000. These bands correspond to electron transfer within the ligands (inner ligand transitions) and to charge transfer from the central metal ion to the ligands or *vice versa* (charge transfer bands). Together they are designated as ligand bands. In most cases, only a strong intensity increase in the range of about 30,000 cm^{-1} can be observed at concentrations convenient for a good determination of the d→d bands.

From the point of view of ligand field theory we are interested only in the d→d bands. Their low intensity is due to the fact that transitions between orbitals of equal parity are formally forbidden by quantum mechanical selection rules. The parity is concerned with the symmetry behaviour of the wave functions with respect to inversion through the origin of the coordinate system; an orbital may be symmetric (g) or antisymmetric (u) [*cf.* Section 4.3.3.]. The d orbitals are inherently of g character and hence d→d transitions are "parity forbidden". But there are mechanisms which afford some relaxation of this rule. They involve the "admixture" of wave functions of other parity (*e.g.* p orbitals which are antisymmetric), which becomes possible if the compound as a whole has no inversion center [4]. In tetrahedral compounds this is always the case, and in fact tetrahedral transition metal complexes have in general relatively intense bands (ε in the range 100–1000). Octahedral complexes, on the other hand, do have an inversion center, but this symmetry may be slightly perturbed by suitable vibrations of the ligands. The term "vibronic coupling" is used to denote the combination of electronic and vibrational wave functions. But the relaxation of the parity forbiddenness by this mechanism is not very effective; octahedral complexes have usually d→d bands with $\varepsilon < 100$. Summarizing we note: the relatively low intensities observed in the d→d bands are indicative of the essentially "forbidden" character of these transitions.

5.2.2. Spectra of d^1 Systems

We shall first consider a d^1 system in an octahedral ligand field, say $[Ti(H_2O)_6]^{3+}$. The spectrum consists of just one broad absorption band at *ca.* 20,400 cm^{-1} (Fig. 5.9). In fact it was in connection with the interpretation of this simple absorption, that Ilse and Hartmann (1951) first indicated the importance of LFT in the study of absorption spectra [5].

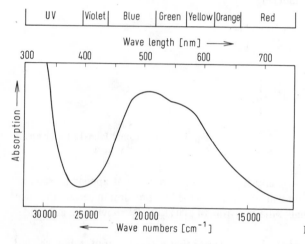

Fig. 5.9. Absorption of $[Ti(H_2O)_6]^{3+}$.

The absorption band covers the blue green region of the visible spectrum. The color of a compound being determined by those wave lengths which are not absorbed, but transmitted, the complex $[Ti(H_2O)_6]^{3+}$ is pale red-purple.

In the ligand field model, this single absorption corresponds to the promotion of the one d electron from the t$_{2g}$ to the e$_g$ level; absorption occurs when $h\nu = \Delta_0$ (*cf.* Fig. 5.7). Evidently tetrahedral d^1 complexes are expected to show also just one absorption, although in this case the transition is e→t$_2$.

Fig. 5.9 reveals a typical aspect of the spectra of transition metal complexes: the absorption bands are generally rather broad. This implies that the energy levels, between which the electron transfer occurs, are spread over a comparable range of energies. The ligand field depends upon the metal-ligand distance. Vibrations modulate this distance. The energies of the fundamental modes of vibrations and their thermally accessible harmonics are of the order of several hundreds of cm^{-1}. Since in most cases the energies of the sublevels of ground state and excited state do not run parallel to one another, a band breadth of the order of 10^4 cm^{-1} may be accounted for by this mechanism [18]. One would expect to see a fine structure in the band, arising from the many possible transitions between the individual vibration levels of ground and excited states. But usually only the envelope of the fine structure lines is observed.

In many cases, and particularly with $[Ti(H_2O)]^{3+}$, another important factor contributes to the breadth of the absorption bands. There is a generally valid theorem of Jahn and Teller [6] which states that many-atom compounds with an orbitally degenerate ground state are unstable. Such systems stabilize themselves by a slight distortion of the ligands such that the symmetry is lowered and the orbital degeneracy is lifted. Octahedral complexes tend to tetragonal or trigonal distortion. A somewhat simplified idea of the driving force of the Jahn-Teller effect may be obtained from a consideration of the splitting of the energy levels when going from a purely octahedral arrangement to a slightly tetragonally distorted one with the axial ligands nearer to the metal than the equatorial ligands (*cf*. Fig. 5.5, left hand side). Usually, the Jahn-Teller splitting is small as compared with Δ_0, but nevertheless there is a definite stabilization of the system, if the single electron of a d^1 complex is placed in the b_{2g} orbital of the distorted arrangement (Fig. 5.10).

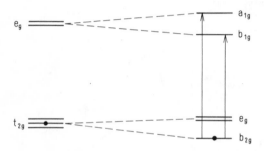

Fig. 5.10. Tetragonal Jahn-Teller distortion of an octahedral d^1 system.

The two possible transitions $b_{2g} \rightarrow a_{1g}$ and $b_{2g} \rightarrow b_{1g}$ will normally not appear as separate lines in the absorption spectrum, but will contribute to the line breadth; in certain instances a shoulder will be observed. (Actually, the spectrum of $[Ti(H_2O)_6]^{3+}$ shows such a shoulder, *cf*. Fig. 5.9.)

In the course of the next section we shall demonstrate that it is quite predictable which d^n configurations are liable to Jahn-Teller deformations and which are not.

5.2.3. The Splitting of Free Ion Terms in Ligand Fields

Whenever we have to deal with transition metal complexes, where the central ion has more than one electron in the d orbitals, the situation is complicated by the interelectronic repulsion. As discussed in Section 3.3, this perturbation causes, even in the free ion, a splitting of the energy levels. Hence, if a d^n ion (with $n > 1$) is placed into a ligand field, there are evidently two different types of perturbation to be considered: one arising from the interelectronic repulsion and the other originating from the electrostatic ligand field. It cannot be said *a priori* which perturbation will have the stronger effect on the energy levels of the d orbitals, because this depends on the strength of the ligand field. Two approaches have been adopted to evaluate the combined influence of both perturbations. The first assumes formally the interelectronic repulsion to be the most important effect. By this route one determines first the terms arising from the given d^n configuration, and in a further step one studies the influence of the ligand field on each of the terms. This method is called the "weak field approach". In the second, the "strong field approach", one proceeds the other way round: the ligand field is considered first, and the interelectronic repulsion is treated as a perturbation of the energy levels arising from the ligand field splitting. If carried out properly, both approaches should lead to the same result [17]. We shall restrict ourselves to a qualitative description of the weak field approach.

The first step has already been carried out in Section 3.3; the results have been summarized in Tables 3.3 and 3.4. We recall that the terms of a free ion are characterized by the Russell-Saunders symbol ^{2S+1}L. We ask now for number and energy ordering of the component terms arising from each free ion term as a consequence of the influence of the electrostatic ligand field.

As with one-electron orbitals (Section 5.1), group theory is the tool for the determination of the number of energy levels. At the end of the chapter "Symmetry and Group Theory" we have already anticipated the splitting of terms in different symmetry environments (Section 4.4.). Making use of Tables 3.4 and 4.7, we shall now study the splitting which the ground terms of all d^n configurations suffer in an octahedral ligand field.

The number and type of components into which a certain ligand field will split a term of a given L is the same regardless of the d^n configuration from which the term arises. Hence the ground terms of d^1, d^4, d^6 and d^9 which are all D terms (Table 3.4), are split by the octahedral field into T_{2g} and E_g (Table 4.7). The terms of the complex ions are characterized by the group theoretical label of the relevant nonreducible representation of the group, using capital letters (as always with terms). An important point concerning the multiplicity has to be mentioned. The chemical environment does not interact directly with the electron spins, and therefore all components into which a particular term is split by the ligand field have the same spin multiplicity as the parent term (see Table 5.3.).

Table 5.3. Splitting of the ground state terms of all d^n configurations in an octahedral field [17].

Configuration	Examples	Splitting		Splitting		Examples	Configuration
d^1	Ti^{3+}	^2D	^2E_g / $^2T_{2g}$	^2D	$^2T_{2g}$ / ^2E_g	Cu^{2+}	d^9
d^2	V^{3+}	^3F	$^3A_{2g}$ / $^3T_{2g}$ / $^3T_{1g}$	^3F	$^3T_{1g}$ / $^3T_{2g}$ / $^3A_{2g}$	Ni^{2+}	d^8
d^3	Cr^{3+} V^{2+}	^4F	$^4T_{1g}$ / $^4T_{2g}$ / $^4A_{2g}$	^4F	$^4A_{2g}$ / $^4T_{2g}$ / $^4T_{1g}$	Co^{2+}	d^7
d^4	Mn^{3+} Cr^{2+}	^5D	$^5T_{2g}$ / ^5E_g	^5D	^5E_g / $^5T_{2g}$	Co^{3+} Fe^{2+}	d^6
d^5	Fe^{3+} Mn^{2+}	^6S	$^6A_{1g}$				

Group theory does not indicate the energy sequence of the new levels. Perturbation theory has to be applied to the quantum mechanical description of the free ion, in order to obtain quantitative data. But again (as in Section 5.1.1.), a common sense consideration may help to find a qualitative answer at least in some cases.

A D term of a free ion is fivefold degenerate. In order to imagine the extension of the five wave functions in space we may rely upon the symmetry of the d orbitals. Hence two of the five components of a D term will have their maximum extension in the direction of the axes, and the other three in the directions between these axes, in the xy, xz and yz planes. The one electron of ^2D originating from d^1 will suffer less repulsion from the ligand electrons, if it is located in one of the three latter component terms, which together form the triply degenerate T_{2g} set. Hence the energy ordering is $T_{2g} < E_g$*).

To understand the situation in the d^9 case, we have to return to the "hole formalism" introduced in Section 3.3.3. Nondegenerate terms, the same as "hydrogen-like orbitals", are able to house two electrons each. If a fivefold degenerate free ion D term contains 9 electrons, there is one "hole" which, from an electrostatic point of view, may be treated as a positron. A positron will be most stable in those regions where an electron is least stable, and *vice versa*. Therefore the energy pattern is reversed for the splitting of the ^2D term arising from d^9: $E_g < T_{2g}$.

The half-filled shell configuration, d^5, gives rise to the nondegenerate, centrosymmetric ^6S ground term of the free ion; it is not split by the field. (Nevertheless the term observed in the field is designated by its group theoretical label, $^6A_{1g}$.) The ^5D ground term of d^4 has one hole as compared with the half-filled shell. Hence the situation is similar to that of the d^9 case: $E_g < T_{2g}$. The ^5D ground term of d^6, on the other hand, has one electron more than the centrosymmetric ^6S term; we may compare the situation with the d^1 case: $T_{2g} < E_g$.

Similar arguments apply to the F terms, although it is more difficult to imagine the spatial extension of the corresponding seven degenerate F wave functions. The splittings of the ground terms of all d^n configurations are summarized in Table 5.3. The table is arranged to emphasize the close correspondence of d^n and d^{10-n} configurations, and the terms arising from them which is, as we have seen, a consequence of the reciprocity of electrons and holes.

So far we have discussed only the splitting of the ground terms in octahedral ligand fields. In a tetrahedral complex, the number and type of the component terms are the same as in the octahedral ones (*cf.* Table 4.7.), but the pattern of energies is reversed. This is quite analogous to, and has the same reasons as, the results for the d^1 ions treated in Section 5.1.3.

Table 5.3. permits us to predict which of the d^n configurations will be liable to Jahn-Teller deformation in an octahedral field. The totally symmetric arrangement of the six ligands is stable only if the ground term is orbitally nondegenerate; this is the case with d^3, d^5 and d^8 ions. All others have T_{2g} or E_g ground terms**), i.e. without Jahn-Teller distortion the ground state would be orbitally degenerate. The symmetry of the ligand arrangement must be lowered to such an extent as to completely lift this degeneracy. The Jahn-Teller theorem indicates neither the direction nor the magnitude of the distortion to be expected. In many instances the deviations from the regular octahedron are large enough to be detected by X-ray crystallographic analysis. The most frequently observed distortion is the tetragonal; for

---

* The same conclusion was, of course, arrived at from the simpler one electron orbital picture of d^1 species in Section 5.1.2.

** The statement as given is valid only for high-spin complexes (*cf.* Section 5.2.4.).

instance nearly all "octahedral" Cu^{2+} complexes (d^9, ground term ^2E_g) and Mn^{3+} complexes (d^4, ground term ^5E_g) have actually D_{4h} symmetry, because of a slight tetragonal distortion.

5.2.4. Energy Level Diagrams

As already mentioned in the preceding section, the determination of the magnitude of the splitting produced by the ligand field, as well as the quantitative ordering of the various energy levels, require quantum mechanical computations of some sophistication.

The perturbing potential caused by the ligand field is a function of the distance between the central ion and the ligands, of the effective charges, and of the dipole moments of the ligands. Usually, none of these data is known with sufficient exactitude. The procedure adopted to overcome this difficulty is the following: the quantum mechanical calculations are carried out maintaining the uncertain magnitudes as parameters in the equations. Actually these parameters which characterize a particular ligand field, are implicitly contained in the magnitude Δ, the ligand field splitting. Hence the computed energy levels are generally given in a diagram, having as ordinate the energy and as abscissa the parameter Δ, measuring the strength of the ligand field. In this form, the energy level diagrams are known as Orgel diagrams [7], Figs. 5.11. and 5.12. show Orgel diagrams for d^2 and d^8 ions in octahedral fields. At the extreme left of the two graphs, at $\Delta_0 = 0$, we have the Russell-Saunders terms of the free ions (V^{3+} and Ni^{2+} respectively). Each of these terms splits up into the components specified in Table 4.7. With increasing field strength the splitting becomes larger. It will be noted that not only the ground term, but also higher terms of the free ion are shown. Some of them, although not all, are important for the interpretation of spectra (Section 5.2.5.). They are marked by heavier lines.

The qualitative aspect of the graphs in Figs. 5.11 and 5.12 is similar. There is, however, a reversal of the energy ordering within the group of components arising from each free ion term, just as already noted for the ground state terms in Table 5.3 (reciprocity of electrons

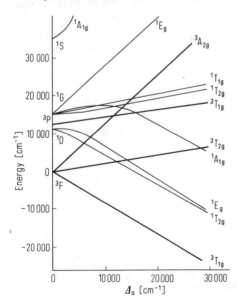

Fig. 5.11. Energy level diagram (Orgal diagram) for a d^2 ion (V^{3+}) in an octahedral ligand field. (Heavier lines mark those levels which are important for spectra).

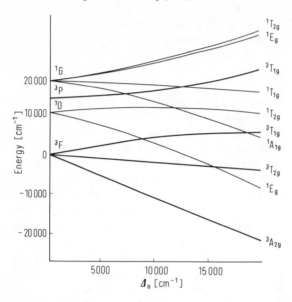

Fig. 5.12. Energy level diagram (Orgel diagram) for a d^8 ion (Ni^{2+}) in an octahedral field.

and holes). The graphs show also that some of the lines are straight and others are curved. The curvature is due to the fact that terms of identical labeling (*i.e.* equal symmetry behavior and multiplicity) interact with each other. The corresponding lines never cross ("non-crossing rule"), instead they suffer a repulsion, *i.e.* they are bent away from each other. This repulsion may even change the energy pattern (compare, for instance, the components of the ^1G terms in Figs. 5.11 and 5.12).

Fig. 5.13 shows the lower energy part of the energy level diagram for a d^4 ion. There is one important difference as compared with Figs. 5.11 and 5.12. In the latter, one and the same term remains the ground term, independent of the strength of the ligand field. In the d^4 case, on the other hand, the ^5E_g term (quintuplet term, 4 unpaired electrons) is the ground term in weaker fields, whereas the $^3T_{1g}$ term (triplet term, two unpaired electrons) becomes lowest in energy in strong fields. Hence, the combined and quantitative consideration of interelec-

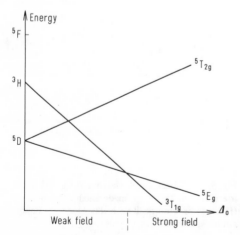

Fig. 5.13. Part of the Orgel diagram for a d^4 ion in an octahedral field.

tronic repulsion and ligand field splitting leads to the same result as the simplified and qualitative consideration of the ligand effect alone, in Table 5.1: a d^2 species has always two unpaired electrons, but a d^4 species is "high-spin" in a weak octahedral field, and "low-spin" in a strong one.

The energy level diagrams for all other d^n configurations are available in the literature (*e.g.* [16, 18, 19]). Frequently a somewhat different representation, due to Tanabe and Sugano [8] is given. Instead of using absolute energy units (for instance cm^{-1}) for the ordinate and abscissa scales, the unit is the interelectronic repulsion parameter (Racah parameter) B. The advantage is evident: the separations of the free ion terms depend in a characteristic way on B (and on $C \simeq 4 B$), as shown in Section 3.3.5., but B is different from metal to metal. If B is used as the energy unit, each d^n diagram becomes generally applicable, and is no longer restricted to just the one particular ion, whose free ion separations have been used in establishing the diagram. A further peculiarity of the Tanabe-Sugano diagrams results from the fact that the energy level of the lowest term is always taken as the zero of energy. This gives a somewhat different aspect to these diagrams, particularly in cases where a change of the ground term occurs. Fig. 5.14 represents a part of the Tanabe-Sugano diagram for d^4 (the same terms as in Fig. 5.13). The diagram is discontinous at the field strength at which another term becomes the ground term.

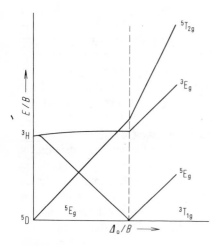

Fig. 5.14. Part of the Tanabe-Sugano diagram for a d^4 ion in an octahedral field.

The energy level diagrams for d^n systems in tetrahedral environment are closely related to those obtained for octahedral fields. We have already seen that number and types of the sublevels into which each free ion term is split by the field are the same in both cases (Table 4.7.). The difference comes from the fact that the energy pattern is just inverted. Thus, for a d^2 ion in a tetrahedral complex, the triplet terms arising from ^3F are $^3A_2 < {}^3T_2 < {}^3T_1$ (refer to Fig. 5.11.); the singlet terms arising from ^1D are $^1E < {}^1T_2$; *etc.*

5.2.5. The Interpretation of Spectra

The visible and near ultraviolet spectra of transition metal complexes are due to electron transitions from the ground term to certain excited terms. The important selection rule says:

only transitions between terms of equal spin multiplicity are allowed, all others are "spin-forbidden".

The assignment of bands is done by fitting the observed spectrum to the corresponding energy level diagram (*e.g.* Fig. 5.12. for a d^8 ion in an octahedral field). The energies of the various terms are expected to lie on a vertical line in the diagram. As an example, consider the two octahedral d^8 complexes $[Ni(H_2O)_6]^{2+}$ and $[Ni(NH_3)_6]^{2+}$ [*]. Their spectra consist of three rather well-defined bands (Table 5.4.).

Table 5.4. The electronic spectra of $[Ni(H_2O)_6]^{2+}$ and $[Ni(NH_3)_6]^{2+}$ [9, 10].

Complex	Observed energy (cm^{-1})	Predicted energy (cm^{-1})
$Ni(H_2O)_6{}^{2+}$	8,500	8,500
	13,500	14,000
	25,300	27,000
$Ni(NH_3)_6{}^{2+}$	10,750	10,750
	17,500	17,680
	28,200	30,410

The energy level diagram (Fig. 5.12.) tells us that the ground term is $^3A_{2g}$. Since only transitions between terms of the same multiplicity are allowed, the bands can be assigned to the transition $^3A_{2g} \rightarrow {}^3T_{2g}$, $^3A_{2g} \rightarrow {}^3T_{1g}$ (F) and $^3A_{2g} \rightarrow {}^3T_{1g}$ (P). In order to see how the predictions of LFT agree with the experimental spectra, we must see if there is any value of Δ which would require absorption bands at the observed energies (*cf.* Fig. 5.12.). In fact if one draws vertical lines into Fig. 5.12 in such a way that the energy gap between $^3A_{2g}$ and $^3T_{2g}$ just fits the lowest energy bands (8,500 cm^{-1} for the water complex, and 10,750 cm^{-1} for the ammonia complex), then the further absorption bands corresponding to spin-allowed transitions are predicted at wave numbers which are in reasonable agreement with experiment (see Table 5.4.). Note that for the d^8 case the abscissa value Δ and the energy gap between the lowest two triplet terms are equal. Hence Δ can be obtained directly from the maximum of the lowest energy band of the spectrum. This is not always so. In the cases of d^2 and of low spin d^4, d^6, d^7 complexes, corrections for configuration interaction and term interaction must be taken into account [17]: the value of Δ differs then by some 20% from the first band maximum.

Table 5.4. gives the additional information that the ammonia ligand causes a stronger ligand field than water does, because its Δ value (10,750 cm^{-1}) is greater than that of water (8,500 cm^{-1}).

Metal ions with the half-filled d-shell (d^5), such as Mn^{2+} or Fe^{3+} are special. Their ground term ^6S is not split by the ligand field (Table 4.7), and there are no further sextuplet terms available for spin-allowed transitions (Table 3.3). Nevertheless complexes of these ions may be weakly colored, and may show absorption spectra, although of very low intensity. They are due to "spin-forbidden" transitions to quadruplet terms which may gain some intensity

---

* For a competent description and interpretation of the spectra of many transition metal complexes see Ref. [21].

by a mechanism involving spin-orbit coupling [17]. These "intercombination bands" are some 10^2 times weaker than those due to spin-allowed transitions. In spectra containing bands from transitions between terms of equal spin multiplicity, the intercombination bands are mostly too weak to be observed. In some instances they are seen as shoulders.

The interpretation of the spectra of tetrahedral complexes does not require any additional knowledge as compared with that of octahedral ones. The same energy level diagrams are valid if the reversed ordering of the levels as well as the reciprocity of electrons and holes are properly taken into account (Section 5.2.4.). Hence, Fig. 5.12 is not only valid for a d^8 ion in an octahedral environment, but also for a tetrahedral d^2 complex*), and Fig. 5.14 may be used for a d^6 ion in a tetrahedral field, etc. As already mentioned earlier (Section 5.2.1.), tetrahedral complexes exhibit in general more intense spectra than octahedral ones.

In square planar complexes, the D and F ground state terms of the different d^n configurations (with the exception of d^5) are split in four sublevels each (D_{4h} symmetry, Table 4.7). Hence, there should be three spin-allowed transitions from the ground state term. In certain cases (e.g. $[PtCl_4]^{2-}$) the three bands are observed, but mostly one or two of them are masked by charge transfer bands of much higher extinction coefficient [11].

5.2.6. Spectrochemical and Nephelauxetic Series of Ligands

In the preceding section we have discussed how the ligand field parameter Δ can be determined from experimental spectra. We are now in a position to compare different ligand environments and metal ions with regard to the ligand field strength. Table 5.5 gives a selection of such data.

Table 5.5. Some Δ values of octahedral complexes ML_6.

Metal ion	Configuration	Δ (cm^{-1})				Ref.
		Cl$^-$	H$_2$O	NH$_3$	CN$^-$	
V(III)	d^2	11,000	17,800	–	22,200	21
Cr(III)	d^3	13,600	17,400	21,600	26,300	12
Cr(II)	d^4	–	13,900	–	–	12
Co(III)	d^6	–	20,760	22,870	32,200	21
Co(II)	d^7	–	9,300	10,100	–	12
Rh(III)	d^6	20,400	27,000	34,000	45,500	12, 21
Ni(II)	d^8	7,300	8,500	10,800	–	12

It turns out that the ligands may be put into a series, with increasing Δ values, and that this series is the same for different central ions. It is called the spectrochemical series. A broad group of ligands has been catalogued in such a series (see e.g. [21, 24]):

$$I^- < Br^- < Cl^- \simeq \underline{S}CN < N_3^- < F^- < (NH_2)_2C\underline{O} < OH^- < oxalate^{2-} \simeq malonate^{2-} <$$
$$H_2O < \underline{N}CS^- < pyridine \simeq NH_3 \simeq PR_3 < \underline{N}H_2CH_2CH_2\underline{N}H_2 \simeq \underline{S}O_3^{2-} < NH_2OH < \underline{N}O_2^- \simeq$$
$$dipyridyl \simeq o\text{-phenanthroline} < H^- \simeq CH_3^- \simeq C_6H_5^- < C\underline{N}^- \simeq \underline{C}O < \underline{P}(OR)_3$$

(In cases of doubt, the atom which coordinates to the metal has been underlined.)

---

* This is only true provided the Racah parameter B is similar; otherwise the graph gives only qualitative information, and one must have recourse to the corresponding Tanabe-Sugano diagram for quantitative relationships.

A "rule of average environment" has been set up empirically, stating that the Δ values of mixed complexes $[MA_nB_{6-n}]$ are approximately given by a linear interpolation between the Δ's of $[MA_6]$ and $[MB_6]$. Replacement of one ligand by another which stands more on the right in the series will cause a shift of the spectrum to higher wave numbers (cf. Table 5.5).

Table 5.5 indicates also some trends concerning the central ion: Δ increases with increasing oxidation state; for the same ligand and oxidation state there is no great difference in Δ within one transition metal series, but Δ increases from one series to the next.

The order of the spectrochemical series of ligands is very difficult to rationalize in terms of a purely electrostatic model, since neutral ligands such as pyridine, ammonia or water exert a stronger ligand field than the negative ions Cl^-, OH^-, etc. The series rather indicates that the ligand field parameter Δ is influenced by contributions from electrostatic effects as well as from covalency. This latter influence will be better understood from the point of view of a molecular orbital description of the metal-ligand bond (see Section 6.4.1.).

There is another possible way of ordering the various ligands into a series, which is known as the nephelauxetic*) series [13]. In order to understand the principle behind this series we must return to the interelectronic repulsion parameter (Racah parameter) B. In Section 3.3.5 we saw that the free ion terms are given as functions of three parameters A, B and C. Among these, B is particularly useful in that the energy separation between terms of maximum spin multiplicity is a function of B alone. In the case of a d^2 ion, represented in Eqs. (3.7), the energy separation between the ground term ^3F and the first excited term ^3P is 15 B. Hence B of the free ion is accessible from an analysis of the emission spectrum of the ion. Free ion values of B (and C) for all important transition metal ions are available in the literature (e.g. [21]).

The energy levels of a metal ion in a ligand field, as represented by the Orgel or Tanabe-Sugano diagrams, are functions not only of A, B and C, but also of Δ. If Δ is determined from the spectrum, B can be obtained by computation [7, 8, 17]. It was found that B values for metal complexes are invariably smaller than those for the corresponding free ions. This implies a decrease in interelectronic repulsion among the electrons in d orbitals which indicates a delocalization of electron density from the metal onto the ligand atoms or, in other words, an expansion of the "electron cloud" belonging to the metal ion (hence the name, nephelauxetic).

Table 5.6 gives some B values of complexes as compared with the corresponding value of the free ion (B_0); it shows the order of magnitude of the effects that can be observed.

Table 5.6. Some B values of octahedral complexes ML_6 [12].

Metal ion	Configu- ration	B_0 (cm^{-1})	B (cm^{-1})			
			H_2O	NH_3	Cl^-	CN^-
Cr(III)	d^3	950	750	670	510	520
Co(III)	d^6	1,050	720	660	–	440
Rh(III)	d^6	800	500	460	400	–
Ni(II)	d^8	1,130	940	890	780	–

---

* Nephelauxetic comes from the greek words νεφελη = cloud, and αυζανειν = amplification, expansion.

The reduction in B is generally represented by the ratio:

$$\frac{B \text{ in complex}}{B \text{ in free ion}} = \frac{B}{B_0} = \beta$$

A number of ligands has been put together in a series of decreasing β, the nephelauxetic series [12]:

$$F^- > H_2O > (NH_2)_2CO > NH_3 > \text{oxalate}^- \approx NH_2CH_2CH_2NH_2 > \underline{N}CS^- > Cl^-$$
$$\approx CN^- > Br^- > S^{2-} \approx I^-.$$

Evidently the smaller β, the greater is the electron cloud expanding effect of the ligand.
The nephelauxetic series runs essentially parallel to the electronegativity of the adjacent ligand atom, and this is in line with the picture of electron expansion through covalent bonding. In fact there are experimental tools able to demonstrate the electron sharing between metal and ligands, such as electron paramagnetic resonance (EPR) or nuclear magnetic resonance (NMR). These methods are, however, better discussed within the framework of the molecular orbital theory (see Section 6.4).

5.3. Magnetic Moments of Transition Metal Complexes

Many transition metal compounds have one or more unpaired electrons (see Table 5.1) which are the cause of a permanent magnetic moment. Such substances are attracted into a magnetic field; they are said to be paramagnetic.

5.3.1. Paramagnetism from Unpaired Spins

Contrarily to the interpretation of the electronic spectra which has to take into account excited states, the discussion of the magnetic behavior of a transition metal compound is primarily based on the ground state. As we have seen in Section 5.1.2, the number $n^{*)}$ of unpaired electrons in the ground state depends on the strength of the ligand field (*cf.* Table 5.1.). The quantum number S of the ground term is given by the sum of the s_i of all unpaired electrons (see Appendix 3.1):

$$S = \sum s_i = \text{n} \cdot \frac{1}{2} \tag{5.1}$$

Just as for a single electron, where the magnetic moment due to its spin, in Bohr magnetons, is given by [see Eq. (2.27)]:

$$\mu_s = 2 \sqrt{s(s+1)}$$

---

* The same letter n is commonly used for the total number of d electrons in a d^n configuration, and for the number of unpaired electrons. In this book we use italics for the former and a roman n for the latter, to keep the two items apart.

so also for a many-electron ion, the magnetic moment due to spin is:

$$\mu_S = 2 \sqrt{S(S+1)} \tag{5.2}$$

Making use of Eq. (5.1), one obtains:

$$\mu_S = \sqrt{n(n+2)} \tag{5.3}$$

Hence, from the determination of the magnetic moment it is possible to evaluate the number, n, of unpaired electrons in a particular ion. This gives the possibility to check the predictions of LFT (Table 5.1). Moreover, the determination of n often provides a way of finding out the valence state of an ion. The magnetic moment of a paramagnetic entity cannot be measured directly, but it can be evaluated from the magnetic susceptibility of the solid or of a solution of the compound. The experimentally accessible volume susceptibility $\varkappa$ is related to the molar susceptibility χ_{mol} by

$$\chi_{mol} = \frac{\varkappa}{c} \tag{5.4}$$

where c is the molar concentration of the paramagnetic compound. On the other hand, χ_{mol} depends on the magnetic moment and on the temperature according to:

$$\chi_{mol} = \frac{N\mu^2}{3\,kT} = C/T \tag{5.5}$$

where N is Avogadro's number and k is the Boltzmann constant. The inverse proportionality between χ_{mol} and T is known as Curie's law. But many substances obey a somewhat different law, known as the Curie-Weiss law:

$$\chi_{mol} = \frac{C}{T - \Theta} \tag{5.6}$$

where Θ is a constant of the dimension of a temperature. It is, however, common practice not to bother which of the two temperature laws is obeyed, but to compute magnetic moments using Eq. (5.5), and call the result the "effective magnetic moment", μ_{eff}, at the specified temperature (generally room temperature):

$$\mu_{eff} = 2.84 \sqrt{\chi_{mol}\,T} \tag{5.7}$$

Table 5.7 shows that in many cases experimental data of μ_{eff} are in good agreement with Eqs. (5.2) or (5.3), and also with the predictions of LFT. But there are also numerous transition metal complexes having magnetic moments which are considerably higher than those expected from Eq. (5.2). This is due to contributions from the orbital angular momentum to the magnetic moment.

Table 5.7. Magnetic moments of octahedral complexes of first row transition metal ions [14].

Configuration	Central ion	Unpaired electrons	$2\sqrt{S(S+1)}$ (BM)	μ_{eff} (BM)
d^1	Ti^{3+}	1	1.73	1.65–1.79
d^2	V^{3+}	2	2.83	2.75–2.85
d^3	Cr^{3+}	3	3.87	3.70–3.90
Low field ("high spin")				
d^4	Cr^{2+}	4	4.90	4.75–4.90
d^5	Fe^{3+}	5	5.92	5.70–6.0
d^6	Co^{3+}	4	4.90	4.3
d^7	Co^{2+}	3	3.87	4.30–5.20
High field ("low spin")				
d^4	Cr^{2+}	2	2.83	3.20–3.30
d^5	Fe^{3+}	1	1.73	2.00–2.50
d^6	Co^{3+}	0	0	diamagn.
d^7	Co^{2+}	1	1.73	1.8
d^8	Ni^{2+}	2	2.83	2.80–3.50
d^9	Cu^{2+}	1	1.73	1.70–2.20

5.3.2. Contributions from Orbital Magnetism

Taking into account spin and orbital magnetism, the magnetic moment, in Bohr magnetons, for one unpaired electron is [cf. Eqs. (2.25 and 2.27)]:

$$\mu_{l+s} = \sqrt{l(l+1) + 4s(s+1)}$$

Correspondingly, the moment of a many-electron ion might be expected to have the value:

$$\mu_{L+S} = \sqrt{L(L+1) + 4S(S+1)} \tag{5.8}$$

provided Russell-Sanders coupling is valid and spin-orbit interactions are negligible. This would mean that μ_{eff} should be greater than expected from Eq. (5.2) in all cases except for ions with $L = 0$ (i.e. with a S ground term). Nevertheless, the "spin-only formula", Eq. (5.2), is a good approximation in many cases, and whenever deviations are found, they are generally smaller than expected from Eq. (5.8). The reason is that the orbital contribution is partially or totally "quenched" by the chemical environment of the metal ion. This is be-cause the electrostatic field of the ligands restricts the orbital motion of the electrons. Pertubation theory is required to find out if, and to what extent, the orbital contribution to μ_{eff} is suppressed in any particular case [17]. It turns out that for metal ions in octahedral and tetrahedral ligand field, having triply degenerate ground terms (T terms), the magnetic moments are expected to deviate considerably from the spin-only formula, whereas for ions with doubly degenerate, or non-degenerate ground terms, the orbital contribution should be completely quenched. This is summarized in Table 5.8 (cf. Table 5.3).

Table 5.8. Expected orbital contribution to the magnetic moments of octahedral and tetrahedral complexes.

Configuration	Octahedral		Tetrahedral[a]	
	Ground term	Orbital contribution	Ground term	Orbital contribution
d^1	$^2T_{2g}$	+	^2E	−
d^2	$^3T_{1g}$ [b]	+	$^3A_2$	−
d^3	$^4A_{2g}$	−	$^4T_1$	+
d^4 high spin	^5E_g	−	$^5T_2$	+
d^4 low spin	$^3T_{1g}$ [c]	+		
d^5 high spin	$^6A_{1g}$	−	$^6A_1$	−
d^5 low spin	$^2T_{2g}$	+		
d^6 high spin	$^5T_{2g}$	+	^5E	−
d^6 low spin	$^1A_{1g}$	−		
d^7 high spin	$^4T_{1g}$	+	$^4A_2$	−
d^7 low spin	^2E_g	−		
d^8	$^3A_{2g}$ [d]	−	$^3T_1$	+
d^9	^2E_g	−	$^2T_2$	+

[a] Tetrahedral complexes are generally of the high-spin type, because Δ_t is usually small compared with the pairing energy; [b] cf. Fig. 5.11; [c] cf. Fig. 5.13 or Fig. 5.14; [d] cf. Fig. 5.12.

Let us consider two examples. A Cr(II) complex (d^4) in a weak octahedral ligand field has four unpaired electrons, and a ground term ^5E_g. No orbital contribution is to be expected (Table 5.8). The spin-only formula requires $\mu_{eff} = 4.90$ BM; values of 4.75 to 4.90 BM have been experimentally found (Table 5.7). The same Cr(II) ion in a strong octahedral field is low-spin, i.e. it has only two unpaired electrons. The ground term $^3T_{1g}$ implies orbital contribution. In fact the experimental values of μ_{eff} (3.2 to 3.3 BM) are markedly higher than that of the spin-only formula (2.83 BM).

But there are also other mechanisms by which deviations from the spin-only formula may occur. Tetrahedral Co(II), for instance, with three unpaired electrons and a $^4A_2$ ground term (Table 5.8) should have no orbital contribution. Nevertheless, magnetic moments of 4.4 to 4.8 BM are observed instead of the spin-only value of 3.87 BM. It has been shown [15] that in this case spin-orbit interaction provides a mechanism by which a certain amount of the first excited state, $^4T_2$, is mixed into the ground state, thus introducing orbital angular momentum, and making μ_{eff} greater than the spin-only value.

In certain cases (especially with ions of the second and third transition series, and with the rare earths) spin-orbit interaction among the various sub-levels of the ground term becomes quite important. As mentioned in Section 3.4., spin-orbit coupling splits the ground terms (and all other terms) into $2S + 1$ states. If the energy separation between states is $> kT$, only the lowest lying state is populated, and hence becomes the ground term. In this particular case the magnetic moment is given by:

$$\mu_{eff} = g \sqrt{J(J+1)} \tag{5.9}$$

where g is the "Landé factor"*:

---

* Landé discovered this relationship empirically before it was deduced quantum mechanically.

$$g = 1 + \frac{J(J+1) + S(S+1) - L(L+1)}{2J(J+1)} \qquad (5.10)$$

But, if the energy gaps between the $2S+1$ states is of the order of kT, the influence of spin-orbit coupling can give rise to μ_{eff} values which are either greater or smaller than the spin-only value. The platinum and palladium groups, for instance, have very low experimental magnetic moments.

Note that for $L = 0$ Eq. (5.9) becomes identical with the spin-only formula, Eq. (5.3), because then $J = S$, and hence $g = 2$.

5.3.3. Diamagnetism, Ferromagnetism and Antiferromagnetism

Diamagnetism is due to a magnetic moment induced in closed shells of electrons by an outer magnetic field. Since the induced moment is opposed to the applied field, diamagnetic substances are repelled from the field. Diamagnetism is a property of all matter, because all substances possess closed shells of electrons. But diamagnetism is usually several orders of magnitude weaker than paramagnetism. With paramagnetic transition metal compounds, corrections for diamagnetism can therefore generally be neglected. Dilute solutions of paramagnetic substances may, of course, be diamagnetic. The paramagnetic susceptibility of the solute is then determined as the difference between the susceptibilities of solution and solvent.

Ferromagnetism and antiferromagnetism are caused by interionic interaction between paramagnetic species. In ferromagnetic substances the magnetic moments of the separate ions tend to align themselves parallel, thus reinforcing each other. In antiferromagnetic compounds, on the other hand, the alignment of the moments is such as to cancel one another. Above a certain temperature (different for each compound, and called Curie- and Neel-temperature for the two types of magnetism respectively), thermal agitation randomizes the orientation; the compounds behave then as normal paramagnetics. Ferromagnetism is uncommon in trasition metal complexes. Its observation in solutions of complexes of iron, cobalt etc. generally indicates decomposition of the complex and reduction to collodial metal.

Antiferromagnetic behavior is found sometimes in polynuclear transition metal complexes with two or three paramagnetic ions; μ_{eff} is then markedly less than that for the isolated ions. In particular, antiferromagnetism is to be expected in bridged dimers, with halide, oxygen, sulphur, *etc.* in the bridge.

References

[1] H. Bethe, Ann. Phys. *3,* 133 (1929).   [2] Ref. [17], p. 22.   [3] F. A. Cotton, *Chemical Applications of Group Theory,* Interscience Publ., 1963, p. 220.   [4] Ref. [17], p. 90.   [5] F. Ilse and H. Hartmann, Z. Phys. Chem., Frankfurt, *197,* 239 (1951).   [6] H. A. Jahn und E. Teller, Proc. Roy. Soc. Ser. *A, 161,* 220 (1937).   [7] L. E. Orgel, J. Chem. Phys., *23,* 1004 (1955).   [8] Y. Tanabe and S. Sugano, J. Phys. Soc. Jap., *9,* 755 (1954). [9] H. B. Gray, J. Chem. Educ., *41,* 2 (1964).   [10] L. E. Orgel, *An Introduction to Transition Metal Chemistry,* Methuen, London, 1960.

[11] H. B. Gray and C. J. Ballhausen, J. Amer. Chem. Soc., *85,* 260 (1963).   [12] C. K. Jørgensen, *Energy Levels of Complexes and Gaseous Ions,* J. Gjellerups Forlag, Copenhagen, 1957.   [13] C. E.

Schäffer und C. K. Jørgensen, J. Inorg. Nucl. Chem., 8, 143 (1958).   [14] B. N. Figgis and J. Lewis in: J. L. Lewis and R. G. Wilkins, Eds., Modern Coordination Chemistry, Interscience Publ. New York, 1960.   [15] W. G. Penny and R. Schlapp, Phys. Rev., 42, 666 (1932).

Suggested Additional Reading

[16] F. A. Cotton and G. Wilkinson, Advanced Inorganic Chemistry, Interscience Publ. 1966, Ch. 26. [17] H. L. Schläfer and G. Gliemann, Einführung in die Ligandenfeldtheorie, Akadem. Verlagsgesellschaft, Frankfurt, Germany, 1967.   [18] B. N. Figgis, Introduction to Ligand Fields, Interscience Publ., 1966.   [19] P. Schuster, Ligandenfeldtheorie, Verlag Chemie, 1973.   [20] C. J. Ballhausen, Introduction to Ligand Field Theory, McGraw Hill, New York, 1962.
[21] A. B. P. Lever, Inorganic Electronic Spectroscopy, Elsevier, 1968.   [22] J. Ferguson, Spectroscopy of 3d complexes, in: S. J. Lippard, Ed. Progress in Inorg. Chem. 12, 159 (1970).   [23] B. N. Figgis and J. Lewis: The Magnetic Properties of Transition Metal Complexes, in: F. A. Cotton, Ed., Progress in Inorg. Chem. 6, 37 (1964).   [24] C. K. Jørgenson, Modern Aspects of Ligand Field Theory, North Holland Publ. Comp., 1971.

6. Molecular Orbital (MO) Theory for Transition Metal Complexes

In the preceding Chapter we have become acquainted with ligand field theory as a convenient tool for exploring the origin and consequence of central metal ion orbital splitting by its chemical environment. Thus, with emphasis at the metal ion, it proved possible to interprete the stereochemistry, the absorption spectra and the magnetic properties of a complex. But LFT considers the metal ion and the ligands as held together by electrostatic forces; that means, it does not explicitly take care of other contributions to the bond. Nevertheless, certain phenomena indicate the existence of electron sharing. Thus, the nephelauxetic effect (Section 5.2.6) implies the expansion of metal electrons as a result of covalent interaction with ligand electrons; furthermore, the spectrochemical series cannot be interpreted on purely electrostatic grounds. Moreover, experimental methods provide direct information concerning the distribution of electron density within a complex, also indicating electron sharing (EPR, NMR, ESCA). These methods will be discussed later in this chapter (Section 6.4).

Molecular orbital (MO) theory provides a framework for understanding this covalency. The theory was originally developed for diatomic molecules and for aromatic hydrocarbons (Mulliken, Hund, Hückel), the first application to coordination compounds being made by van Vleck [1]. Although quantitative MO calculations for transition metal complexes are far beyond the scope of this book, a qualitative discussion of the theory and its results will provide a good conceptual picture of the chemical bonding in these compounds.

We shall first introduce language, concepts and methods of MO theory in the treatment of small molecules, selecting preferentially potential ligands for transition metal complexes and substrates for catalytic processes. Subsequently we shall proceed to a qualitative MO description of the complexes themselves.

6.1. Concepts of MO Theory

Quantum chemistry has taught us to treat electrons in atoms no longer as particles, but to describe them by wave functions (atomic orbitals) which are solutions of the Schrödinger Equation [*cf.* Chapter 2]:

$$\mathcal{H} \, \Psi = E \, \Psi \tag{6.1}$$

Consequently, we have to consider the process of bond formation between atoms as an alteration of the wave functions of the individual atoms. Evidently the atomic orbitals will overlap somewhat if the atoms are brought within bonding distance, whence they will influence each other. The Schrödinger Equation embodies also this quantum mechanical problem and it is, in principle, possible to write down the corresponding Hamiltonian taking into account the interactions between all electrons and nuclei involved. However, as with the higher atoms (Chapter 3), the mathematical difficulties in solving the wave equation are impossible to overcome (except for the simple case of the hydrogen molecule ion with just one electron and two nuclei).

MO theory provides a device for formulating molecular wave functions which are approximate solutions of the Schrödinger Equation*⁾.

6.1.1. Linear Combination of Atomic Orbitals (LCAO)

The basic trade of MO theory may be summarized as follows: Take all atoms involved in forming the molecule under consideration, strip them of their electrons, fixing the nuclei in their bonded positions; construct molecular wave functions (molecular orbitals, MO's) embracing all nuclei, and fill them according to Aufbau principle, Pauli's principle and Hund's rules with the available electrons. In other words, MO theory attempts to describe a molecule in terms of the basic concepts of atomic structure. The most obvious difference is that the MO's are polycentric.

For constructing the MO's it is assumed that an electron in the vicinity of a certain nucleus A behaves similarly as if it would belong only to this nucleus, *i.e.* as if it would "move" in any one of the atomic orbitals (AO's) of atom A, and the same is assumed for the electron in the vicinity of nucleus B, *etc.* The mathematical expression for this assumption is a linear combination of the atomic orbitals involved, *i.e.* a sum where each AO has a coefficient:

$$\Psi = c_1 \varphi_1 + c_2 \varphi_2 + c_3 \varphi_3 + \ldots \tag{6.2}$$

(we use the symbol Ψ for MO's, and φ for AO's in this chapter).

Two important rules must be taken into account when selecting AO's of the individual atoms which are to be combined into MO's**⁾:

1. The energies of the orbitals to be combined must not differ too much.
2. The combining AO's must have the same symmetry with respect to the molecular framework, and the resulting MO has also this symmetry.

The energy requirement may be rationalized considering a transition metal-carbon bond. Carbon offers only 2s and 2p orbitals in its valence shell. The bond formation, however, will not be with the metal 2s or 2p orbitals, although they evidently would have the same symmetry; but the 2s and 2p electrons of the metal are heavily attracted by the higher nuclear charge and hence are too low in energy. The LCAO−MO will rather contain some of the valence orbitals of the metal (*e.g.* 3d, 4s and 4p in case of a first row transition metal). The symmetry requirement is, of course, most rigorously covered by group theory, and we shall make use of it when dealing with transition metal complexes. With simple diatomic molecules, the symmetry behaviour of the AO's is easily seen by inspection, using the intuitive boundary surface pictures of the AO's (*cf.* Fig. 2.4). Two s orbitals have evidently the same symmetry (Fig. 6.1a). But an s orbital at one atom and a p_x or p_y orbital at the other (taking the bond axis as z direction) cannot be combined effectively into a MO (Fig. 6.1c): the "positive overlap" of the positive lobe of the p orbital is canceled by the "negative over-

---

* Valence bond theory is another approach to the problem. It will not be treated here, because it has several defects when applied to transition metal complexes, *e.g.* not taking into account the splitting of d energy levels and hence not permitting the interpretation or prediction of spectra, and not accounting for detailed magnetic properties.
** Actually no error would be made neglecting these rules; they merely reduce the computational work. AO's that do not fit the rules would emerge with vanishing coefficients in the course of the calculation.

lap" of the negative lobe. The exact mathematical significance of positive and negative overlap will become clear in Section 6.1.3. For the moment remember the dual purpose character of the graphical description of the orbitals: the squares of the wave functions are represented, but the changes of sign refer to the wave functions themselves, *cf.* Section 2.2.3. The combination s-p_z, on the other hand, is possible, and so is d_{yz}-p_y or d_{xz}-p_x (Fig. 6.1b and d).

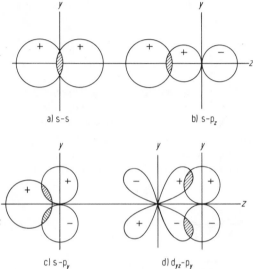

a) s-s   b) s-p_z

c) s-p_y d) d_{yz}-p_y Fig. 6.1. Overlap of atomic orbitals.

6.1.2. The Variation Method

The problem of finding an approximate molecular wave function has by now been reduced to the problem of evaluating suitable coefficients c_i in Eq. (6.2) which make the LCAO–MO fit the Schrödinger Equation. This is done by the variation method. Its application is based on the plausible postulate that the ground state of a molecule represents the most stable electron distribution ever possible. If this distribution is described by a wave function Ψ, and its energy is E, then any trial wave function Ψ_i we may imagine will, on application of Eq. (6.1), result in an energy value $E_i > E$, unless we have been lucky enough to choose the one exact Ψ. Evidently that trial function which gives the lowest value for the energy will be the best approximation. LCAO–MO's [Eq. (6.2)] are particularly convenient trial functions, because they have several adjustable parameters, namely the coefficients c_1, c_2, c_3 *etc.* The variation method consists in varying these coefficients systematically, so that a combination is found that minimizes E. This is a defined mathematical problem; the following conditions have to be met:

$$\left[\frac{\partial E}{\partial c_1}\right]_{c_2,\, c_3 \ldots} = 0; \qquad \left[\frac{\partial E}{\partial c_2}\right]_{c_1,\, c_3 \ldots} = 0; \quad etc. \qquad (6.3)$$

6.1.3. Secular Equations and Secular Determinant

From Eq. (6.1) it would appear that E is simply given by $(\mathscr{H}\,\Psi/\Psi)$. For mathematical reasons, however, Eq. (6.1) must be transformed before the partial differentiations, Eqs. (6.3), can be carried out. Both sides of Eq. (6.1) are multiplied by Ψ, and then integrated over all space*):

$$\int \Psi \mathscr{H} \Psi d\tau = E \int \Psi^2 \, d\tau$$

Hence:

$$E = \int \Psi \mathscr{H} \Psi d\tau \, / \int \Psi^2 \, d\tau \tag{6.4}$$

Fortunately we must not be concerned with the detailed form of the Hamiltonian for obtaining suitable wave functions. We just leave it, abbreviated as $\mathscr{H}$, in the equation.

We now introduce the LCAO−MO, Eq. (6.2), into Eq. (6.4). For making the calculation less bulky, we confine ourselves to a diatomic molecule. It will be easy to extend the result to many-atom problems. The two AO's, φ_1 and φ_2, to be combined into a MO must not be defined at this stage, but it is assumed that they satisfy the two rules outlined in Section 6.1.1. Eq. (6.4) then gives:

$$E = \frac{\int (c_1\varphi_1 + c_2\varphi_2)\,\mathscr{H}\,(c_1\varphi_1 + c_2\varphi_2)\,d\tau}{\int (c_1\varphi_1 + c_2\varphi_2)^2\,d\tau} =$$

$$= \frac{c_1^2 \int \varphi_1 \mathscr{H}\varphi_1 d\tau + c_2^2 \int \varphi_2 \mathscr{H}\varphi_2 d\tau + 2\,c_1 c_2 \int \varphi_1 \mathscr{H}\varphi_2 d\tau}{c_1^2 \int \varphi_1^2 d\tau + c_2^2 \int \varphi_2^2 d\tau + 2\,c_1 c_2 \int \varphi_1\varphi_2 d\tau} \tag{6.5}$$

(All operators in quantum mechanics are "hermitian", that is $\int \varphi_1 \mathscr{H}\varphi_2 \, d\tau = \int \varphi_2 \mathscr{H}\varphi_1 \, d\tau$.)

Two types of integrals appear in Eq. (6.5); they are abbreviated as follows:

$$H_{rs} = \int \varphi_r \, \mathscr{H}\varphi_s \, d\tau$$
$$S_{rs} = \int \varphi_r \varphi_s \, d\tau$$

This brings Eq. (6.5) into the form:

$$E = \frac{c_1^2 H_{11} + c_2^2 H_{22} + 2\,c_1 c_2 H_{12}}{c_1^2 S_{11} + c_2^2 S_{22} + 2\,c_1 c_2 S_{12}} \tag{6.6}$$

Carrying out now the partial differentiations, Eqs. (6.3), one obtains the following two equations for the problem including two AO's:

$$(H_{11} - E S_{11})\,c_1 + (H_{12} - E S_{12})\,c_2 = 0 \tag{6.7}$$

$$(H_{21} - E S_{21})\,c_1 + (H_{22} - E S_{22})\,c_2 = 0 \tag{6.8}$$

---

* This corresponds to averaging the equation over all space. Note that $\Psi \mathscr{H} \Psi \neq \mathscr{H} \Psi^2$. All integrals run from $-\infty$ to $+\infty$.

These two equations together are referred to as the secular equations (for the present problem). They are of the general form:

$$ax + by = 0$$
$$cx + dy = 0$$

and if we solve this set of linear homogeneous equations, we obtain

$$(ad - bc)\,x = 0$$
$$(ad - bc)\,y = 0$$

One trivial solution would be $x = y = 0$ which evidently makes no sense since translated into the set of equations (6.7) and (6.8), it would mean that the coefficients c_1 and c_2 are zero. The second solution is $ad - bc = 0$. This may conveniently be written in form of a determinant:

$$\begin{vmatrix} a & b \\ c & d \end{vmatrix} = 0$$

In terms of the secular equations we obtain the secular determinant[*]

$$\begin{vmatrix} H_{11} - ES_{11} & H_{12} - ES_{12} \\ H_{21} - ES_{21} & H_{22} - ES_{22} \end{vmatrix} = 0 \tag{6.9}$$

Solution of the determinant will give E (as a function of the various integrals H_{rs} and S_{rs}) which then, in turn, can be used to determine c_1 and c_2 from the secular equations (6.7) and (6.8), see next section.

For the more general case that n AO's are involved in the formation of a MO, the secular determinant takes the form:

$$\begin{vmatrix} H_{11} - ES_{11} & H_{12} - ES_{12} & H_{13} - ES_{13} & \dots & H_{1n} - ES_{1n} \\ H_{21} - ES_{21} & H_{22} - ES_{22} & H_{23} - ES_{23} & \dots & H_{2n} - ES_{2n} \\ H_{31} - ES_{31} & H_{32} - ES_{32} & H_{33} - ES_{33} & \dots & H_{3n} - ES_{3n} \\ \cdot \\ \cdot \\ \cdot \\ H_{n1} - ES_{n1} & H_{n2} - ES_{n2} & H_{n3} - ES_{n3} & \dots & H_{nn} - ES_{nn} \end{vmatrix} = 0 \tag{6.10}$$

---

[*] The name comes from the Latin *saeculum* = century. The mathematical device of secular equations and determinants was used much earlier in astronomy, where the periodicity of perturbations of orbits may be in the order of magnitude of centuries.

The various integrals require some comments. The terms $H_{rr} = \int \varphi_r \mathcal{H} \varphi_r d\tau$ are called Coulomb integrals; they give approximately the energy of an electron in the atomic orbital φ_r. The terms $H_{rs} = \int \varphi_r \mathcal{H} \varphi_s d\tau$ are called exchange integrals; they represent the energy of interaction of the two atomic orbitals φ_r and φ_s. All integrals $S_{rr} = \int \varphi_r^2 d\tau$ are unity if normalized AO's are considered (*cf.* Section 2.1.2). The terms $S_{rs} = \int \varphi_r \varphi_s d\tau$ are named overlap integrals. This designation becomes evident if we consider that the integral is different from zero only in those regions of space where neither of the two atomic orbitals involved has zero electron density, *i.e.* in the zone of overlap (*cf.* Fig. 6.1; in the case of s-p$_y$, the positive and the negative part of the integral cancel). It is common practice in first approximation MO theory to neglect the overlap, *i.e.* to set all $S_{rs} = 0$. This may seem to be a rather drastic measure since it is fairly obvious that AO's will overlap when a bond is formed. The justification comes from the fact that qualitative conclusions from MO calculations are practically not influenced by this neglect, and even quantitative results often change very little if overlap is included. On the other hand the mathematics are generally greatly simplified by the "zero-overlap approximation".

6.1.4. Energies and Coefficients of the LCAO−MO's

Let us first consider the most simple case of a homonuclear diatomic molecule where the two AO's forming the MO are identical. The secular determinant (6.9) is considerably simplified, because $H_{11} = H_{22}$, $H_{21} = H_{12}$ and $S_{12} = S_{21} \equiv S$. With $S_{11} = S_{22} = 1$, we obtain:

$$\begin{vmatrix} H_{11} - E & H_{12} - ES \\ H_{12} - ES & H_{11} - E \end{vmatrix} = 0$$

$$(H_{11} - E)^2 - (H_{12} - ES)^2 = 0$$

The two roots of this quadratic equation are:

$$E_a = \frac{H_{11} - H_{12}}{1 - S} \; ; \quad E_b = \frac{H_{11} + H_{12}}{1 + S} \tag{6.11}$$

which, in the approximation of zero overlap, gives

$$E_a = H_{11} - H_{12} \quad \text{and} \quad E_b = H_{11} + H_{12} \tag{6.12}$$

Thus, the combination of two AO's leads to two MO's, one lower and the other higher in energy than the starting atomic orbitals. The number of MO's is always equal to the number of AO's involved.

It is useful to construct a diagram showing the relative orbital energies (Fig. 6.2). The orbital energies of the combining AO's are given on the left and on the right of the diagram, and the energy of the MO is in the middle. The exchange integral H_{12} is a negative quantity, hence the MO with the energy $H_{11} + H_{12}$ is more stable than the AO's. Such a MO is said to be bonding (Ψ_b). The MO with energy $H_{11} - H_{12}$, on the other hand, is less stable; it is said to be antibonding (Ψ_a). Actually, the phenomenon "bonding" is by now defined as the fact that an orbital (MO) is formed which is energetically more favorable than the starting AO's. Consider a case where φ_1 and φ_2 contain one electron each (*e.g.* 1 s in the hydrogen atom). According to Pauli's principle, Ψ_b can house both electrons. Hence an energy gain of $2 H_{12}$ (minus the pairing energy) is effected if two hydrogen atoms form a molecule. Electrons in the antibonding orbital, on the other hand, cancel the bonding effect of electrons in Ψ_b. Therefore helium $(1 s^2)$, for which four electrons would have to be accommodated in the scheme (Fig. 6.2), does not form a stable molecule He_2.

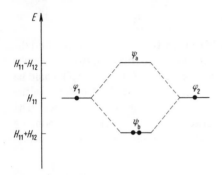

Fig. 6.2. Energy level diagram for LCAO–MO's made up from two identical AO's. (Zero-overlap approximation.)

We are still left with the problem of finding the coefficients c_i in the two LCAO−MO's. We shall make use of the secular equations (6.7) and (6.8). But we cannot expect to solve a total of three unknowns (E, c_1, c_2) from only two equations. Normalization, however, introduces an additional equation. As with atomic orbitals, the MO's must satisfy the condition $\int \Psi^2 \, d\tau = 1$ (*cf.* Section 2.1.2).

Hence, starting with Ψ_b $(E_b = H_{11} + H_{12})$, and making the same simplifications as above $(\varphi_1$ and φ_2 identical and normalized; zero-overlap), it follows from Eq. (6.7) or (6.8) that:

$$c_1 = c_2 \tag{6.13}$$

and from the normalization condition[*]:

$$\int (c_1 \varphi_1 + c_2 \varphi_2)^2 \, d\tau = c_1^2 \int \varphi_1^2 \, d\tau + c_2^2 \int \varphi_2^2 \, d\tau + c_1 c_2 \int \varphi_1 \varphi_2 \, d\tau = 1$$
$$c_1^2 + c_2^2 = 1 \tag{6.14}$$

The combination of (6.13) and (6.14) gives:

$$c_1 = c_2 = 1/\sqrt{2}$$

For Ψ_a one obtains, along the same lines,

$$c_1 = 1/\sqrt{2}; \quad c_2 = -1/\sqrt{2}$$

---

[*] More generally, for LCAO−MO's made up from several atomic orbitals φ_r $(\Psi = \sum_r c_r \varphi_r)$, the normalization condition in the zero-overlap approximation is:
$$\int (\sum_r c_r \varphi_r)^2 \, d\tau = \sum_r c_r^2 = 1.$$

This gives us now the final normalized LCAO−MO's:

$$\Psi_b = \frac{1}{\sqrt{2}} \, (\varphi_1 + \varphi_2) \tag{6.15}$$

$$\Psi_a = \frac{1}{\sqrt{2}} \, (\varphi_1 - \varphi_2) \tag{6.16}$$

6.1.5. Symmetry of Molecular Orbitals

In deriving the normalized LCAO−MO's, Eqs. (6.15) and (6.16), we have not specified the atomic orbitals φ_1 and φ_2. It could have been two s orbitals, two p_x orbitals, two p_z orbitals, *etc.* Evidently, however, the spatial distribution of electron density within the MO would be different in each case. Fig. 6.3 shows boundary surface pictures of LCAO−MO's. In the upper part, the two MO's formed by combination of two s orbitals are represented. The bonding MO possesses increased electron density between the two nuclei, the antibonding MO has zero electron density in this range, and a change of sign of the wave function (a node).

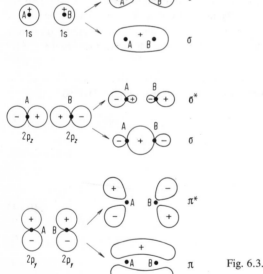

Fig. 6.3. Boundary surface pictures of LCAO−MO's.

The boundary surface pictures of the two MO's resulting from two p_z orbitals are represented in the center, those for two p_y orbitals in the lower part of Fig. 6.3.

The MO's shown are not all equal concerning symmetry with respect to the molecular axis. Those MO's which do not change sign on rotation about this axis are called σ MO's. Two electrons in a bonding σ MO make a σ bond. MO's which change sign on rotation by 180° are called π MO's. Two electrons in a bonding π MO make a π bond. Evidently, σ bonds are formed from s or p_z orbitals, whereas p_x or p_y orbitals lead to π bonds. The antibonding MO's are generally starred: σ^*, π^*. Their characteristic is a nodal plane (change of sign) between the two atoms. Considering the apparently unconnected parts of such MO's, we have to forget about the particle concept of electrons and keep in mind that the concepts of classical physics break down in quantum mechanical description of electrons.

If a bond is formed between nonidentical atoms having different effective nuclear charge, the electron cloud is attracted more strongly by one of the nuclei. Such an asymmetric bond is said to be more or less polar. The coefficients of the LCAO−MO's are unequal. In the bonding orbital $c_1 > c_2$, if φ_1 belongs to the atom that attracts the electron cloud more; in the antibonding orbital $c_2 > c_1$ (this refers to the numerical value, apart from the sign). Under these conditions, the bonding MO has more the character of φ_1, the antibonding more that of φ_2.

6.2. MO Description of Relevant Ligands and Substrates

6.2.1. Homonuclear Diatomic Molecules

With the basic knowledge of MO concepts developed in Section 6.1, we are now in a position to discuss molecular orbitals and energy level diagrams of simple molecules.

The hydrogen molecule has been treated in Section 6.1. Its two MO's are represented in Fig. 6.3 (upper part); its energy level diagram corresponds to Fig. 6.2, with φ_1 and φ_2 equal to 1 s, and with two electrons in Ψ_b.

The nitrogen molecule is made up from two atoms with the electronic configuration $1s^2 2s^2 2p^3$. In constructing an energy level diagram (see Fig. 6.4) we must take into account the two rules given in Section 6.1.1. Although 1 s and 2 s evidently have the same symmetry, they are too different in energy (different cores) for effective bonding. Hence, MO's will form between the two 1 s and the two 2 s orbitals separately. From the p orbitals, the p_z give a σ MO, whereas p_x and p_y have π symmetry, giving a degenerate pair of π MO's. The p_z orbitals, when brought into bonding distance, overlap considerably more than the p_x or p_y orbitals, thus giving rise to a stronger bond (lower energy). Correspondingly, the σ antibonding MO (σ_p^*) is higher in energy than the π antibonding MO [*cf.* Eq. (6.12)].

If the available electrons are fed into the MO scheme in Fig. 6.4, all orbitals up to the degenerate π level are doubly occupied. We see that the bonding effect of 2 electrons in σ_{1s}

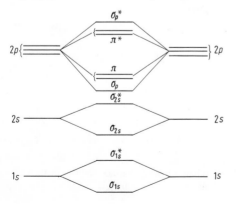

Fig. 6.4. Energy level diagram for a homonuclear diatomic molecule of the first short period (no $\sigma_s - \sigma_{p_z}$ interaction).

is cancelled by two electrons in the antibonding σ_{1s}^*. This shows that inner core orbitals do not contribute to the bonding in a molecule; therefore they are generally omitted in energy level diagrams. Among the MO's arising from valence AO's, σ_{2s} and σ_{2s}^* also appear to cancel. This leaves us with one σ bond (σ_p) and two degenerate π bonds.

According to this simple scheme, the highest occupied level in the nitrogen molecule should be the degenerate pair of π MO's. But experimental evidence, from a detailed analysis of the electronic spectrum of N_2 [2], has demonstrated that the highest occupied level is a σ MO, indicating that the MO energy level diagram is not as straightforward as represented in Fig. 6.4. The experimental finding has been traced back to the fact that electrons in orbitals having equal symmetry (either σ or π), and comparable energy, interact. Roughly speaking they repel each other, bringing themselves into orbitals which, energywise, are more separated. This happens with the two σ MO's σ_{2s} and σ_p, the former being lowered, the latter increased in energy. If the interaction is strong (and this is evidently the case with "dinitrogen"), the upper σ MO can be lifted above the degenerate π level. This situation is represented in Fig. 6.5. The σ orbitals have no longer pure s or p character, but are mixed. The subscripts s and p have therefore been omitted in the orbital labeling in Fig. 6.5.

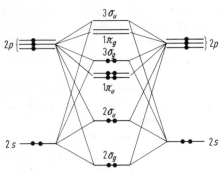

Fig. 6.5. Qalitative MO energy level diagram for dinitrogen, with $\sigma_s - \sigma_{p_z}$ interaction (qualitative with respect to distances of energy levels).

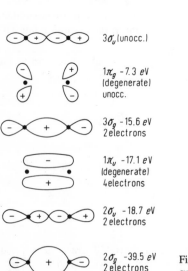

Fig. 6.6. Approximate boundary surface pictures of the molecular orbitals of dinitrogen (*cf.* Fig. 6.5).

The same has been done with the stars indicating antibonding orbitals, since bonding and antibonding properties are no longer sharply separated in this scheme. The labeling rather takes care of the symmetry properties g and u (*cf.* Chapter 4, and Fig. 6.6). The different symmetry species (σ_g, σ_u, π_g, etc.) are numbered separately. The labeling starts with $2\,\sigma_g$; $1\sigma_g$ and $1\sigma_u$, resulting from the 1s atomic orbitals (core orbitals) are not included in the scheme.

Fig. 6.6 represents the approximate boundary surfaces of the molecular orbitals, taking into account some mixing of s and p in the σ orbitals, as indicated in Fig. 6.5. In the nitrogen molecule the four lower orbitals are filled, the degenerate pair of $1\pi_g$ being the lowest energy vacant orbitals. The orbital energies given in Fig. 6.6 are equal to the negative ionization potentials (Koopman's theorem).

This description of the molecule may produce some conceptional difficulties to the chemist, "knowing" this molecule to have a triple bond between the two nitrogen atoms, and a lone pair of electrons on each. Which is the triple bond, and where are the lone pairs? A linear combination of the two MO's $2\,\sigma_u$ and $3\,\sigma_g$ brings the whole picture nearer to the chemical concepts [3]*):

$$\Psi' = (1/\sqrt{2})\,(\Psi_{2\sigma_u} - \Psi_{3\sigma_g}) \tag{6.17}$$

$$\Psi'' = (1/\sqrt{2})\,(\Psi_{2\sigma_u} + \Psi_{3\sigma_g}) \tag{6.18}$$

Inspection of Fig. 6.6 shows that the summation results in an orbital predominantly localized at the left hand nitrogen atom, and directed 180° away from the nitrogen-nitrogen bond (the lobes on the left of the orbitals are additive, whereas the lobes on the right cancel). Subtraction leads to an equivalent MO, localized at the right hand atom. This leaves us with three bonding, filled orbitals, one σ MO ($2\,\sigma_g$) and two degenerate π MO's ($1\,\pi_u$), and with two orbitals for the lone pairs (Ψ' and Ψ''):

$$(\uparrow\downarrow N \equiv N \uparrow\downarrow)$$

The different ways of representing the MO's of the nitrogen molecule, as well as the manipulation of the wave functions $2\,\sigma_u$ and $3\,\sigma_g$ in Eqs. (6.17) and (6.18) may, at first sight, seem rather confusing. In general a chemist, whose theoretical background is not fairly extensive, has a strong tendency to attach an excessively concrete physical significance to a calculated set of MO's, or to the qualitative picture of boundary surfaces deduced from these MO's, and to forget that another set may represent an equally correct solution. In a very rough manner one may visualize this situation in the following way: the one and definite "electron cloud" of a given molecule can be formally divided into different regions by mathematical procedures, so that different calculation methods may give different patterns. For the treatment of certain molecular or chemical properties of the molecule, the one or the other MO description might then be more advantageous.

---

* It should be noted that the wave functions themselves are often designated by their relevant symmetry symbols, so that Eqs. (6.17) and (6.18) would read as

$$\Psi' = (1/\sqrt{2})\,(2\,\sigma_u - 3\,\sigma_g)$$
$$\Psi'' = (1/\sqrt{2})\,(2\,\sigma_u + 3\,\sigma_g)$$

The oxygen molecule is represented by the same qualitative MO energy level diagram as dinitrogen (Fig. 6.5), but there are two more electrons to be placed into the MO's. Since the next empty level ($1\pi_g$) is degenerate, each of the electrons corresponds to one of the two π orbitals (Hund's rule). Therefore, the oxygen molecule is paramagnetic with two unpaired electrons. The two electrons in the strongly antibonding π_g level cancel the bonding effect of two of the four electrons in the $1\pi_u$ orbitals. Hence the two oxygen atoms are held together by one σ and only one π bond.

The $O-O$ bond lengths in the oxygen molecule, and in cations and anions thereof, represent convincing evidence for the antibonding character of the highest occupied $1\pi_g$ level, and hence of the MO concept as a whole:

$$O_2: 1.2074\,\text{Å} \qquad O_2^-: 1.26\,\text{Å}$$

$$O_2^+: 1.1227\,\text{Å} \qquad O_2^{2-}: 1.49\,\text{Å}$$

If an electron is removed from the $1\pi_g$ level, the bonding becomes stronger. On the other hand, more electrons in this level weaken the bond.

6.2.2. Heteronuclear Diatomic Molecules

Heteronuclear diatomic molecules such as CO or NO also may be described by an energy level diagram similar to that of Fig. 6.5, except that all AO's of one of the atoms (namely the one with the greater effective nuclear charge) are lower in energy than those of the other. In CO, for instance, the oxygen p_z level is low enough to give strong bonding overlap with the carbon 2 s orbital. As a consequence the oxygen p_z and carbon 2 s orbitals are more involved in the σ bonding than are the oxygen 2 s and carbon p_z orbitals and the lone pair orbital at oxygen has more 2 s, and that on carbon more p_z character. That means that the carbon lone pair has a larger lobe (protrudes further into space) than the oxygen lone pair. Furthermore, the electron cloud of all bonding orbitals is shifted towards the oxygen atom, whereas the antibonding orbitals have larger lobes at carbon. Qualitative boundary surface pictures of the MO's of CO are shown in Fig. 6.7. Since CO is isoelectronic with dinitrogen,

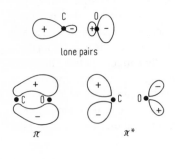

Fig. 6.7. Approximate boundary surface pictures of the molecular orbitals of the CO molecule. (One of the degenerate π and π^* MO's each represented.)

the molecule is also held together by one σ bond and two degenerate π bonds. The particular form of the MO's is important for the properties of CO as a ligand (see Section 6.3.1).

6.2.3. Olefins and Conjugated Diolefins (Hückel MO Theory)

Olefins are the most important substrates in homogeneous catalysis, therefore we shall summarize briefly the relevant features of their electronic structure. We can restrict ourselves to their π-electron system, since the electrons in σ bonds are too low in energy to play a part in the interaction of an olefin with a transition metal center, or in the relevant chemical reactions of olefins[*]. The most simple, but nevertheless highly efficient and instructive treatment of π-electron systems is that of Hückel [4]. The Hückel MO theory is based on the quantum mechanical background outlined briefly in Section 6.1, but it operates with some additional approximations, apart from the introduction of LCAO−MO's and that of zero overlap. It assumes that the Coulomb integrals H_{rr} involved in a conjugated π system have all the same value which is designated $α$. This assumption is not unreasonable since $H_{rr} = \int φ_r \mathscr{H} φ_r d\tau$ is approximately the energy of an electron in the atomic orbital $φ_r$, and all atomic orbitals involved in a conjugated π system are carbon 2p orbitals. Furthermore, all exchange integrals H_{rs} are assumed to have the same value, designated $β$, as long as they refer to neighbouring carbon atoms. For nonneighbouring carbon atoms this integral is taken as zero. These approximations result in a considerable simplification of the secular determinants, and hence of the computational work.

We shall not go into detailed Hückel calculation which is described excellently in a number of books [5]. We do, however, consider the most significant features of important substrate molecules such as ethylene, propylene, butadiene, and of the allyl group.

Fig. 6.8 lists the relevant data for the ethylene molecule. In Hückel MO theory, the energy levels of the molecular orbitals are given in terms of $α$ and $β$, which are both negative energy quantities. Thus the energy level $α + β$ is more negative (more stable) than the energy of an electron in a single carbon 2 p orbital ($α$), and the two π electrons of the ethylene molecule are located in the bonding π MO with this energy.

The coefficients for the two MO's made up from two identical carbon 2p orbitals, are of course the same as those evaluated in Section 6.1.4 for homonuclear diatomic molecules

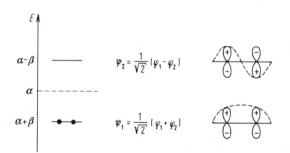

E

$α-β$ —— $ψ_2 = \frac{1}{\sqrt{2}} (φ_1 - φ_2)$

$α$ - - - - - - -

$α+β$ •—•—• $ψ_1 = \frac{1}{\sqrt{2}} (φ_1 + φ_2)$

Fig. 6.8. Energy levels, wave functions and graphic representations of Hückel MO's for ethylene.

---

[*] Actually it is a somewhat gross simplification to consider the π electrons independently, since they are under the electrostatic influence of the total electron cloud. More elaborate calculations than those we can discuss in this context, do take this influence into account (*e.g.* [23]).

[*cf.* Eqs. (6.15) and (6.16)]. The boundary surface pictures of the two MO's are given in Fig. 6.3 (lower part). Because of their π symmetry (with respect to the $C-C$ bond axis) the two ethylene MO's Ψ_1 and Ψ_2 are often referred to as the π and π^* MO of this molecule. Sometimes it is convenient to represent the Hückel MO's by drawing separately the $2p$ atomic orbitals, each orbital with height and sign corresponding to its coefficient in the LCAO$-$MO (Fig. 6.8, right hand side; see also Figs. 6.9 and 6.10). This representation shows that Ψ_1 has no change of sign between the two carbon atoms (bonding MO), whereas Ψ_2 changes sign (antibonding MO). Fig. 6.9 gives the same kind of data for the allyl group. Besides the bonding MO (Ψ_1) and the antibonding MO (Ψ_3), there is also one MO (Ψ_2) which has the same energy as the starting carbon $2p$ orbital. An electron in such a MO does not contribute any stabilization to the system; the orbital is therefore said to be nonbonding. The allyl radical has one electron in the nonbonding level, the allyl anion two, and the allyl cation none.

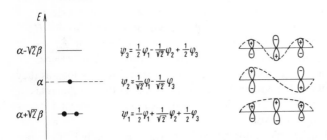

Fig. 6.9. Energy levels, wave functions and graphic representations of Hückel MO's for the allyl radical.

$$\psi_3 = \tfrac{1}{2}\varphi_1 - \tfrac{1}{\sqrt{2}}\varphi_2 + \tfrac{1}{2}\varphi_3$$
$$\psi_2 = \tfrac{1}{\sqrt{2}}\varphi_1 - \tfrac{1}{\sqrt{2}}\varphi_3$$
$$\psi_1 = \tfrac{1}{2}\varphi_1 + \tfrac{1}{\sqrt{2}}\varphi_2 + \tfrac{1}{2}\varphi_3$$

It should be reminded that the allyl group is not linear as represented, for the sake of simplicity, in Fig. 6.9, but has a $C-C-C$ bond angle of $120°$. Note that the nonbonding MO has a nodal plane (change of sign) at the site of the central carbon atom. Ψ_3 is, of course, antibonding between C^1 and C^2 as well as between C^2 and C^3.

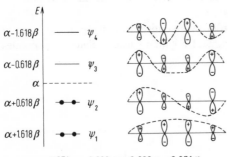

$$\psi_4 = 0.371\,\varphi_1 - 0.600\,\varphi_2 + 0.600\,\varphi_3 - 0.371\,\varphi_4$$
$$\psi_3 = 0.600\,\varphi_1 - 0.371\,\varphi_2 - 0.371\,\varphi_3 + 0.600\,\varphi_4$$
$$\psi_2 = 0.600\,\varphi_1 + 0.371\,\varphi_2 - 0.371\,\varphi_3 - 0.600\,\varphi_4$$
$$\psi_1 = 0.371\,\varphi_1 + 0.600\,\varphi_2 + 0.600\,\varphi_3 + 0.371\,\varphi_4$$

Fig. 6.10. Energy levels, wave functions and graphic representations of Hückel MO's for butadiene.

Butadiene is represented in Fig. 6.10. In its ground state the molecule has two electrons each in Ψ_1 and Ψ_2. Ψ_1 is bonding between all four carbon atoms; Ψ_2 is bonding between C^1 and C^2, antibonding between C^2 and C^3 and bonding between C^3 and C^4. But the bonding properties of Ψ_1 between C^2 and C^3 are more pronounced than the antibonding properties of Ψ_2. Hence a net π bonding between C^2 and C^3 results. Such qualitative observations have been formulated quantitatively by Coulson [6], expressing the π bond order p_{rs} between adjacent carbon atoms as

$$p_{rs} = \sum_j n_j c_{jr} c_{js}$$

in which n_j is the number of electrons in the j th MO, and the c's are the coefficients of the AO's in the LCAO – MO. For the butadiene molecule the π bond orders are (*cf.* Fig. 6.10):

$$p_{12} = p_{34} = 2 \times 0.371 \times 0.600 + 2 \times 0.600 \times 0.371 \quad\quad = 0.894$$
$$p_{23} = \quad\quad 2 \times 0.600 \times 0.600 + 2 \times 0.371 \times (-0.371) = 0.447$$

Because of the partial double bond character of the bond between C^2 and C^3, butadiene and related molecules in solution are present in two conformations, named *s-trans* and *s-cis* (*trans* and *cis* with regard to the central "single" bond):

$$
\begin{array}{ccc}
\underset{\text{H}}{\overset{\text{H}_2\text{C}}{\diagdown}}\text{C} =\!=\!= \text{C}\underset{\text{CH}_2}{\overset{\text{H}}{\diagup}}
&\rightleftharpoons&
\underset{\text{H}}{\overset{\text{H}_2\text{C}}{\diagdown}}\text{C} =\!=\!= \text{C}\underset{\text{H}}{\overset{\text{CH}_2}{\diagup}}
\\[2mm]
\textit{s-trans} & & \textit{s-cis}
\end{array}
$$

In butadiene, the *trans* conformation is energetically favored, the energy difference between the two conformations amounting to 9.6 kJ/mol [7]. For isoprene, on the other hand, the *cis* conformation is predominant.

The important feature of the propylene molecule is caused by the methyl group which does not belong to the π system but nevertheless has an influence on it. Similarly as in the heteronuclear diatomic molecules discussed in Section 6.2.2, the electron cloud in the π MO is not symmetrically distributed among the carbon atoms on either side of the double bond. The methyl group has the effect of "driving" the electron cloud away from the carbon atom to which it is attached, onto the β-position. It has been calculated [8] that the π electron density distribution*) in the propylene molecule is the following:

$$
\begin{array}{c}
0.972 \;\; 1.043 \\
CH_3 - C^\alpha H = C^\beta H_2
\end{array}
$$

(In ethylene and butadiene the relevant numbers are all unity [5].) The approximate boundary surface pictures are shown in Fig. 6.11.

A similar dissymmetry exists also in the π bond of the higher olefins with terminal double bonds.

---

* The π electron density at an atom r, q_r, is the sum of electron densities contributed by each electron in each MO; hence $q_r = \sum_j n_j c_{jr}^2$.

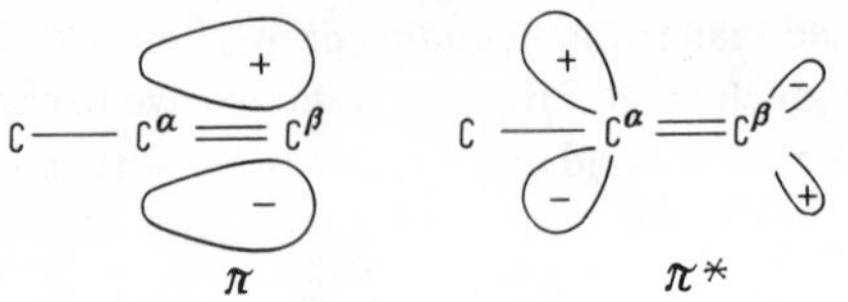

C —— C^α = C^β

π

C —— C^α = C^β

π^*

Fig. 6.11. Approximate boundary surface pictures for the bonding and the antibonding π MO's of propylene.

6.3. MO Description of Transition Metal Complexes

6.3.1. The Coordinative Bond

In the "classical" chemical bond, each of the two atoms involved in bond formation contributes one electron. The coordinative bond, which plays an important role in transition metal complexes, is characterized by the fact that both electrons come from one of the partners. A bonding and an antibonding MO are formed from a filled orbital of one and an empty orbital of the other partner. The bonding MO takes up the two electrons, the antibonding remains empty. The molecule is held together, because this situation is energetically more favorable than that with the two electrons in the isolated orbital of the donor partner (the bonding MO is always lower in energy than any one of the constituent orbitals). The transition metal center as well as the ligands may function as the donor partner. Therefore we have to consider as "valence orbitals" not only the highest occupied orbitals of metal and ligands, but also the lower lying empty orbitals. In the particular case of a transition metal center, $(n-1)$d, ns and np orbitals are generally taken into account.

Let us consider an octahedral complex, say $[Co(NH_3)_6]^{3+}$. The ammonia ligand has three very stable $N-H$ σ MO's, which are too low in energy to take any part in the bonding to the metal. The important ligand orbital is a lone pair orbital at nitrogen. Fig. 6.12 shows the six ligand lone pairs arranged octahedrally in a coordinate system, at the origin of which we have to imagine the nucleus of the metal (compare Fig. 4.1).

Fig. 6.12 gives also a numbering system for the six ligand orbitals. Let us focus for the moment at one of them, say number 1, located on the x axis (Fig. 6.13a). Among the d orbitals of the metal, only $d_{x^2-y^2}$ has electron density in this direction (*cf.* Fig. 2.4). The ligand orbital as well as $d_{x^2-y^2}$ have σ symmetry with respect to the bond axis metal-ligand (no

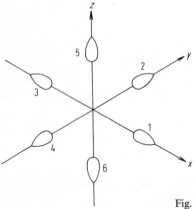

Fig. 6.12. A set of six ligand σ orbitals in an octahedral complex.

change of sign on rotation). Hence, a σ bond may be formed, provided the d orbital is empty, since the lone pair orbital contributes the two electrons.

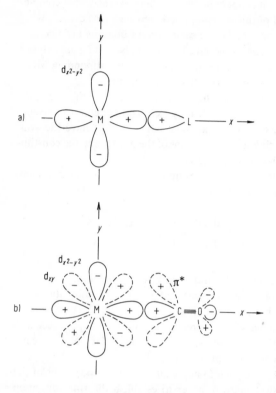

Fig. 6.13. Metal-ligand bonding: a) σ bonding between the metal $d_{x^2-y^2}$ and a ligand lone pair orbital; b) σ bonding and π back donation between metal d orbitals and a CO molecule.

Now replace the ammonia ligand at site number 1 by a CO molecule. This time we have two lone pairs at the ligand molecule, but Fig. 6.7 indicates already that bond formation is more favorable with the carbon lone pair. Moreover, there are low lying empty orbitals available at the ligand, namely the degenerate pair π*. One of these orbitals has the right symmetry properties to match with the d_{xy} orbital (see Fig. 6.13b), and to form a π MO (change of sign on rotating by 180° around the bond axis). The longer lobes at C are particularly suited for this purpose. The other antibonding π* orbital of the CO molecule evidently matches in the same way with d_{xz}. Hence a CO ligand in position 1 may in principle form a coordinative σ bond and one or two coordinative π bonds with a transition metal, provided $d_{x^2-y^2}$ is empty, and d_{xy} and/or d_{xz} are occupied. In the σ bond, electron density is donated from the ligand to the metal, in the π bond *vice versa* ("electron back-donation"). It should be noted that in the particular case of CO the π interaction is much more important than the σ interaction. The CO molecule as a ligand is said to be a typical π acceptor.

If we review the other positions (2–6, Fig. 6.12) in the octahedral complex as we have done with position 1, we shall see that the metal orbitals $d_{x^2-y^2}$ and d_{z^2} are able to form σ MO's with suitable ligands, whereas, d_{xy}, d_{xz} and d_{yz} are available only for π bonding. It is easily remembered (although not as simply vizualized) that in tetrahedral complexes it is *vice versa*: d_{xy}, d_{xz}, d_{yz} are σ orbitals, and $d_{x^2-y^2}$, d_{z^2} participate in the π bonding to ligands [9, 20–22].

6.3.2. Ligand Symmetry Orbitals

The discussion of localized metal-ligand bonds, as made in the preceding section, is sometimes helpful as a first approximation, but it is not quite correct, because the whole complex
must be dealt with as a unity. The pitfall of this oversimplified view becomes evident when
taking into account that each d orbital could make the same type of bonding arrangement
with more than one ligand orbital; $d_{x^2-y^2}$, e.g., could give four σ bonds with ligand orbitals,
numbers 1, 2, 3 and 4, and this would result in four bonding and four antibonding MO's,
starting from only five constituent orbitals – which is evidently not possible.

Group theoretical considerations provide a solution to the problem. One of the important
statements of group theory says: if a transition metal complex (or in general a molecule)
belongs to a certain symmetry group, every molecular wave function must have the symmetry properties of any one of the irreducible representations of the group, and the constituent metal and ligand orbitals must have the same symmetry.

The character table of the O_h group (Table 4.6) indicates immediately the symmetry labels
of the nine valence orbitals of the metal (cf. Chapter 4):

s:	a_{1g}	$d_{z^2}, d_{x^2-y^2}$:	e_g
p_x, p_y, p_z:	t_{1u}	d_{xy}, d_{xz}, d_{yz}:	t_{2g}

None of the isolated ligand orbitals alone (Fig. 6.12) transforms as any one of the irreducible
representations of the group. But the orbitals can be combined into linear combinations, so-
called "symmetry orbitals" which have the required symmetry properties. The procedure is
as follows: Use all the ligand orbitals as a basis for a representation of the group, reduce it to
the constituent irreducible representations, and set up linear combinations of the ligand
orbitals transforming as the resulting irreducible representations.

To find first the reducible representation, we have to apply all symmetry operations of the
group O_h to the six σ orbitals of the six ligands in order to establish the transformation
matrices. Let us take arbitrarily one of the symmetry operations belonging to the class $6\,C_4$,
say rotation by 90° about the z axis, as an example. This rotation brings ligand orbital 1 in
position 2 $(1\rightarrow2)$, $2\rightarrow3$, $3\rightarrow4$, $4\rightarrow1$; the ligand orbitals 5 and 6, on the other hand, remain
in their position. The transformation matrix is then (cf. Section 4.3):

$$C_4: \begin{bmatrix} 0 & 1 & 0 & 0 & 0 & 0 \\ 0 & 0 & 1 & 0 & 0 & 0 \\ 0 & 0 & 0 & 1 & 0 & 0 \\ 1 & 0 & 0 & 0 & 0 & 0 \\ 0 & 0 & 0 & 0 & 1 & 0 \\ 0 & 0 & 0 & 0 & 0 & 1 \end{bmatrix} \quad \chi(C_4) = 2$$

The character of this matrix is $1 + 1 = 2$. The other five symmetry operations of the class
$6\,C_4$ give, of course, the same result with respect to the character (cf. Section 4.3). Only
rows of the transformation matrix having values $\neq 0$ in the main diagonal contribute to the
character. These rows represent ligand orbitals which during the considered symmetry

operation remain unvaried. This observation leads to a simplified procedure of determining the character of each symmetry operation in the reducible representation: simply sum up those ligand σ orbitals which are invariant, e.g.:

$$
\begin{array}{lll}
E & i & \sigma_h \\[6pt]
\sigma_1 \longrightarrow \sigma_1 & \sigma_1 \longrightarrow \sigma_3 & \sigma_1 \longrightarrow \sigma_1 \\
\sigma_2 \longrightarrow \sigma_2 & \sigma_2 \longrightarrow \sigma_4 & \sigma_2 \longrightarrow \sigma_2 \\
\sigma_3 \longrightarrow \sigma_3 & \sigma_3 \longrightarrow \sigma_1 & \sigma_3 \longrightarrow \sigma_3 \\
\sigma_4 \longrightarrow \sigma_4 & \sigma_4 \longrightarrow \sigma_2 & \sigma_4 \longrightarrow \sigma_4 \quad\quad etc. \\
\sigma_5 \longrightarrow \sigma_5 & \sigma_5 \longrightarrow \sigma_6 & \sigma_5 \longrightarrow \sigma_6 \\
\sigma_6 \longrightarrow \sigma_6 & \sigma_6 \longrightarrow \sigma_5 & \sigma_6 \longrightarrow \sigma_5 \\[6pt]
\chi(E) = 6 & \chi(i) = 0 & \chi(\sigma_h) = 4
\end{array}
$$

It will be easy for the reader to verify that the six ligand σ orbitals form the basis for the representation $\Gamma(6\sigma)$ given below. This representation is not irreducible (Table 4.6), hence it has to be reduced, making use of Eq. (4.3).

O_h	E	$8\,C_3$	$6\,C_2$	$6\,C_4$	$3\,C_2'$	i	$6\,S_4$	$8\,S_6$	$3\,\sigma_n$	$6\,\sigma_d$
$\Gamma(6\sigma)$	6	0	0	2	2	0	0	0	4	2

$\Gamma(6\sigma) = a_{1g} + e_g + t_{1u}$

This result tells us that, in octahedral symmetry, the six ligand σ orbitals may be combined to form six symmetry orbitals, one of a_{1g}, a degenerate pair of e_g, and three degenerate orbitals of t_{1u} symmetry. These may form MO's with metal valence orbitals of the corresponding symmetry.

To find the linear combinations of ligand orbitals of a_{1g}, e_g and t_{1u} symmetry, we are allowed to have recourse to a simple, straightforward commonsense method, considering the shape of the metal orbital with which a particular symmetry orbital has to match. (A more rigorous procedure may be found in the literature [20, 22].) Let us begin with one of the metal t_{1u} orbitals, say p_z. A ligand symmetry orbital matching with p_z must have a positive coefficient at $+z$, and an equal but negative coefficient at $-z$. Hence, the corresponding linear combination is $c_5\sigma_5 - c_6\sigma_6$. For symmetry reasons, c_5 and c_6 are equal. Applying the normalization condition [compare Eq. (6.14)]

$$\sum_r c_r^2 = 2c^2 = 1 \tag{6.19}$$

this gives the following ligand symmetry orbital of t_{1u} symmetry, and matching with the p_z metal orbital:

$$\Psi_L(t_{1u}, p_z) = (1/\sqrt{2})\,(\sigma_5 - \sigma_6)$$

The same procedure applied to all relevant symmetry species leads to the symmetry orbitals summarized in Table 6.1. Together with the metal orbitals of corresponding symmetry, they form the six σ MO's linking the six ligands to the metal. Evidently the t_{2g} metal orbitals do not find a symmetry match, they are nonbonding in an octahedral complex with only σ bonding ligands. (We saw already in Fig. 6.13b that the metal t_{2g} orbitals come into play if the ligands have suitable orbitals for π bonding.)

Table 6.1. The symmetry orbitals formed from 6 ligand σ orbitals in an octahedral complex.

Symmetry species	Metal orbital	Ligand symmetry orbital	
a_{1g}	s	$(1/\sqrt{6})$	$(\sigma_1 + \sigma_2 + \sigma_3 + \sigma_4 + \sigma_5 + \sigma_6)$
t_{1u}	p_x	$(1/\sqrt{2})$	$(\sigma_1 - \sigma_3)$
	p_y	$(1/\sqrt{2})$	$(\sigma_2 - \sigma_4)$
	p_z	$(1/\sqrt{2})$	$(\sigma_5 - \sigma_6)$
e_g	$d_{x^2-y^2}$	$(1/2)$	$(\sigma_1 - \sigma_2 + \sigma_3 - \sigma_4)$
	d_{z^2}	$(1/\sqrt{12})$	$(2\sigma_5 + 2\sigma_6 - \sigma_1 - \sigma_2 - \sigma_3 - \sigma_4)^{*)}$

* Remember that the analytical function for d_{z^2} is proportional to $2z^2 - x^2 - y^2$ [Eq. (2.22)].

6.3.3. MO Energy Level Diagrams for Transition Metal Complexes

The determination of the MO energies for the complex is possible only by quantum mechanical computation, and even then one has to rely on very rough estimates of the interaction energies (integrals H_{rs}, cf. Section 6.1.4). For a qualitative MO energy level diagram, however, it suffices to estimate the relative ordering of the metal and ligand orbitals. Such a diagram is shown in Fig. 6.14. The order of increasing energy for the metal valence orbitals in transition metal complexes is $nd < (n+1)\,s < (n+1)\,p$. The ligand σ valence orbitals are generally more stable than any one of the metal valence orbitals. Under these conditions, the ligand orbitals make a larger contribution to the σ bonding MO's (a_{1g}, t_{1u} and e_g) than the metal orbitals, whereas the corresponding antibonding MO's (e_g^*, t_{1u}^* and a_{1g}^*) have more the character of metal orbitals. The nonbonding t_{2g} metal orbitals have not changed their energy in the process of complex formation.

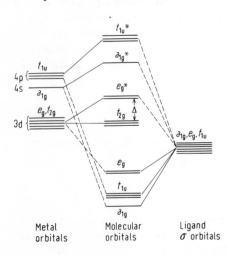

Fig. 6.14. Schematic MO energy level diagram for an octahedral complex ML_6, in which the ligands, L, have no π interaction with the metal, M (e.g. $[Ti(H_2O)_6]^{3+}$ or $[Co(NH_3)_6]^{3+}$).

Metal orbitals   Molecular orbitals   Ligand σ orbitals

The ordering of the σ MO's ($a_{1g} < t_{1u} < e_g$) may be rationalized in first approximation taking into account that these MO's are predominantly ligand orbitals. For a set of symmetry orbitals made up from identical ligand orbitals, the energy increases with the number of nodal planes (and hence with the number of changes of sign) in the symmetry orbital[*]. But, as we have seen, there is also a certain metal contribution in the σ MO's. As a consequence the t_{1u} MO's may be a little below rather than above the a_{1g} orbital in certain cases.

An important feature of the energy level diagram in Fig. 6.14 is the occurence of an e_g^* set and a t_{2g} set of orbitals, with the former higher in energy, which consist predominantly of metal d orbitals. This is qualitatively the same result as that given by LFT (*cf.* Section 5.1). But MO theory interprets this fact as a result of bonding between the metal and the environment (in the present case with the six octahedrally oriented σ bonding ligands).

In the course of filling the MO's with the available electrons there will be as many electrons in the t_{2g} and e_g^* level as there have been d electrons in the starting metal ion. In the example of [Co(NH$_3$)$_6$]$^{3+}$, there are $2 \times 6 = 12$ lone pair electrons from the ligands and six d electrons from the Co^{3+}; hence all levels up to and including t_{2g} are completely filled. The consequences concerning spectra and magnetic properties (high and low spin complexes) are, of course, the same as predicted by LFT. But the magnitude Δ of the energy difference (*cf.* Fig. 5.3 and 6.14) is no longer conditioned by the intensity of electrostatic interaction, but by the stability of the metal-ligand bonds: the more stable the bonding MO's a_{1g}, t_{1u} and e_g, the higher in energy are the antibonding MO's e_g^*, a_{1g}^*, and t_{1u}^*, and hence the greater is Δ.

The MO picture becomes somewhat more complicated if π bonds between metal and ligands have to be taken into account, as for instance with Cr(CO)$_6$ (*cf.* Fig. 6.13b). The procedure followed in setting up the π MO's is, however, much the same as that employed in the preceding section for σ MO's. We shall focus here only on one important feature of π bonding. From Fig. 6.13b we can deduce that the three t_{2g} metal orbitals (d$_{xy}$, d$_{xz}$, and d$_{yz}$) may be involved in the π bonding. As a consequence, there is a bonding (t_{2g}) and an antibonding (t_{2g}^*) level, as shown in Fig. 6.15a.

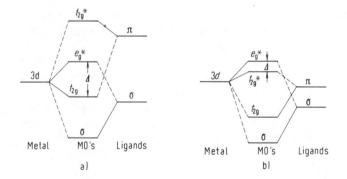

Fig. 6.15. Schematic MO energy level diagram for an octahedral complex ML$_6$, including some aspects of π bonding. a) effect of empty antibonding ligand π* orbitals (*e.g.* CO in [Cr(CO)$_6$]). b) effect of occupied ligand π orbitals (*e.g.* Cl in [IrCl$_6$]$^{2-}$).

---

[*] Compare simple Hückel MO theory for organic π systems [5], *e.g.* Fig. 6.10. The number of nodal planes in the ligand symmetry orbitals can easily be judged from Table 6.1.

The lowering of the t_{2g} level by π bonding increases the energy difference Δ between t_{2g} and the noninvolved e_g^*. We conclude that a ligand that, besides a σ bond, can also form a π bond (using an occupied metal orbital and a vacant π ligand orbital), is a "stronger ligand" than it would be if it could not form a π bond.

The situation is different, if a ligand possesses occupied orbitals having π symmetry with respect to the metal-ligand bond axis, such as the lone pair orbitals in chlorine (atomic $3\,\mathrm{p}$ orbitals). Since the energy of such π orbitals is in general considerably lower than that of, e.g., the antibonding π^* orbitals of CO, the t_{2g} and t_{2g}^* MO's are lower than shown in Fig. 6.15a, the t_{2g}^* level coming even below the e_g^* level (see Fig. 6.15b). More electrons have now to be incorporated into the scheme, and the t_{2g}^* and e_g^* levels will end up containing as many electrons as the original metal d level. The energy difference Δ, now between t_{2g}^* and e_g^*, can be – and in the case of halogen ligands actually is – very small. Thus, the halogens are "weak ligands".

We have restricted ourselves in this chapter to the highly symmetric octahedral ML_6 complexes. For the purpose of a qualitative discussion of bonding, the schematic energy diagrams Figs. 6.14 and 6.15 serve also if the six ligands are not identical. For complexes of other symmetries (the most important being tetrahedral and square planar), the corresponding MO diagrams have also been established, along the same lines as set up in this chapter; the reader is referred to the literature for details [9, 20–22].

6.4. Evidence for Covalent Bonding

6.4.1. The Spectrochemical Series in the Light of MO Theory

We saw in Section 5.2.6 that the spectrochemical series which compares empirically measured values of the energy splitting Δ, shows several inconsistencies if seen from the point of view of purely electrostatic interaction between metal and ligands. Thus it is difficult to see why neutral ligands such as CO, NH_3, H_2O should cause stronger electrostatic fields than ions as Cl^- or I^-. However, if covalent bonding is taken into account, much of the spectrochemical series may be rationalized in terms of the influences of σ and π bonding. We have seen in the last section that good "π acceptor" ligands such as CO (or CN^-) cause large splittings by lowering the t_{2g} level. Ligands such as amines, NH_3 or H_2O, with no π bonding capability, have intermediate values of Δ; in these cases, the better the "σ donor" capacity of the ligand, the higher it rises the e_g^* level, thus increasing Δ (amines $> NH_3 > H_2O$). Finally the "π donor" ligands (Cl^-, I^-) cause particularly small "ligand field splittings" by a mechanism pointed out at the end of Section 6.3.3. Seen as a whole, the experimental spectrochemical series turns out to be a reasonable corroboration of the MO interpretation of covalent bonding in transition metal complexes.

6.4.2. Electron Paramagnetic Resonance (EPR)

One of the most direct evidences that electrons in transition metal complexes are in molecular orbitals is provided by Electron Paramagnetic Resonance. The EPR spectra of certain complexes, containing transition metal ions with unpaired electrons, show a hyperfine pattern due to the interaction of the unpaired electrons with magnetic nuclei in the ligands, clearly indicating that the unpaired electrons are to some extent delocalized over the ligands. The main ideas underlying the EPR experiment are relatively simple, but the full elaboration of the theory, particularly in context with transition metal complexes, has become a sophisticated mathematical task. All magnetic interactions of the electron(s) with the nuclei and with the chemical environment are taken care of by a Spin Hamiltonian; the reader is referred to the specialist literature [24–26]. The purpose of this section is merely to point out briefly what kind of information one may expect from an EPR spectrum.

Let us consider a paramagnetic complex with one unpaired electron. Due to its spin, the electron has associated with it a magnetic moment [compare Eq. (2.27)]; so it may be treated as a tiny bar magnet. In the absence of a magnetic field the unpaired electrons are aligned at random, but the presence of an external magnetic field forces them to align themselves either parallel or antiparallel to the field. There is a difference in energy between these two states[*] which is proportional to the strength of the applied field, H_0:

$$\Delta E = g\beta H_0 \tag{6.20}$$

The proportionality constant g is called the spectroscopic splitting factor, or just the g factor; β is the Bohr magneton. Transitions between the two energy levels can be induced, applying electromagnetic radiation of the correct frequency ($\Delta E = h\nu$) to the paramagnetic sample in a magnetic field. Hence the EPR method is essentially a method of absorption spectroscopy. The appropriate radiation is in the microwave range (9,000–34,000 MHz), for magnetic fields available in the laboratory. In principle the magnetic field could be maintained constant and the frequency varied until resonant absorption takes place. For technical reasons, however, the reverse procedure is adopted: at constant microwave frequency the magnetic field is varied through the appropriate range.

The primary result of the EPR experiment is the determination of the g factor, according to Eq. (6.20), and substituting ΔE by $h\nu$:

$$g = \frac{h\nu}{\beta H_0} \tag{6.21}$$

For one unpaired electron in a S state, the g factor has the value of 2 (exactly 2.0023, if relativistic corrections are taken into account). For a free ion, with quantum numbers S, L, J, the g factor is given by Eq. (5.10), but for a transition metal ion in a ligand field it is much more complicated. Unsymmetric ligand fields give directional properties to the g factor. Essentially all ligand fields are unsymmetric from the point of view of EPR; remember that spin-orbit coupling and the Jahn-Teller effect remove the degeneracy of all levels which would appear degenerate if judged only from the rough macrosymmetry of a complex. The g value in different directions can be determined from EPR on single crystals and, in favorable cases, also from frozen solutions in solvents giving a glass on solidification (see e.g. Figs. 11.7 and 11.8). The directional properties of the g factor depend in a characteristic way on the ligand field splitting constant Δ and on the spin-orbit coupling constant λ. Thus, EPR measurements may in principle be used to determine λ (if Δ is known from optical spectra)

---

[*] For more than one unpaired electron there are more energy levels, resulting from the fact that the total angular momentum has quantized orientations with respect to the magnetic field; see Section 3.4.

[24–26], or to check an assumed ground state configuration [10]. In particular, it is sometimes possible to estimate in which of the d orbitals an unpaired electron is located [10, 11]. The evidence for covalent bonding in transition metal complexes mentioned at the beginning of this section comes from a secondary effect in EPR: the magnetic moment of the unpaired electron(s) interacts with magnetic moments of neighbouring nuclei, resulting in a splitting of the otherwise single line of the EPR signal. In the case of n equivalent ligand atoms of nuclear spin I, a characteristic pattern of $2nI + 1$ lines is observed. The EPR theory says that, in solution, such magnetic interaction is possible only if the unpaired electron of the central metal ion has a definite probability to be found at the site of the nucleus of the ligand atom. Translated into the language of MO theory this means, that the unpaired electron is described by a wave function which is a linear combination of the metal d orbital originally housing the electron, and s orbitals*) of the relevant ligands with magnetic nuclei:

$$\Psi = c_1 \varphi_{d,metal} + c_2 \varphi_{s,ligand} + \sum_i c_i \varphi_i$$

The last term represents other atomic orbitals which might be involved in the MO.

Besides demonstrating covalency, EPR spectra can, of course, provide structural information on transition metal complexes concerning number and identity of certain ligand atoms. However, unfortunately only paramagnetic central ions, and of those only ions with an odd number of unpaired electrons, are generally suitable for EPR measurements [25], and only a restricted number of possible ligand atoms has magnetic nuclei.

6.4.3. Nuclear Magnetic Resonance (NMR) and Mössbauer Spectroscopy

These two methods are grouped together here because both are nuclear resonance techniques, and in both atomic nuclei are used as source of information concerning the electron cloud.

The formalism of the primary effect in NMR is similar to that in EPR. The magnetic moment of a magnetic nucleus (*e.g.* H with nuclear spin $I = 1/2$; $P: I = 1/2$; $N: I = 1$) has quantized orientations in an external magnetic field. The different orientations have different energy; the energy difference ΔE is proportional to the strength of the magnetic field H_0. Transitions can be induced applying electromagnetic radiation. The energy requirement is 2–3 orders of magnitude lower than with EPR, therefore NMR operates with radio waves (10–300 MHz). The frequency is maintained constant; the magnetic field is varied until absorption is observed, or *vice versa*.

The relevant information comes from secondary effects. The magnetic field at a magnetic nucleus is not exactly equal to H_0, because the surrounding electron cloud shields the nucleus. Hence different chemical environments require slightly different external magnetic fields for achieving resonant absorption. This phenomenon is called "chemical shift". Variations in chemical shift of ligands in complexes as compared with free ligands indicate electron delocalization and hence covalent bonding.

In compounds with unpaired electrons an additional effect is observed. The magnetic field originating from the spin moment of the unpaired electrons adds to the external magnetic field, the contribution of the former being proportional to the unpaired electron density at

---

* Remember that only s orbitals have electron density at the nucleus; Section 2.2.3.

the nucleus under consideration [12]. An instructive example is provided by a comparison of the NMR spectra of the acetylacetonates of the paramagnetic vanadium(III) and the diamagnetic cobalt(III) [13]:

$$M \left(\begin{array}{c} O-C \diagdown \,^{CH_3} \\ \quad\quad CH \\ O-C \diagup \,_{CH_3} \end{array} \right)_3$$

The proton resonances of the CH group as well as that of the CH_3 groups are considerably shifted to lower fields in the case of M = vanadium, as compared with that of the cobalt complex. This shows that the d electrons are delocalized into the π system of the ligands, and have even a certain probability to be found at the protons.

There is, of course, also magnetic interaction between neighbouring magnetic nuclei in the same molecule, giving rise to a splitting of NMR absorption lines. Metal-hydrogen interaction has been useful in determining whether or not an alkyl group is bonded directly to a metal. A rhodium complex, for instance, which analyzes as $RhCl(PPh_3)_2(CH_3I)_2$, shows two methyl group resonances [14]. One is a well-resolved triplet, further split into doublets. This fine structure is consistent with that expected for a methyl group directly bound to a rhodium atom (^{103}Rh; $I = 1/2$), to which are symmetrically bound two phosphine groups (^{31}P; $I = 1/2$). The other methyl resonance has no fine structure and is somewhat broad; it is clearly not due to a methyl group bound to the rhodium, hence indicating the coordination of one of the methyl iodide molecules to rhodium through the iodine atom. The following structure has been deduced from these results [14]:

In Mössbauer Spectroscopy the excitation energy is provided by a γ-radiating nucleus. Gamma emission occurs when a nucleus drops from an excited state to one of lower energy. If these γ-rays fall on a nucleus of the same isotope which is in the lower state, resonant absorption can occur, and the second nucleus becomes excited. Condition for resonance is, however, that the nucleus in the "source" and that in the "absorber" are in identical chemical environment. Otherwise the radiation is not absorbed, but transmitted. But resonance may be restored imparting a velocity to the absorber relative to the source. Due to the Doppler effect this motion changes the energy of the incident quanta. The shift in the resonance absorption due to chemical environment (isomer shift) is therefore expressed in units of velocity (mm/s).

The isomer shift arises from the fact that the nucleus is surrounded and penetrated by electronic charge with which it interacts electrostatically. Although it depends directly only on the s electron density at the nucleus, it is indirectly influenced by changes in density of other electrons participating in the chemical bonding because of their screening effect.

Mössbauer Spectroscopy is restricted to only some 28 isotopes of certain elements; most work has been done with ^{57}Fe and ^{119}Sn. Relevant information on chemical bonding in transition metal complexes has been obtained with ^{57}Fe [15, 16]. In the particular case of transition metal complexes the isomer shift depends on the tendency of the ligands to donate electrons to (or withdraw from) the metal which, evidently, is synonymous with covalent bonding.

6.4.4. Photoelectron Spectroscopy

This relatively new analytical method is also known under the names ESCA (electron spectroscopy for chemical analysis) and IEE (induced electron emission). It permits the direct observation of electrons in atoms or molecules (complexes). An X-ray beam ejects electrons from the various orbitals of the substrate under investigation. The kinetic energy, E_{kin}, of these "photoelectrons" is measured. The binding energy E_b (= ionization potential) of an ejected electron is given by the difference between the energy of the incident X-ray photon, $h\nu$, and the kinetic energy of the electron, which latter is measured with an ingenious device [17]:

$$E_b = h\nu - E_{kin}$$

The analysis of photoelectron energies enables the determination of the binding energies of electrons with high precision. With X-ray beams of sufficient energy, electrons from all levels may in principle be ejected, *i.e.* inner shells as well as valence molecular orbitals can be investigated [17].

The most important applications in our context are based on the fact that the orbital energies depend on the chemical environment. If there is a certain deficiency in the electronic cloud of an atom, the remaining electrons will be more strongly bounded to the positive nucleus, than would be the case if the electronic clouds were complete. Measurements with transition metal complexes actually have confirmed that even the coordination of neutral ligands may well lead to a considerable charge transfer from the metal to the ligand or *vice versa* [17]. The method is also used with increasing frequency as a check on validity and limitations of quantitative or semiquantitative MO calculations on transition metal complexes [18, 19].

References

[1] J. van Vleck, J. Chem. Phys., *3*, 803, 807 (1935).   [2] H. B. Gray, *Electrons and Chemical Bonding*, W. A. Benjamin Inc., New York, 1965.   [3] K. G. Caulton, R. L. DeKock and R. F. Fenske, J. Amer. Chem. Soc., *92*, 515 (1970).   [4] E. Hückel, Z. Phys., *76*, 628 (1932).   [5] e.g.: A. Streitwieser, *Molecular Orbital Theory for Organic Chemists*, John Wiley & Sons, Inc., New York, 1961; E. Heilbronner and H. Bock, *Das HMO-Modell und seine Anwendung*, Verlag Chemie, Weinheim, 1968.
[6] C. A. Coulson, Proc. Roy. Soc., Ser. A, *169*, 413 (1939).   [7] R. G. Parr and R. S. Mulliken, J. Chem. Phys., *18*, 1338 (1950).   [8] J. A. Pople and M. Gordon, J. Amer. Chem. Soc., *89*, 4253 (1967).
[9] e.g.: H. B. Gray, J. Chem. Educ., *41*, 2 (1964).   [10] A. H. Maki, N. Edelstein, A. Davison and R. H. Holm, J. Amer. Chem. Soc., *86*, 4580 (1964).
[11] e.g.: B. A. Goodman, J. B. Raynor and M. C. R. Symons, J. Chem. Soc. A, *1969*, 2572.   [12] H. L. Schläfer and G. Gliemann, *Einführung in die Ligandenfeld-Theorie*, Akademische Verlagsgesellschaft, Frankfurt, 1967.   [13] A. Forman, J. N. Murrell and L. E. Orgel, J. Chem. Phys., *31*, 1129 (1959).

[14] D. N. Lawson, J. A. Osborn and G. Wilkinson, J. Chem. Soc. A, *1966*, 1733. [15] R. H. Herber, Mössbauer Spectroscopy. Progr. in Inorg. Chem., *8*, 1 (1967). [16] K. Burger, Inorg. Chim. Acta Rev., *6*, 31 (1972). [17] C. Nordling, Angew. Chem. Internat. Edit., *11*, 83 (1972). [18] Ch. K. Jørgensen, Chimia, *27*, 203 (1973). [19] R. F. Fenske, 167th ACS National Meeting, Los Angeles, April 1974; Abstr. of Papers, Inorg. 50.

Suggested Additional Reading

[20] M. Orchin and H. H. Jaffé, *Symmetry, Orbitals, and Spectra,* Wiley-Interscience, New York, 1971, Ch. 6. [21] C. J. Ballhausen and H. B. Gray, *Molecular Orbital Theory,* W. A. Benjamin, Inc., New York, 1964. [22] B. N. Figgis, *Introduction to Ligand Fields,* Interscience Publ., New York, 1966, Ch. 8. [23] D. A. Brown, W. J. Chambers and N. J. Fitzpatrik, *Molecular Orbital Theory of Transition Metal Complexes,* Inorg. Chim. Acta Rev., *6*, 7 (1972). [24] N. M. Atherton, *Electron Spin Resonance,* Ellis Horwood Ltd., Chichester, 1973. [25] B. A. Goodman and J. B. Raynor, *Electron Spin Resonance of Transition Metal Complexes,* Advan. Inorg. Chem. Radiochem., *13*, 135 (1970). [26] A. Carrington and A. McLachlan, *Introduction to Magnetic Resonance,* Harper and Row, New York, 1967. [27] W. Jolly, *The Application of X-Ray Photoelectron Spectroscopy to Inorganic Chemistry,* Coord. Chem. Rev., *13*, 47 (1974).

7. General Aspects of Catalysis with Transition Metal Complexes

The word catalysis (from the Greek καταλυσισ, meaning destruction, dissolution) was introduced by Berzelius [1] in 1836 to describe the phenomenon whereby certain chemical reactions, in particular decomposition processes, were enhanced by the presence of substances which themselves remained unchanged. At that time, catalyzed processes were regarded as taking place under the influence of an undefined "catalytic force". Although the general concept of catalysis has developed considerably in the meantime, and the scope of reactions susceptible to catalytic influence has exceeded by far the original range of decompositions, the term catalyst has been retained and is now generally employed for a substance capable of enhancing the rate of a chemical reaction without itself being consumed in the process. As knowledge of catalytic processes increased, it became clear that the catalyst enters into the chemical reaction providing an alternative and speedier route to the desired product. This can usually be explained in terms of normal chemical reactions between the catalyst and the reactants (also called substrates), to give intermediates which finally yield the products with regeneration of the catalyst. The catalyzed reactions proceed faster because the activation energy is lowered and/or because of changes in the preexponential factor in the Arrhenius equation for the rate constant. Since the catalyst is unchanged at the end of the process, it imparts no energy to the system; according to thermodynamics therefore, it can have no influence on the equilibrium position of a particular reaction.

Frequently, an extremely small amount of catalyst will cause a considerable increase in the reaction rate. The effectiveness of a catalyst is sometimes expressed in terms of its turnover *i.e.* the number of moles of substrate reacted per mol of catalyst.

During the past few decades, soluble transition metal complexes have become very important as catalysts for a wide range of reactions. The starting point of this development was probably the discovery by O. Roelen [2] in 1938 of the reaction of olefins with carbon monoxide and hydrogen to form aldehydes (the "oxo process" or hydroformylation, *cf.* Chapter 10), which takes place in the presence of a soluble carbonylcobalt complex. The solubility of this and of other active transition metal complexes permits the bringing together of catalyst and reactants in solution (homogeneous catalysis), whereas with the classical insoluble metal and metal oxide catalysts there is always a gross interface, generally of the solid/gas type (heterogeneous catalysis).

Soluble catalysts often operate under milder conditions of pressure and temperature than do heterogeneous catalysts, thus offering considerable advantages in specificity and selectivity. The homogeneous systems are also very attractive from the point of view of mechanistic studies, since they allow one to bypass serious problems inherent in heterogeneous systems, such as particle size and surface quality of the catalytic species, dosification, stirring, mass transport phenomena, etc. The prospects of elucidating the mechanism of a particular type of catalysis is certainly greater with soluble systems. Moreover, the massive advance of organometallic chemistry in recent years has provided methods for preparing a wide variety of transition metal complexes, thus permitting systematic investigation of catalytic properties as a function of metal, valency, and ligand environment.

Soluble transition metal complexes will certainly not replace the heterogeneous metal oxide and metal catalysts in all areas, at the present stage of chemical development. While the solid catalysts are usually not too selective, this disadvantage is counterbalanced by a number of practical advantages such as versatility, resistance to heat, ease of continuous operation, separation at the end of reaction, and, finally, facile regeneration of the catalyst.

One would expect to find parallels between the functioning of soluble and solid transition metal catalysts. Actually, a trend exists whereby the catalytic behavior of heterogeneous systems may be discussed no longer only in terms of "active surfaces", but also from the point of view of individual active metal centers. In this sense, the results obtained with soluble systems also facilitate the understanding of heterogeneous catalysis.

7.1. Transition Metal Ions in Catalysts

There are several reasons why the transition metals play such an important catalytic role. In Chapter 3 we have defined transition metals as those which have partially filled d orbitals in any of their chemically important oxidation states. The $(n-1)$d orbitals have energies similar to the corresponding ns and np orbitals, and therefore form part of the valence shell.

The availability of empty d orbitals makes possible the coordinative bonding of suitable neutral molecules to the central metal ion (*cf.* Section 6.3.1). Since the most important substrates for the catalysis under consideration are neutral molecules (olefins, conjugated diolefins, carbon monoxide), the significance of this particular type of bonding for fixation of these molecules to the catalyst becomes evident (*cf.* Section 7.2).

However, the transition metals can also establish covalent σ bonds with hydrogen, alkyl or aryl groups such as, for instance, the metal-carbon bond between titanium and the ethyl group in the complex represented in Fig. 7.14. The intermediate formation and opening of such metal-carbon σ bonds is also crucial for catalysis (*cf.* Sections 7.3 and 7.4).

Depending upon number and type of the surrounding ligands and on the symmetry of the ligand arrangement, all d orbitals, or only some of them, participate in bond formation. We saw this when comparing (Chapter 6) octahedral complexes containing only σ bonding ligands (Fig. 6.14) with complexes having σ and π bonding ligands (Fig. 6.15). In the first case the three t_{2g} orbitals remain nonbonding, whereas in the latter they participate in π bonding. That the d orbitals are able to accommodate different numbers of electrons is important; as a consequence, several stable oxidation states of the same metal are generally available, and mostly they are accompanied by different preferential coordination numbers. Frequently, one observes several changes of oxidation state and/or coordination number in the course of one and the same catalytic cycle (*cf.* Section 7.4.3).

It is not yet possible, in general, to predict which metal is most effective for which catalysis situation. However, certain trends are evident. Soluble complexes of Rh, Ir, Ru, and Co are particularly important in the hydrogenation of olefins (*cf.* Section 8.2); Co and Rh have proved especially suitable for hydroformylation (Section 10.2), whereas Ni is the preferred catalyst at present for the cyclooligomerization of butadiene (Section 9.5). Dimerizations of olefins require group VIII catalysts; transition metals situated further to the left in the Periodic System (Ti, V, Cr), on the other hand, have proved particularly suitable for the oligomerization and polymerization of α-olefins (Section 8.4). Some reasons for this metal complex "specialization" will be discussed in the following sections and in the course of subsequent chapters.

7.2. Coordinative Bonding and Catalysis

7.2.1. Carbon Monoxide and Olefins

The coordination of CO to a transition metal ion has been discussed in Section 6.3.1 (*cf.* Fig. 6.13). The CO molecule is a "strong" ligand, being at the head of the spectrochemical series of ligands (*cf.* Section 5.2.6). Qualitative MO theory tells us that this is due to the particularly strong π acceptor capacity of the pair of antibonding π orbitals (*cf.* Sections 6.3.3 and 6.4.1). Electrons in antibonding orbitals cancel the bonding effect of electrons in the corresponding bonding orbitals. Hence, if the π* orbitals of CO accept electron density donated by the metal, the C−O bond is weakened. This corollary has been confirmed experimentally by comparing the vibrational spectra of free and coordinated CO. The stretching frequency of a bond (measured by infrared spectroscopy) is higher, the more stable the bond. The free molecule has a stretching frequency $\nu_{CO} = 2143$ cm^{-1}, whereas terminal CO groups in carbonyl complexes {such as $Cr(CO)_6$ or $Fe(CO)_3$ $[P(C_6H_5)_3]_2$} have ν_{CO} in the range 2100–1800 cm^{-1}.

Presumably, weakening of the CO bond on coordination plays a role in catalysis, making the molecule more susceptible to reaction, provided there is a suitable reaction partner present in the coordination sphere of the transition metal.

The coordinative bonding of an olefin molecule has a certain similarity to that of CO, although the orientation of the molecule with respect to the metal is different in the two cases. As is known from X-ray structural analyses of relevant complexes, metal carbonyls have a linear M−C−O arrangement, whereas in ethylene complexes both carbon atoms are equidistant from the metal atom:

$$M\text{----}\overset{\displaystyle C}{\underset{\displaystyle C}{\|}} \qquad M\text{---}C\equiv O$$

Fig 7.1 shows the orbitals involved in the coordinative bonding of an olefin to a transition metal ion. Only the π bonding and the π* antibonding MO's of the olefin have to be taken into consideration. The C−C and C−H σ bonds are too stable, i.e. the energy levels of the corresponding MO's are too low for effective interaction with the metal d orbitals.

The bonding π orbital of the olefin, which contains two electrons, has σ symmetry with respect to the metal-olefin bond (no change in sign of the molecular wave function occurs

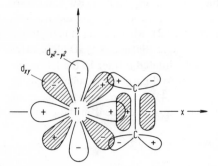

Fig. 7.1. Coordination of an olefin (ethylene) to transition metal (titanium); schematic representation of the relevant orbitals in the *xy* plane of an octahedral complex (occupied orbitals are shaded).

on rotation through 180° about the bond axis; cf. Section 6.1.5). It can overlap with an empty dσ orbital of the metal, with formation of a coordinative σ bond. Ligand field theory tells us that, among the d orbital group, the dσ orbitals represent always the highest energy levels. Therefore we can generally expect the relevant dσ orbital to be empty and hence available for coordinative σ bonding. Strictly speaking, the other σ orbitals of the metal (s and p) also participate in this σ bond (cf. Fig. 6.14); but this needs not trouble us for the moment.

The empty π* antibonding MO of the olefin has π symmetry with respect to the metal-olefin bond (a single change of sign on rotation through 180° about the bond axis). If the dπ orbital of the metal which lies in the same plane is occupied, a coordinative π bond is also established. Hence the metal can, in principle, form a double bond with the olefin. This description of the metal and olefin orbital interaction is known as the Dewar-Chatt-Duncanson model [3, 4] for the metal-olefin bond.

Electron density passes via the σ bond from the olefin to the metal, and via the π bond from the metal into the π* antibonding orbital of the olefin. The latter interaction is named electron back-donation or retrodative π-bonding. Both interactions tend to reduce the strength of the carbon-carbon bond in the olefin. The extent of this effect is reflected in the increased C−C bond distance observed in X-ray structures of olefin complexes, and can be quite remarkable, in particular if the olefin possesses electron withdrawing substituents such as in fumaronitrile and tetracyanoethylene (cf. Table 7.1). For TCNE complexes, the carbon-carbon bond distance is nearly equal to that of a single bond (compare ethane).

Table 7.1. Carbon-carbon distances in free and coordinated olefins.

Compound	C−C bond length (Å)	Ref.
$H_2C = CH_2$	1.337	[5]
(CN)HC = CH(CN) (FUMN)	1.34*)	[6]
$(CN)_2C = C(CN)_2$ (TCNE)	1.34	[7]
$[Pt(C_2H_4)Cl_3]K$	1.375	[8]
$Ni(C_2H_4)[P(C_6H_5)_3]_2$	1.43	[9]
IrH(CO) (FUMN) $[P(C_6H_5)_3]_2$	1.43	[6]
Pt(TCNE) $[P(C_6H_5)_3]_2$	1.49	[10]
Ir(TCNE)Br(CO) $[P(C_6H_5)_3]_2$	1.51	[11]
$CH_3 - CH_3$	1.54	[12]

X-ray structural analysis has also revealed that, in the case of extreme bond lengthening, the olefin is no longer planar but that the electron withdrawing substituents are bent away from the metal [10]. It appears tempting therefore to describe the bonding in these extreme cases in terms of two σ bonds from the metal to two approximately sp³ hybridized carbon atoms, as shown in Fig. 7.2.

However, this would require a highly strained three-membered ring. The great thermal stability of such complexes (e.g. Pt(TCNE) $[P(C_6H_5)_3]_2$ dec. 268–270 °C) does not agree with

---

* Although no structural data are available for the FUMN molecule, there seems to be no reason to suppose that the olefin bond length differs much from that in ethylene and TCNE [6].

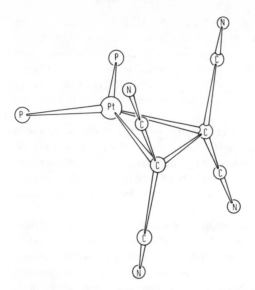

Fig. 7.2. "Platina-cyclopropane" structure for the complex $Pt(TCNE)[P(C_6H_5)_3]_2$.

such a strained structure. Hence the Dewar-Chatt-Duncanson model (Fig. 7.1) remains the best description of bonding in these extreme cases also.

An increase in bond length is equivalent to activation of the molecule. In other words, a coordinated olefin may be more liable to reaction, provided there is a suitable partner present in the coordination sphere of the metal center. If this is not the case, a strong double bond between the metal and the olefin will impart stability to the complex, so that it can be isolated and crystallized as shown in the examples given above.

It has been found empirically, by X-ray analysis of stable complexes, that coordinated olefins tend to be oriented perpendicular to the molecular plane in square-planar compounds, and in the trigonal plane in trigonal and trigonal bipyramidal complexes:

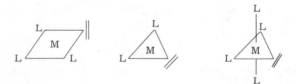

In octahedral complexes this question is irrelevant since parallel or perpendicular orientations with respect to one main axis are *vice versa* with respect to another (*cf.* Fig. 4.1). Nevertheless, other ligands in the complex may compel the olefin to adopt a certain orientation, for instance as a result of steric influence.

In solution, however, one should not vizualize the olefin as being fixed in a certain orientation. It has been shown using ^1H NMR, particularly for ethylene complexes of Rh(I) and Os(I), that the olefin molecules rotate. This was first observed with the complex $(C_5H_5)Rh(C_2H_4)_2$, shown in Fig. 7.3 [13]. The complex has two times four equivalent ethylenic protons, marked i (inner) and o (outer), which would be expected to give two separate NMR signals.

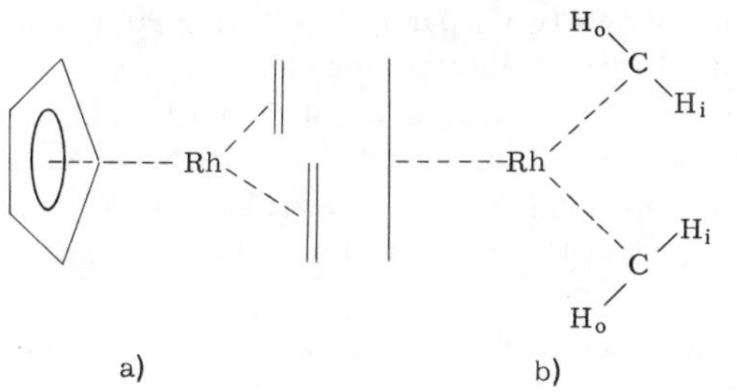

Fig. 7.3. The complex $(C_5H_5)Rh(C_2H_4)_2$. a) perspective sketch; b) view parallel to the C−C double bonds.

Actually the NMR spectrum at low temperature ($-20\,°C$) exhibits three sharp signals with intensity ratio $5:4:4$, which are due to the cyclopentadienyl ring protons and the two types of hydrogens respectively. But the temperature dependence of the NMR spectrum is characteristic for so-called fluxional behavior, i.e. for a rapid transition between two configurations which are nondistinguishable as regards structure and bonding. At room temperature, the two ethylene peaks are strongly broadened showing that the nonequivalent protons exchange at an intermediate rate in the NMR time scale; at $57\,°C$ the two peaks coalesce, indicating that the exchange is now so fast that NMR can no longer distinguish between the nonequivalent protons. (The peak due to the cyclopentadienyl protons remains sharp over the whole temperature range.)

Two rotation modes for the olefin molecule would satisfy the variable temperature NMR data, namely rotation about the C−C axis of the ethylene molecule, or a propeller-like movement about the metal-olefin bond axis:

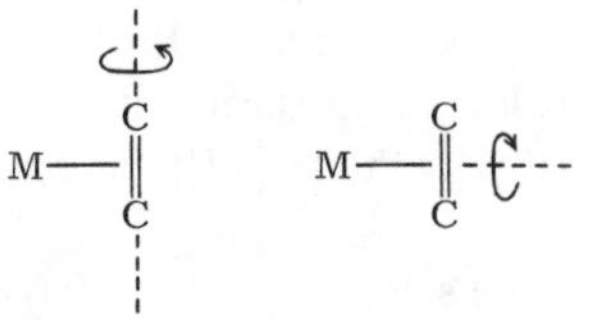

This question was settled by means of the complex $\{Os(CO)\,(NO)\,(C_2H_4)\,[P(C_6H_5)_3]_2\}PF_4$ (Fig. 7.4) [14] which contains two pairs of equivalent hydrogens, H' and H''.

Because of the particular ligand arrangement, the two protons on either side of the ethylene double bond are equivalent, but the H's are different from the H''s. Rotation about the C−C axis would not alter this situation, but the propeller-like movement would exchange

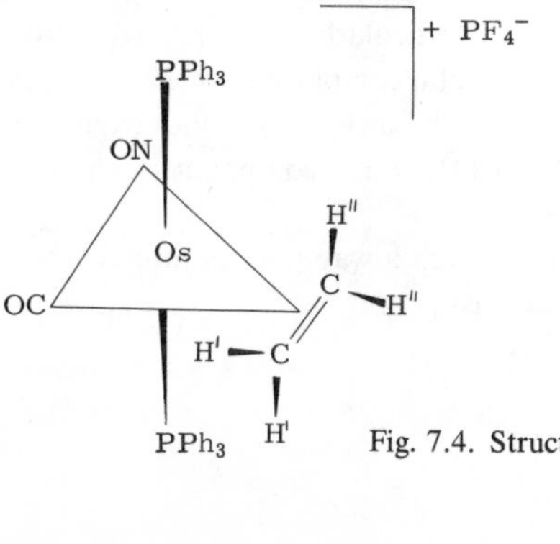

Fig. 7.4. Structure of the complex $\{Os(CO)\,(NO)\,(C_2H_4)\,[P(C_6H_5)_3]_2\}PF_4$.

non-equivalent protons. The NMR spectrum shows two separate peaks at $-90\,°C$ and coalescence at $-65\,°C$, hence proving that the olefin movement is propeller-like.

From these and similar experiments an energy barrier to rotation of 50–60 kJ/mol has been estimated, for ethylene itself, as well as for monosubstituted ethylene [15]. No rotation of the olefin has been observed in complexes containing $F_2C = CF_2$ or $(CN)_2C = C(CN)_2$, and this has been ascribed to the stronger π bond between metal and these olefins, as compared with ethylene.

With regard to catalysis, the relative coordinating ability of different olefins is also of interest. From equilibrium measurements of the type:

$$M - L \ + \ \text{olefin} \ \underset{\rightleftharpoons}{K} \ M - \text{olefin} \ + \ L$$

for Pd(II) [16], Ni(0) [17], and Rh(I) [18] complexes the following series were found for the coordination ability: butene-1 > *cis*-butene-2 > *trans*-butene-2; hexene-1 > hexene-2 > 2-methylpentene-1 > 2-methylpentene-2.

These two series evidently reflect increasing steric hindrace to coordination; the following series, however, show the importance of electronic effects: *viz.* ethylene > propylene > buten-1 ≃ hexene-1; and NCCH = CHCN > $CH_2 = $ CHCN > $CH_2 = $ CHCOOCH$_3$ > $CH_2 = $ CHCOCH$_3$ > $CH_2 = CH - C_6H_5$ > hexene-1 > $CH_2 = $ CHO(CH$_2$)$_3$CH$_3$.

The last two series have been established with a Ni(0) complex [17], and show that the stability of the nickel-olefin bond is markedly enhanced by electron withdrawing substituents such as cyano or carboxyl, and is reduced by electron donating groups (the donor ability increases in the order, methyl < ethyl < alkoxyl). This behavior indicates that electron back-donation from filled metal d to empty olefin π^* orbitals (π-bonding, *cf.* Fig. 7.1) is the predominant part of the metal-olefin bond, and this is plausible with a Ni(0) d^{10} center which has all d orbitals occupied. The electronic influence is less pronounced with metal centers having a lower d electron density {Ni(0) > Fe(0) > Rh(I) > Pt(II) [17]}, and is probably negligible with titanium centers.

Differences in coordination ability can sometimes provide selectivity for a catalytic process in the sense that, from a mixture of terminal and internal olefins, predominantly the former are converted into product. Examples of this behavior will be found in the following chapters.

7.2.2. The Allyl Group

Allyl complexes of transition metals play an important role as intermediates in the catalytic reactions of 1,3-diolefins (Chapter 9), and also in certain isomerizations (Chapter 8).

The metal-allyl bond has been investigated extensively, particularly by X-ray structural analysis, NMR and infrared spectroscopy of available allylmetal complexes [19, 20]. Only in very rare cases is the allylgroup fixed to the transition metal by a simple σ bond; examples are $[CH_2 = CHCH_2Co(CN)_5]^{3-}$ and $CH_2 = CHCH_2Mn(CO)_5$. Such compounds exhibit a characteristic C = C stretching frequency at *ca.* 1620 cm^{-1}.

Far more frequent is the attachment of the allyl group in such a way that its three carbon atoms are all bonded to the metal center, generally written as:

$$\text{HC}\begin{smallmatrix}\text{CH}_2 \\ \\ \text{CH}_2\end{smallmatrix} \text{—— ML}_n \quad \text{or} \quad \text{—— ML}_n$$

where L_n stands for the other ligands in the complex. Instead of having two electrons in a localized π bond and one in a σ orbital forming the σ bond to the metal, the allyl group behaves as a delocalized π system as shown in Fig. 6.9. All three electrons are now valence electrons with respect to the allylmetal bond, and we have to consider the three π orbitals as valence orbitals.

The two types of attachment are generally called σ allyl and π allyl bonding, respectively. Cotton [21] has proposed a different nomenclature which is widely applicable to compounds containing organic residues bound to a metal. The prefix hapto (from the Greek ηαπτειν = to fasten) is placed before the organic moiety and the number of carbon atoms which are attached to the metal atom is specified by the appropriate Greek numerical prefix, mono, bi, tri, tetra, etc. In this nomenclature a σ allyl complex reads, for instance, monohaptoallyl-pentacarbonyl-manganese, which is abbreviated $(\eta^1\text{-}C_3H_5)Mn(CO)_5$. An example of a π complex is $[(\eta^3\text{-}C_3H_5)PdCl]_2$.

A schematic representation of the structure of $[(\eta^3\text{-}C_3H_5)PdCl]_2$ is given in Fig. 7.5. X-ray measurements show that for this and for a number of other π allyl transition metal complexes in which the metal is joined to an allyl group and to two other ligands, the plane of the allyl group is always somewhat tilted; the dihedral angle between this plane and that defined by the metal and the other two ligands lies in the 106–125° range. The coordination plane of the metal and the other ligands (in Fig. 7.5, the PdCl₂Pd plane) cuts the π allyl skeleton at a line *ca.* two thirds of the distance from the central towards the terminal carbon atoms. The carbon-carbon distances are generally 1.4–1.43 Å, *i.e.*, intermediate between those of carbon-carbon double bonds (1.34 Å) and single bonds (1.54 Å). The allyl group is considered as a bidentate ligand occupying two coordination sites, so that the metal atom is surrounded by an approximately square-planar ligand arrangement, in the dimeric complex of Fig. 7.5 as well as in monomeric complexes such as those represented in Figs. 7.6–7.8, where the bridging chloride ions are replaced by other ligands L' and L''.

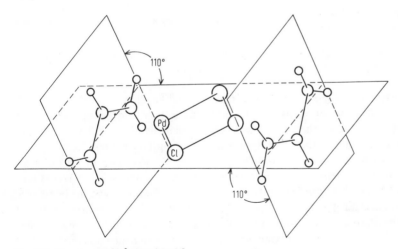

Fig. 7.5. Structure of $[(\eta^3\text{-}C_3H_5)PdCl]_2$.

The π allylmetal bond may be described in the following qualitative way [22]. The occupied molecular orbital Ψ_1 of the π allyl group (2 electrons, *cf.* Fig. 6.9) has the correct symmetry for overlap with a metal orbital which is a linear combination of the empty s, p_x and p_y valence orbitals (sp^2). This interaction is shown in Fig. 7.6a. A σ bond is established, in which electron density is donated from the allyl ligand to the metal.

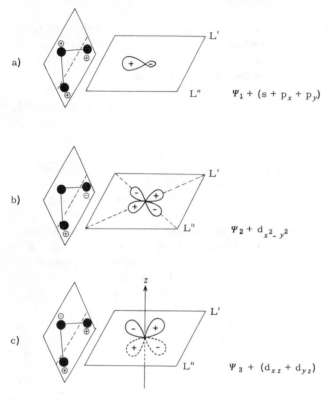

Fig. 7.6. Coordination of a trihapto-allyl group to a metal in a square-planar complex; interactions of metal orbitals and allyl orbitals. (After van Leeuven and Praat [22].)

The nonbonding allyl MO, Ψ_2, is able to form a π bond with $d_{x^2-y^2}$ (Fig. 7.6b). Ψ_2 of the free ligand contains one electron, hence this MO may act as an acceptor or as a donor. Whether it does so or not, will depend on the number of electrons available in the metal, and on the donor or acceptor character of the other ligands L' and L" in the complex. If Ψ_2 donates electron density to the metal (as it does in $[\eta^3\text{-}C_3H_5PdCl]_2$ according to MO calculations [23]) the allyl group behaves more as a cation; if the electron flow is *vice versa* (as in $(\eta^3\text{-}C_3H_5)_2Pd$ [23]), the allyl ligand behaves more as an anion. Nevertheless, the bonding is best described as covalent. Finally, there is a relatively weak [23] interaction between the empty antibonding Ψ_3 of the allyl ligand and a suitable combination of metal orbitals, namely d_{xz} and d_{yz} which, in square-planar symmetry, are low-lying and hence occupied (*cf.* Fig. 5.5). This interaction results in an additional π bond, this time with electron back-donation, i.e. electron flow from the metal to the ligand (Fig. 7.6c). This interaction is weak, presumably, because the orbitals involved are rather different in energy (energy criteria for bonding, *cf.* Section 6.1.1).

This qualitative MO picture permits a plausible interpretation of the experimentally determined geometry of the allyl group. The interaction of Ψ_1 and Ψ_3 with the relevant metal orbitals is evidently most effective, if the plane defined by the metal and the ligands L′ and L″ intersects the molecular plane of the allyl group at a line through the center of gravity of the latter. The interaction of Ψ_2 with $d_{x^2-y^2}$, on the other hand, would be maximal, if the terminal carbon atoms were located in the PdL′L″ plane, not below it. The overlap is, however, clearly improved by tilting the allyl plane, which evidently brings the relevant parts of the orbitals involved closer together.

η^3-Allyl groups, just as coordinated olefin molecules, are not always fixed rigidly to the metal. From the temperature dependence of their NMR spectra it has been deduced that many such complexes show fluxional behavior, i.e. there is a rapid interchange between two configurations which are equivalent as regards structure and bonding. Fig. 7.7 gives again a sketch of a η^3-allylmetal complex, this time including numbered hydrogen atoms.

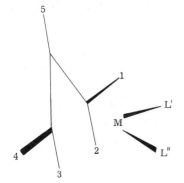

Fig. 7.7. Sketch of a square-planar η^3-allylmetal complex.

It is customary to call H^1 and H^4 *syn*, H^2 and H^3 *anti*. (This designation defines their position relative to the substituent on the central carbon atom of the allyl group [24].) The two *syn* protons are equivalent, and so are the *anti* protons, the latter being somewhat more shielded (nearer to the metal). Hence for a symmetric complex (*i.e.* L′ = L″) one would expect to find two separate signals in the NMR spectrum, the *anti* signal somewhat more upfield. At low temperature this is indeed observed. At higher temperatures, however, the two signals sometimes broaden, and finally coalesce indicating a rapid exchange of H^1 and H^2 as well as of H^3 and H^4 (*syn-anti* exchange) [19].

Unlike the olefinic proton exchange described in the preceding section, this allylic exchange is not produced by rotation of the ligand. Today it is generally agreed that a short-lived σ intermediate is involved. The main evidence comes from the NMR spectra of unsymmetric allyl complexes (L′ ≠ L″). In the absence of fluxional movement one would expect H^1, H^2, H^3, and H^4 to be nonequivalent, because of the differing chemical environment. It has been found, for instance, that a methallylpalladium complex (η^3-C_4H_7)PdL′L″, with L′ = $CH_3CO_2^-$ and L″ = $(C_6H_5)_3P$, shows the four expected signals at low temperature. At $T \approx 0\,°C$, however, H^3 and H^4 become equivalent (coalescence of the two peaks), whereas the H^1 and H^2 peaks remain sharp. Only at $T > 45\,°C$ do H^1 and H^2 also start to interchange [22]. This observation is best explained in the following way [19, 22]. Donor ligands have usually a labilizing influence on the metal-ligand bond *trans* to their own position (*trans* influence, *cf.* Section 7.6.2). Reinforcing a normal thermal vibration, the phosphorus *trans* influence can lead to an opening of the metal-carbon bond *trans* to the phosphine. Two of the π electrons of the allyl group then localize in a double bond, and the third in a σ C−Pd bond (*cf.* Fig. 7.8). In this situation, free rotation about the C−C σ bond is possible. After restoration of the η^3-complex, H^3 and H^4 have exchanged their position relative to the substituent on the central carbon atom (*syn-anti* exchange), whereas H^1 and H^2 retain their former relationship.

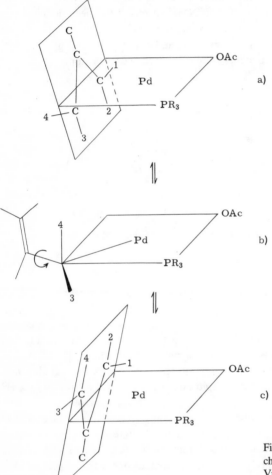

a)

b)

c)

Fig. 7.8. The mechanism of *syn-anti* interchange of $(\eta^3\text{-}C_4H_7)Pd(PR_3)(OAc)$, after Vrieze [19]; $(OAc = CH_3CO_2^-)$.

The difference in *trans* influence of the phosphine and the acetate ligand allows the π→σ→π rearrangement with H^3 and H^4 interchange occuring at a lower temperature than that for interchanging H^1 and H^2. In symmetric complexes $(L' = L'')$, evidently, either side has the same chance to undergo the π→σ→π transformation.

Usually, the molecule is in the σ allyl form for a very short time only, but in the presence of an excess of phosphine or other donor molecule, the η^1-form of several allylpalladium complexes may attain sufficient lifetime to be observable in the infrared region (weak band at 1610–1650 cm^{-1}), presumably *via* the equilibrium [25]:

$$HC\underset{CH_2}{\overset{CH_2}{\Big\langle}}\!\!-\!\!Pd\underset{L''}{\overset{L'}{\Big\langle}} + PR_3 \rightleftharpoons CH_2{=}CHCH_2PdL'\,L''\,(PR_3)$$

Such π→σ interconversions are assumed to play an important role in catalytic reactions involving conjugated diolefins (*cf.* Chapter 9). Interestingly, the fluxional movement discussed above is promoted also by olefins [26].

For the sake of completeness we note that in some asymmetric η^3-allyl complexes H^1 and H^4 exchange and concurrent H^2 and H^3 exchange (*syn-syn* and *anti-anti*, *cf.* Fig. 7.7) have also been observed. An exchange of this type occurs, for instance, on addition of PR$_3$ to solutions of $[(\eta^3$-2-methallyl)PdCl(PR$_3$)]$ [27]. This process is less well understood, and several mechanisms may be operating [19]. The most frequent route appears to involve a pentacoordinate intermediate in which the chlorine ligand occupies the apical position of an (approximate) square pyramid, and an extra phosphine molecule P*R$_3$ occupies the equatorial position. Evidence for such an intermediate comes from the disappearance, upon addition of an excess of phosphine, of an absorption band in the range 300–250 cm^{-1}, indicative of an "in-plane" Pd−Cl bond [27]. Hence, this pathway for *syn-syn* and simultaneous *anti-anti* exchange may be written as follows:

$$\Big\langle\!\!-\!\!Pd\underset{Cl}{\overset{PR_3}{\Big\langle}} \qquad \Big\langle\!\!-\!\!Pd\underset{P^*R_3}{\overset{Cl}{\Big\langle}}$$

$$\pm\,P^*R_3 \diagdown \qquad \diagup \pm\,PR_3$$

$$\Big\langle\!\!-\!\!\overset{Cl}{\underset{P^*R_3}{Pd}}\!\!\overset{PR_3}{}$$

This mechanism actually involves a stationary allyl ligand, and an exchange of the other ligands. No evidence has ever been found for a propeller-like rotation of the allyl group [19].

7.3. Transition Metal-Carbon σ Bonds

Transition metal-carbon σ bonds are relatively stable from a thermodynamic point of view. Bond dissociation energies which are in principle available from calorimetric measurements (although not always with great accuracy) have been determined in several cases. Table 7.2 shows that values in the range 125–380 kJ/mol have been obtained, depending upon the metal itself and on the other ligands in the complex.

Nevertheless, metal-carbon bonds are involved in most catalytic processes at transition metal centers, as illustrated in the following chapters. Evidently this testifies to a certain

Table 7.2. Calorimetric bond dissociation energies (BDE) of some metal-carbon σ bonds.

Bond	Compound	BDE (kJ/mol)	Ref.
Ti −C	Ti[CH$_2$C(CH$_3$)$_3$]$_4$	170	[119]
Ti −C	Ti[CH$_2$Si(CH$_3$)$_3$]$_4$	250	[119]
Ti −C	Ti(CH$_2$C$_6$H$_5$)$_4$	240	[119]
Ti −C	(η^5-C$_5$H$_5$)$_2$Ti(CH$_3$)$_2$	250	[28]
Ti −C	(η^5-C$_5$H$_5$)$_2$Ti(C$_6$H$_5$)$_2$	350	[28]
Zr −C	Zr[CH$_2$C(CH$_3$)$_3$]$_4$	220	[119]
Zr −C	Zr[CH$_2$Si(CH$_3$)$_3$]$_4$	325	[119]
Zr −C	Zr(CH$_2$C$_6$H$_5$)$_4$	380	[119]
Mn−C	Mn(CH$_3$)(CO)$_5$	125	[29]
Re −C	Re(CH$_3$)(CO)$_5$	220	[29]
Pt −C	Pt(C$_6$H$_5$)$_2$[P(C$_2$H$_5$)$_3$]$_2$	250	[30]
Pt −C	(η^5-C$_5$H$_5$)Pt(CH$_3$)$_3$	160	[31]

lability of these bonds, making them accessible to chemical reaction. The labilization required for catalysis comes about in the course of concerted processes, *i.e.* under conditions where simultaneous bond making and bond breaking make unnecessary the full dissipation of the M−C bond dissociation energy. The most frequently encountered concerted process is the so-called β hydrogen transfer. It comprises abstraction of a hydrogen from the carbon β to the metal, with formation of a metal hydride and an olefin:

$$L_nMCH_2CH_2R \rightarrow L_nMH + CH_2 = CHR$$

This reaction, which leads to the destruction of the metal-carbon bond, will be described in detail in Section 7.4.2, because of its importance in catalysis.

Another relatively frequent decomposition involves two alkylmetal moieties, and leads to the expulsion of alkene and alkane, with simultaneous reduction of the metal. The mechanism of this reaction appears to be related to β-hydrogen elimination. For an ethyltitanium-(IV) complex, a bimolecular reaction has been formulated, with β hydrogen transfer from one Ti unit to the other, followed by donation of the hydrogen atom to the alkyl group of the second (formation of ethane) and liberation of ethylene at the first [123] (*cf.* Eq. 7.24):

$$L_nTi(IV)C_2H_5 + L_nTi(IV)C_2H_5 \rightarrow 2L_nTi(III) + C_2H_4 + C_2H_6$$

With an alkylcopper(I) complex, a two step mechanism has been proposed [32]:

$$CH_3CH_2CH_2CH_2CuP(n\text{-}Bu)_3 \rightarrow CH_3CH_2CH = CH_2 + HCuP(n\text{-}Bu)_3$$
$$CH_3CH_2CH_2CH_2CuP(n\text{-}Bu)_3 + HCuP(n\text{-}Bu)_3$$
$$\rightarrow CH_3CH_2CH_2CH_3 + 2\,Cu^0 + 2\,P(n\text{-}Bu)_3$$

It should be noted that no free radicals are involved. In general, it turns out that homolysis, *i.e.* thermal bond breaking to give a carbon radical and a reduced metal species, is quite uncommon with transition metal alkyl or aryl complexes; it has been proposed occasionally in speculative discussion of decomposition pathways, but where more detailed studies have been made it has often turned out to be of relatively little importance [119].

Several other concerted decomposition pathways for transition metal-carbon bonds have been studied in great detail, but are less important for catalysis. They are amply described in the literature (e.g. [33, 34]).

The kinetic lability of metal-carbon bonds, provided by concerted processes, although evidently essential for catalysis, has handicapped for many years preparative transition metal organic chemistry. Many unsuccessful attempts to prepare simple transition metal alkyls and aryls even led to the belief that transition metal-carbon σ bonds are inherently unstable. Only very recently it was suggested, independently, by two groups [35, 36] that such metal-carbon bonds may be made kinetically stable provided concerted decompositions are inhibited, in particular β hydrogen elimination. One such method is to use carbon ligands which do not possess hydrogen in β position. Many very stable complexes with methyl, benzyl, neopentyl, and similar ligands have been prepared, for instance $Zr(CH_2$-$C_6H_5)_4$ m.p. $133\,°C$, $Mn(CO)_5CH_3$ m.p. $95\,°C$, or $W(CH_3)_6$ which is stable below $50\,°C$ [37]. Another possibility is to have all coordination sites occupied by strongly bonded ligands*). Thus it proved possible to isolate, e.g., $[Rh(NH_3)_5(C_2H_5)]^{2+}$ as a perchlorate or bromide, despite the presence of a β hydrogen in the alkyl group [38].

Nowadays an extensive preparative organometallic chemistry of transition metals exists [118, 119]. Very often "stabilizing" π bonding ligands such as CO, phosphines or phosphites, or $η^5$-cyclopentadienyl ligands are present in the σ alkyl and σ aryl derivatives of the metals. Their main function probably is to block coordination sites, an important feature being their strong attachment to the metal. However, a certain degree of electronic stabilization cannot be ruled out either, as will be discussed in more detail in Section 7.6.3.

Apart from transition metal organic complexes containing simple σ bonded alkyl (or aryl) groups, carbene complexes have started to play a role in catalysis also (cf. Chapter 8.5). These novel compounds were described first by E. O. Fischer [39, 40]. The term carbene refers to a divalent carbon species, i.e., one in which a carbon atom forms two two-electron bonds to adjacent atoms, with the two remaining electrons localized on carbon. From spectroscopic and crystallographic data [40] it has been concluded that the carbenoid carbon in the relevant complexes is trigonally hybridized (sp^2), with the two electrons paired in one of the sp^2 orbitals (singlet state), and with one vacant p orbital:

The lone pair orbital forms a strong σ bond with suitable unoccupied σ orbitals of the metal. Although one would expect the p_z orbital of the carbenoid carbon to be available for π bonding with a suitable occupied metal π orbital, the observed metal-carbon distances are generally close to that of metal-carbon single bonds.

---

* β Hydrogen transfer requires an empty coordination site cis to the alkyl group, see Section 7.4.2.

7.4. Key Reactions in Catalysis

The catalytic processes involving transition metal complexes are composed generally of several partial reactions. Only a few basic steps have to be considered which, in varying number and sequence, make up the different catalytic processes such as isomerization, hydrogenation, dimerization and polymerization of olefins, *etc.* These key reactions are described in this section.

7.4.1. The Insertion Process

A partial step which forms part of most catalytic cycles is designated as an insertion. It comprises reaction of a ligand R, covalently σ bonded to a metal center, with a substrate molecule coordinated to the same metal. The ligand R is generally alkyl or hydrogen; the substrate molecule may be, for instance, an olefin or CO. In the course of reaction the substrate molecule inserts between the metal and R:

$$(L_n)M \cdots \overset{R}{\underset{}{|}} \quad \overset{C}{\underset{C}{\|}} \quad \longrightarrow \quad (L_n)M-\overset{|}{C}-\overset{|}{C}-R \tag{7.1}$$

$$(L_n)\overset{R}{\underset{}{M}} \cdots C\equiv O \quad \longrightarrow \quad (L_n)M-\overset{O}{\underset{}{\overset{\|}{C}}}-R \tag{7.2}$$

$$(L_n)\overset{R}{\underset{}{M}} \quad \longrightarrow \quad (L_n)MCH_2CH=CHCH_2R \tag{7.3}$$

$$(L_n)M —$$

In these formulae, (L_n) represents all other ligands in the complex. Their importance will be discussed later (Section 7.6). Reaction (7.1) could be, for instance, the growth process in ethylene polymerization (where R = the growing polymer chain). After the insertion step, one coordination site remains free, and the cycle coordination-insertion could be repeated. With R = H it could be a partial step in the hydrogenation of an olefin, *etc.* Note that CO inserts in a different way as does an olefin, and that after insertion of a conjugated diolefin the allylic end of the ligand stabilizes generally in the η^3 form.

It is assumed commonly that insertions take place by a concerted reaction path, *i.e.*, *via* a more or less polar cyclic transition state involving simultaneous bond breaking and bond making, *e.g.*:

$$\overset{R}{\underset{M}{|}} \overset{C}{\underset{C}{\|}}{}^{C} \quad \rightleftharpoons \quad \left[\begin{array}{cc} \delta^- & \delta^+ C \\ R----C & \\ | & | \\ \delta^+ M----C & \delta^- \end{array} \right] \quad \longrightarrow \quad M-C-\underset{C}{\overset{|}{C}}-R \tag{7.4}$$

(For clarity we have omitted other ligands, L_n, attached to the metal.) Concerted reactions are characterized by relatively low activation energies, which are generally lower than the bond dissociation energies of the weakest bonds involved (indicative of concerted bond breaking and bond making). Activation entropies are either very small or negative (indicative of the restriction in motion resulting from the formation of a cyclic transition state) [41]. For the insertion of an ethylene molecule into the metal-carbon bond of an ethylrhodium-(III) complex the following activation parameters have been obtained [42]: Arrhenius activation energy $E_a = 72.0$ kJ/mol, $\Delta H^* = 69.5$ kJ/mol, $\Delta F^* = 95.0$ kJ/mol, $\Delta S^* = -81.4$ J/mol·deg. Experimental data concerning metal-carbon bond strengths are relatively scarce, but whenever such data have been estimated they lie in the range 125–380 kJ/mol (cf. Table 7.2). The strength of the coordinative rhodium-ethylene bond has been estimated as ≤ 130 kJ/mol*) [43]. Although this is an upper limit, it indicates that metal-olefin bonds are not weak. Finally, the opening of a carbon-carbon double bond requires 264 kJ/mol (cf. Table 12.1). Thus, the overall activation energy is considerably lower than the dissociation energies of all bonds involved.

Activation parameters have also been reported for the insertion of CO into a $Mn-CH_3$ bond [44]: $E_a = 61.9$ kJ/mol; $\Delta H^* = 59.4$ kJ/mol, $\Delta F^* = 86.2$ kJ/mol, $\Delta S^* = -88.3$ J/mol·deg. Both examples form a solid basis for the assumption of a concerted mechanism.

Some discussion centres upon whether the insertion takes place at the coordination site of the σ bonded ligand, or whether this ligand migrates to the site of the coordinated substrate molecule:

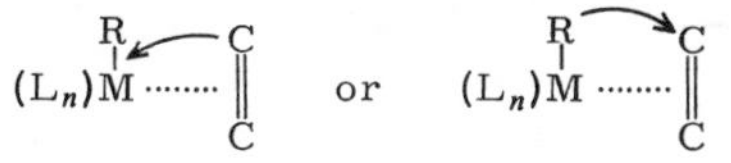

In one case at least, this question appears settled by experimental evidence, namely in a CO insertion [45]. In pentacarbonyl methylmanganese the methyl group is carbonylated under the influence of an excess of CO. It has been shown, using ^{13}CO, that the carbonylation involves a CO molecule originally present in the coordination sphere of the metal, whereas an incoming CO is located cis to the newly formed acyl group. The stereochemistry of the products was examined by infrared spectroscopy.

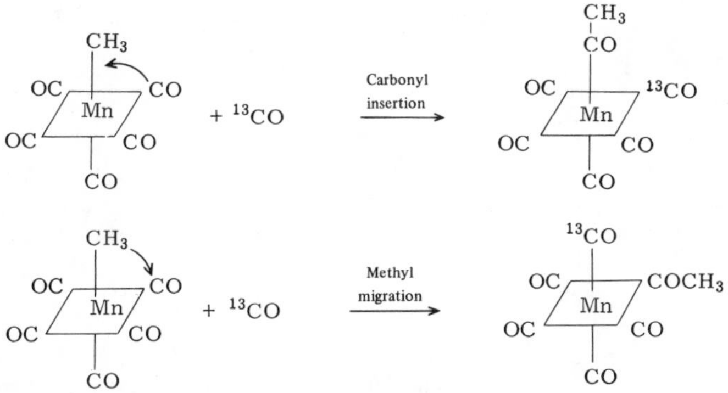

---

* Bond strength should not be confused with the barrier to rotation (Section 7.2.1).

From this experiment alone it is not possible to choose between the two mechanisms, since both lead to the same product, cis-$CH_3COMn(^{13}CO)(CO)_4$. From a study of the stereochemical changes in the carbonylation of cis-$CH_3Mn(^{13}CO)(CO)_4$, however, Noack and Calderazzo [45] decided in favor of the cis migration of the methyl group. Only this mechanism leads to the observed $1:2:1$ ratio of the product pentacarbonylacylmanganeses exhibiting $^{13}COCH_3$, cis, and $trans$ stereochemistry respectively.

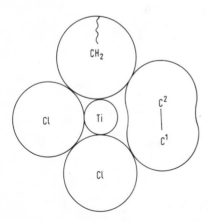

On the basis of commonsense considerations the cis migration mechanism is usually assumed also for ethylene insertion into metal-carbon bonds, particularly with titanium catalysts [46–48]. Fig. 7.9 shows the xy plane of a titanium complex, comprising an alkyl ligand, a coordinated ethylene molecule and two chlorine ligands (*cf., e.g.,* Fig. 7.14). The dimensions of metal and ligands are approximately drawn to scale.

This representation emphasizes that a small "in-plane" displacement of the methylene group, which may be possible within the normal vibrational modes, brings this group into a

Fig. 7.9. xy Plane of an alkyltitanium complex containing a coordinatively bonded ethylene in cis-position to the alkyl ligand (after Cossee [46]).

position of considerable overlap with the olefin [47]. Furthermore, it can be presumed generally that metal-carbon (and also metal-hydrogen) bonds are polar in the sense that the metal atom has a positive, and the ligand atom a negative charge. Actually, a MO calculation of the charge distribution within a relevant titanium complex [48] has shown that the alkyl carbon is strongly negative. The olefin molecule as a whole, on the other hand, is slightly positive because electron back donation does not play a role in this case [Ti(IV), 3 d^0; *cf.* Fig. 7.1]. This initial situation predisposes the alkyl-carbon and the C^2 carbon atom of the ethylene molecule towards stronger interaction. As soon as the alkyl group migration commences, its σ orbital starts to overlap with the metal d_{xy} orbital (*cf.* Figs. 7.1 and 7.9). Note that this overlap is forbidden in the octahedral complex, but becomes allowed, because of the changed symmetry, as soon as the alkyl group moves away from its position on the y axis. Hence there is no significant loss in bonding energy, and consequently no serious activation energy barrier to migration. Moreover, the overlap of the alkyl σ bonding orbital with d_{xy} transfers electron density to this metal orbital, and hence into the antibonding π* orbital of the olefin, thus additionally activating the substrate molecule.

These considerations, although not constituting a proof, make the *cis* migration mechanism a very attractive model also for olefin insertion. Nevertheless, the conclusion should not be generalized without caution. Unless stated otherwise, we shall confine the term "insertion" to merely describing the outcome of the reaction, without attributing any mechanistic significance to it.

The formulation (7.4) of the transition state during olefin insertion raises the question of the orientation of the olefin molecule in this step, which evidently bears on the structure of the product. Markownikoff's rule is frequently cited in conjunction with this problem. The rule states that during the addition of H^+X^- to an asymmetrically substituted olefin the negative ion becomes attached to the unsaturated carbon carrying the smaller number of hydrogen atoms*).

Originally this was an empirical rule, but quantum chemistry and MO theory have since provided a solid theoretical background to it. In the particular case of propylene, for instance, (*cf.* Section 6.2.3), the π electron density distribution is calculated as follows:

$$\begin{array}{cc} 0.972 & 1.042 \\ \end{array}$$
$$CH_3 - CH = CH_2$$

i.e. the double bond is polarized and one expects an anion to attack at the central carbon atom. We define then as "Markownikoff mode" of insertion that one which is expected from an electronic point of view**).

In general, $M^{\delta+} - C^{\delta-}$ and $M^{\delta+} - H^{\delta-}$ bonds are also polarized, in the sense indicated [124]. This polarization evidently favors the transition state shown in (7.4), where the negatively charged carbon (or hydrogen) ligand reacts with the unsaturated carbon next to the methyl group, whereas the metal cation attacks the less substituted carbon of the double

---

* The original "Markownikoff rule" (1875) reads: Lorsqu'à un hydrocarbure non saturé, renfermant des atomes de carbone inégalement hydrogenés, s'ajoute un acide haloidhydrique, l'élément électronégatif se fixe sur le carbone le moins hydrogené" [49].
** This original definition is generally used in the German literature, whereas several English authors have reversed the nomenclature, calling the reaction sequence (7.4) the "anti-Markownikoff mode", and (7.5) the "Markownikoff mode".

bond. This "Markownikoff mode" of insertion is found, for instance, in the coordinative polymerization of propylene with titanium and vanadium catalysts [50].

Frequently, however, the olefin inserts in the opposite way, i.e. the metal cation attacks the more substituted carbon. This behavior, referred to as the "anti-Markownikoff mode" of insertion, is found particularly with group VIII transition metals of the first two rows, in cases where (L_n) are strongly electron donating ligands [124]. (Typical examples will be found in Section 8.4.2.) This behavior requires that these metal centers, with their high d electron population, provoke inversion of olefin double bond polarization*), e.g.:

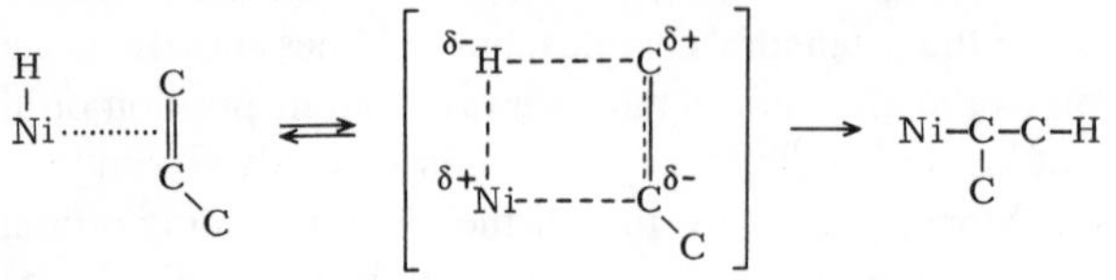

$$
\begin{matrix}
\text{H} & & \text{C} \\
| & & \| \\
\text{Ni} \cdots\cdots & & \text{C} \\
& & \diagdown \text{C}
\end{matrix}
\;\rightleftharpoons\;
\left[
\begin{matrix}
{}^{\delta-}\text{H} \text{------} \text{C}^{\delta+} \\
| \quad\quad | \\
{}^{\delta+}\text{Ni} \text{-----} \text{C}^{\delta-} \\
\quad\quad\quad \diagdown\text{C}
\end{matrix}
\right]
\;\longrightarrow\;
\begin{matrix}
\text{Ni--C--C--H} \\
| \\
\text{C}
\end{matrix}
\tag{7.5}
$$

This hypothesis may be rationalized on the basis of electron back-donation from the metal to the olefin, according to the Chatt model for the metal-olefin double bond (Fig. 7.1), taking into account the unsymmetric nature of the π-orbitals of propylene (Fig. 6.11). Upon co-ordination to a metal, the olefin π orbital loses electron density in a metal-olefin σ bond whereas π^*, in the case of back-donation, gains electron density in a π bond (see Fig. 7.10).

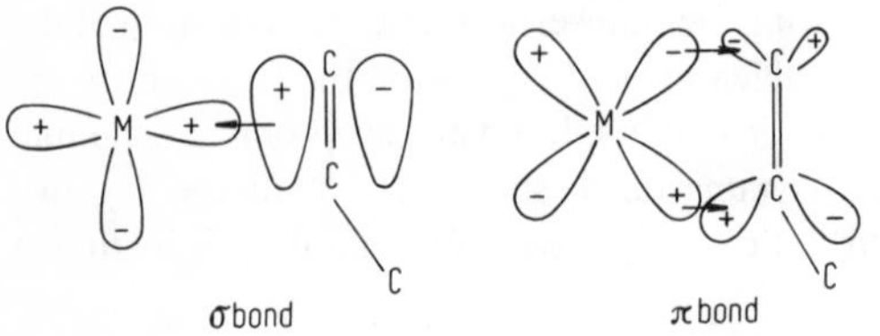

σ bond   π bond

Fig. 7.10. Coordination of propylene to a transition metal center.

Evidently, this effect tends to counterbalance the polarization of the double bond of propylene (and of other unsymmetrically substituted olefins). A strong σ bond combined with effective electron back-donation (high d population) may then reverse the polarization.

7.4.2. β Hydrogen Transfer (β Elimination)

Transition metal complexes with σ bonded organic ligands having hydrogen attached to the β carbon tend to undergo C−H bond rupture forming a metal hydride. The organic ligand leaves the complex with an olefinic end group.

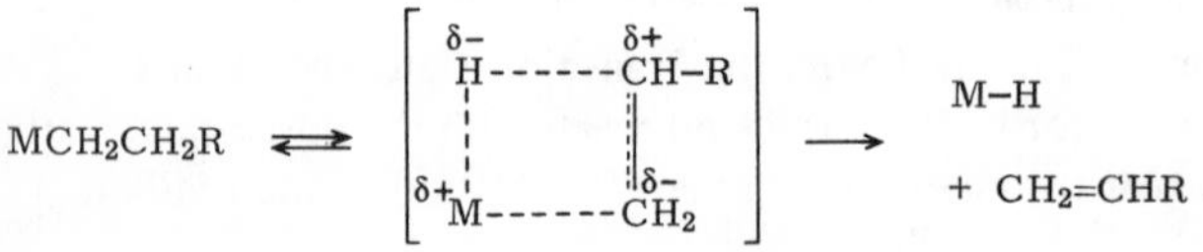

$$
\text{MCH}_2\text{CH}_2\text{R} \;\rightleftharpoons\;
\left[
\begin{matrix}
{}^{\delta-}\text{H} \text{------} \text{CH--R}^{\delta+} \\
| \quad\quad\quad | \\
{}^{\delta+}\text{M} \text{-----} \text{CH}_2^{\delta-}
\end{matrix}
\right]
\;\longrightarrow\;
\begin{matrix}
\text{M--H} \\
\\
+ \text{CH}_2\text{=CHR}
\end{matrix}
\tag{7.6}
$$

---

* An alternative explanation would be the inversion of the polarization of the M−H or M−R bond which would require a shift of the two σ bonding electrons towards the metal ion. The particular conditions (high d electron density, strong electron donor ligands) make such a shift less probable. However, in complexes with several strongly electron accepting ligands such as CO, and in powerfully solvating solvents such as water or alcohols, one may have to take into consideration a reversal of the polarization of M−H bonds, and also of certain M−C bonds [124].

In coordinative polymerizations, for instance, the β hydrogen transfer determines the molecular weight of the polymer; if β elimination takes place after only two growth steps, dimerization of the olefin occurs. The tendency towards β hydrogen abstraction depends upon the metal (group VIII metals > metals from the left hand side of the Periodic Table), on the valency state [*e.g.* Ti(IV) > Ti(III)], and on ligand environment (*cf.* Section 7.6.4).

The reaction does not occur unless an additional site is available within the coordination sphere of the metal, indicating that hydrogen has to approach the metal in a cyclic transition state such as formulated in (7.6). In fact, β elimination is the reverse of insertion (7.1) or (7.4), if we consider a metal hydride as the starting complex, and actually the reaction is often reversible.

A kinetic isotope effect has been observed during β hydrogen transfer in the complex [51]:

$$C_6H_{13}CH(D)CH_2Ir(CO)[P(C_6H_5)_3]_2$$

From the ratio of deuterated to non-deuterated octene, a value of $k_H/k_D = 2.28$ was deter-determined. It implies a primary kinetic hydrogen deuterium isotope effect, involving hydrogen (or deuterium) in a β position to the metal center; the relatively low value*[)] again accords with a cyclic transition state with simultaneous bond breaking and bond making [53].

In the presence of an excess of olefin, a somewhat different mechanism has been suggested for the β hydrogen transfer in titanium based catalysts [54, 55]. From kinetic measurements (oligomerization of ethylene, *cf.* Section 8.4.3), it was concluded that the β hydrogen transfer is a bimolecular process, including the alkyl metal species and an olefin molecule. A six-center cyclic transition state was formulated therefore [54]:

$$
\begin{array}{c}
\raisebox{0.5em}{\scriptsize $>$}C=C\raisebox{0.5em}{\scriptsize $<$} \\
\underset{\mathrm{CH_2CHR}}{\overset{}{\mathrm{Ti}}}\ \mathrm{H}
\end{array}
\rightleftharpoons
\left[
\begin{array}{c}
\overset{\delta-}{\mathrm{H_2C}}\text{\textemdash}\overset{\delta+}{\mathrm{CH_2}} \\
\delta+\mathrm{Ti} \qquad \mathrm{H}\delta- \\
\underset{\delta-}{\mathrm{H_2C}}\text{----}\underset{\delta+}{\mathrm{CHR}}
\end{array}
\right]
\longrightarrow
\begin{array}{c}
\mathrm{TiCH_2CH_3} \\
+\ \mathrm{CH_2{=}CHR}
\end{array}
\qquad (7.7)
$$

According to this mechanism, no metal hydride is formed as an intermediate, but hydrogen is transferred to the olefin generating an alkyl group. A lose interaction Ti...H can, of course not be excluded, and even appears probable. (For further discussion of this mechanism see Section 7.6.4). Theoretical work [55] on the transition states represented in Eqs. (7.6) and (7.7) actually indicates that the olefin-assisted β hydrogen transfer is highly favored energy-wise, at least in some titanium catalysts. It might also play a role with other transition metal alkyls, however. The monomolecular reaction, Eq. (7.6), would require clear-cut first-order kinetics. Where such a rate law has in fact been found, the β hydrogen transfer was not the rate determining step [56]. On the other hand, induction periods and autocatalysis have been reported for the β elimination from alkylmetals in the absence of added olefin [57], which might be due to a slow initial reaction according to (7.6) with progressively increasing participation of (7.7) as free olefin becomes available.

---

* The theoretical value for a linear transition state is $k_H/k_D \simeq 7$ at 25 °C. There exist reasons other than cyclic transition states for lower values [52].

If β elimination forms part of a catalytic reaction cycle it is, from a kinetic point of view, a chain transfer reaction because the metal hydride [Eq. (7.6)], or the metal alkyl [Eq. (7.7)] will continue the kinetic chain by further reaction with substrate molecules.

7.4.3. Oxidative Addition and Reductive Elimination

The term "oxidative addition" is used for the addition of neutral XY molecules to transition metal complexes having coordination sites available. Formally, the XY molecule is reductively dissociated to give two anionic ligands X^- and Y^-, and the metal is simultaneously oxidized. The prototype of this rather widespread class of reactions was detected in 1962 by Vaska and DiLuzio who added hydrogen to a square-planar iridium complex [58]:

$$
\begin{array}{ccc}
\begin{array}{c} L \diagup\; \diagup\, CO \\ \quad \boxed{Ir(I)} \\ X \diagup\quad\diagup L \end{array}
& \xrightarrow{\;H_2\;} &
\begin{array}{c} H \\ L\diagup |\, \diagup H \\ \quad \boxed{Ir(III)} \\ X \diagup\;|\,\diagup L \\ CO \end{array}
\end{array}
\tag{7.8}
$$

[X = Cl, Br, I;   L = $P(C_6H_5)_3$]

Actually the newly formed metal-hydrogen bonds are covalent σ bonds. According to a generally accepted convention, the "oxidation number" of a transition metal in a complex is defined, however, as the charge remaining on the metal when each electron pair shared between the metal and a ligand has been assigned formally to the ligand [122]. In the left hand complex of reaction (7.8) the only shared electron pair is that of the Ir-X bond; the other three bonds are coordinative bonds and the ligand contributes both electrons. Hence the iridium atom in this square-planar complex has the oxidation number (I). In the right hand octahedral complex of reaction (7.8) there are three shared electron pairs, one in Ir-X, and two in the two Ir-H bonds; hence the oxidation number of the metal is (III).

It is a characteristic for oxidative additions that the increase in oxidation number is accompanied by an increase in coordination number. Addition is particularly frequent in (but not restricted to) square-planar d^8 complexes, where the acquisition of two more ligands completes the 18 electron (or "noble gas") configuration (*cf.* Section 7.5.1).

The range of XY molecules which undergo oxidative addition encompasses hydrogen, the halogens, and hydrogen halides [122], as well as acyl chlorides [59], carboxylic acid anhydrides [60], and various organic halides, particularly methyl iodide [61, 122].

The oxidative addition of a neutral molecule to a transition metal center is often equivalent to an activation of such a molecule for further reaction. This is evident particularly for the H_2 molecule, so it is not surprising that oxidative addition is often a key step in the mechanism of catalyzed reactions involving hydrogen, such as hydrogenation or hydroformylation of olefins (*cf.* Chapters 8 and 10).

Frequently (but not generally) oxidative addition is reversible. The term "reductive elimination" is used for the reverse reaction. In catalysis, it often happens that one of the added ligands undergoes further reactions, *e.g.* the insertion of an olefin, before reductive elimination regenerates the catalyst as, for instance, in the following cycle (L = tertiary phosphine; *cf.* Section 8.2):

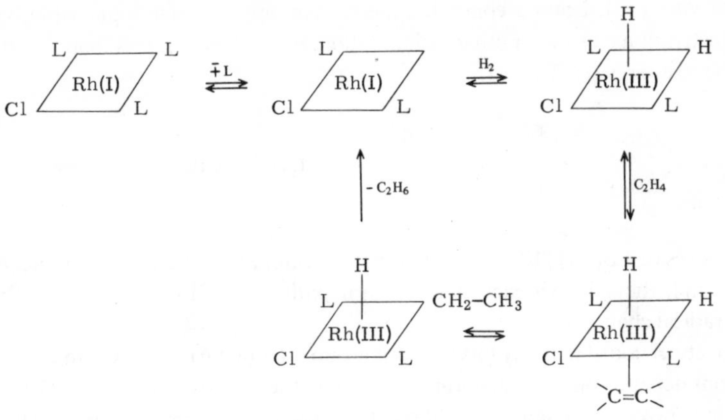

$$\text{(7.9)}$$

The addition often occurs in a stereospecific manner, both *cis* and *trans* modes having been observed depending on the metal, the neutral molecule, and the ligand environment [120, 122]. This raises the question of the detailed mechanism for oxidative addition, notably whether it is a concerted or a stepwise process. In the particular case where $XY = H_2$, the addition is generally believed to be a concerted three-center process with simultaneous bond breaking and bond making, so that the energetically favorable formation of two metal-hydrogen bonds contributes to the dissociation of the $H-H$ bond (435 kJ/mol):

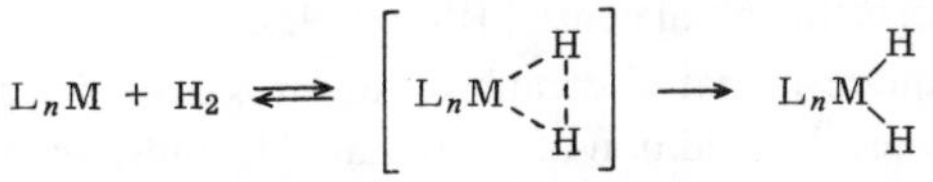

$$L_nM + H_2 \rightleftarrows \left[L_nM \begin{smallmatrix} H \\ \vdots \\ H \end{smallmatrix} \right] \longrightarrow L_nM \begin{smallmatrix} H \\ \\ H \end{smallmatrix} \qquad \text{(7.10)}$$

The addition of alkyl halides also appears to be a concerted reaction, in most relevant cases. For some Ir(I) complexes this is indicated by the observation that the addition of methyl iodide, in the presence of an excess of Cl^-, gives only $L_nIr(CH_3)I$ (no exchange of I^- for Cl^-), and by the fact that an optically active bromoalkane adds with retention of configuration [61]. Low activation energies and large negative activation entropies complete this picture [62].

One would expect *cis* addition for such a concerted process, but it has been pointed out by Pearson and Muir [61] that *trans* addition might also be possible *via* interaction of antibonding empty σ^* orbitals of the incoming XY molecule with occupied d_{xz} or d_{yz} orbitals of the metal (*cf.* Fig. 7.11).

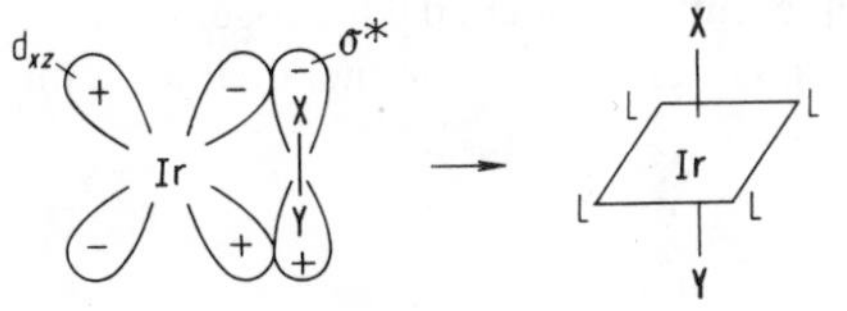

Fig. 7.11. Concerted *trans* addition of an XY molecule to an iridium center [61].

The reductive elimination of ethane from the complex $PtX(CH_3)_3(L-L)$ where $X = Cl$, Br; and $L-L = 1,2$-*bis*(diphenylphosphine)-ethane has also been investigated from a mechanistic point of view [63]: again a concerted mechanism appears highly probable because the measured activation energy for ethane elimination, 69.0 kJ/mol, is less than half the methyl-platinum bond energy (*cf.* Table 7.2).

$$L_n Pt \overset{CH_3}{\underset{CH_3}{}} \rightleftharpoons \left[L_n Pt \overset{CH_3}{\underset{CH_3}{}} \right] \longrightarrow L_n Pt + CH_3 CH_3 \tag{7.11}$$

With complexes of Co(III) [122], as well as with certain Pt(0) and Pd(0) complexes [64], on the other hand, there is evidence that a major pathway for the addition of alkyl halides involves a radical chain process.

The propensity of coordinatively unsaturated group VIII metal complexes to undergo oxidative additions depends on the metal ion itself and on the ligands associated with it.

Donor ligands favor the oxidative addition; thus, the equilibrium constant for the addition (7.12) is 40 atm^{-1} for L = P(p-CH$_3$-C$_6$H$_4$)$_3$, and only 18 atm^{-1} for L = P(C$_6$H$_5$)$_3$ [65].

$$RhClL_3 + H_2 \overset{K}{\rightleftharpoons} Rh(H)(H)ClL_3 \tag{7.12}$$

Evidently high electron density at the metal center is favorable as is also indicated by the recent observation [66] that the oxidative addition of methyl iodide to Rh(CO)IL$_2$ where L = As(C$_6$H$_5$)$_3$ or Sb(C$_6$H$_5$)$_3$ is strongly accelerated by iodide ions. These ions displace one of the neutral L ligands producing an anionic Rh(I) complex, [Rh(CO)I$_2$L]$^-$, which is much more prone to add methyl iodide than is the initial neutral Rh complex.

The very high reactivity of certain square-planar d^8 complexes towards oxidative addition is reflected in the occurence of intramolecular addition of coordinated ligands, opening C$-$H bonds, as in the following two examples. The iridium(I) complex IrCl[P(C$_6$H$_5$)$_3$]$_3$ isomerizes on heating in solution to give an octahedral hydridoaryl complex of Ir(III) [67]:

$$
\begin{array}{c}
(C_6H_5)_3P \diagdown \quad Cl \\
\quad\quad Ir(I) \\
(C_6H_5)_3P \diagup \quad P(C_6H_5)_3
\end{array}
\longrightarrow
\begin{array}{c}
(C_6H_5)_2P \quad\quad Cl \\
\quad\quad Ir(III) \\
H \diagup \quad P(C_6H_5)_3 \\
P(C_6H_5)_3
\end{array}
\tag{7.13}
$$

A ruthenium complex with stoichiometry [((CH$_3$)$_2$PCH$_2$CH$_2$P(CH$_3$)$_2$]$_2$Ru(0) [68] was found to undergo a similar albeit intermolecular process, forming a dimer [69], (Fig. 7.12).

Finally it should be noted that some authors [33, 70] recently formulated β hydrogen transfer (*cf.* Section 7.4.2) in terms of oxidative addition of a C$-$H bond to the metal, at least for square-planar d^8 complexes, *e.g.*:

$$
L_n IrCH_2CH_2R \rightleftharpoons
\left[
\begin{array}{c}
L_n Ir \text{------} H \\
\quad \\
H_2C \text{-----} CHR
\end{array}
\right]
\longrightarrow
\begin{array}{c}
L_n IrH \\
\diagup \diagdown \\
H_2C \text{---} CHR
\end{array}
$$

$$\longrightarrow L_n IrH + CH_2 = CHR$$

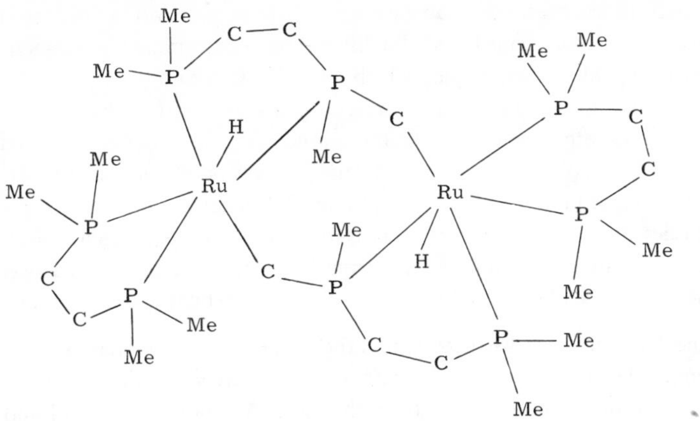

Fig. 7.12. X-ray structure of bis(tetramethyldiphosphinoethane)ruthenium dimer: intermolecular addition of a C−H bond to the metal atom (after Cotton *et al.* [69]).

It should, however, be born in mind that this mechanism is probably not generally applicable because it requires two empty coordination sites for establishing the transition state.

7.4.4. Ligand Dissociation and Ligand Exchange

Catalytically active transition metal species are sometimes formed from precursor complexes by dissociation of a ligand (making available a required coordination site). An example is shown in the reaction sequence depicted in Eq. (7.9). The rhodium complex originally applied is $RhCl[P(C_6H_5)_3]_3$ (Wilkinson's hydrogenation catalyst [71]). This compound loses a phosphine ligand to form the active species $RhClL_2$ which, although present only in very low concentrations, acts as the main catalyst (*cf.* Section 8.2). In the analogous iridium complex $IrClL_3$ on the other hand, the phosphine ligands are bound more tightly. This complex is very reactive for the oxidative addition of hydrogen, but it is ineffective as a hydrogenation catalyst, mainly because the dihydride remains hexacoordinate in solution, and no site is available for the olefin [72]. Another indication of the importance of ligand dissociation in catalytic processes is provided by pentacarbonyliron which acts as a hydrogenation catalyst above 160 °C. Presumably the high temperature is required in order to expel CO and make available coordination sites for the oxidative addition of H_2 and for the substrate molecule [120].

Ligand exchanges are very frequent if a ligand is present in an excess in solution. This is particularly true for phosphines [73] and olefins [13]. For example, the complex (acac)Rh-$(C_2H_4)_2$ (acac = the bidentate acetylacetonate ligand) is thermodynamically very stable; in the absence of an excess of ethylene it does not dissociate noticeably even at 80 °C (in glycol solution). That means that the equilibrium

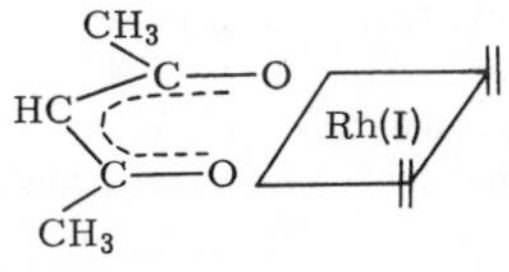

$$\text{(acac)Rh(C}_2\text{H}_4) + \text{C}_2\text{H}_4 \qquad (7.14)$$

is shifted strongly to the left. In the presence of an excess of ethylene, however, a very rapid exchange of ethylene molecules occurs. This kinetic lability has been determined experimentally from the proton NMR signal of the complex which, at $-58\,°C$, exhibits two well defined absorptions corresponding to the inner and outer hydrogen atoms of the ethylene molecules (*cf.* Fig. 7.3). If, however, ethylene is added to the solution at $-58\,°C$, only one broad absorption is present indicating rapid exchange of ethylene molecules, already at this low temperature. It has been estimated that at $25\,°C$ the lifetime of coordinated ethylene is less than 10^{-4} sec. [13]. The relevant corollary to catalysis is as follows: Substrate molecules such as olefins may enter and leave an active catalyst center many times before further reaction takes place, whereas insertions or oxidative additions are more likely to be rate determining.

Two extreme reaction mechanisms, one dissociative and the other associative, may be considered for ligand exchanges. The dissociative exchange proceeds in two steps, the first being a slow, unimolecular dissociation of the ligand X from the metal M, succeeded by a rapid second step in which the new incoming ligand X′ reacts with the metal:

$$L_n MX \xrightarrow{\text{slow}} L_n M + X$$

$$L_n M + X' \xrightarrow{\text{fast}} L_n MX' \tag{7.15}$$

The coordination number is lowered by one in the first, rate determining, step. First order kinetics are expected for such a process, *i.e.* the reaction rate does not depend on the concentration of the incoming ligand X′:

$$-\frac{d[L_n MX]}{dt} = k_1 [L_n MX] \tag{7.16}$$

The rate determining step of the associative ligand exchange is bimolecular:

$$L_n MX + X' \longrightarrow \left[L_n M \overset{X}{\underset{X'}{\diagup}} \right] \longrightarrow L_n MX' + X \tag{7.17}$$

Incoming and outgoing ligands are equally bound to the metal in the transition state in which the coordination number of the metal is increased by one. The rate of this bimolecular process is given by:

$$-\frac{d[L_n MX]}{dt} = k_2 [L_n MX][X'] \tag{7.18}$$

Frequently it is not possible to identify an exchange clearly as dissociative or associative[*], because there are intermediate situations where the incoming ligand is bound only loosely in the transition state. Following Langford and Gray [75], the mechanism of such reactions is

---

* The Ingold S_N1, S_N2 nomenclature, borrowed from organic chemistry, is sometimes used instead of the terms dissociative and associative, although it is now generally recognized as being inadequate when describing ligand exchanges [74].

termed interchange, and may be considered as taking place within a preassembled "encounter complex", where $M - X'$ bond formation occurs before M has "lost memory" of X, that is before the second coordination sphere has had time to relax. More complicated rate laws are expected [74].

7.4.5. External Nucleophilic Attack

Although less documented, and possibly not as important for transition metal catalysis, the direct nucleophilic attack of strong anionic reactants at metal-coordinated olefins, in protic solvents, should be mentioned:

$$R^- + \quad \overset{\diagup \text{C} \diagdown}{\underset{\diagup \text{C} \diagdown}{\|}} \cdots M \quad \longrightarrow \quad \overset{R}{\underset{\diagup}{\overset{\text{\tiny ''''}}{\text{C}}}} - \overset{\text{\tiny ''''}}{\underset{M}{\text{C}}} \tag{7.19}$$

The occurrence of such external attack (*i.e.* without prior coordination of the reactant R^- to the metal) has been deduced from the stereochemistry of the products resulting from the methoxypalladation of olefins, and similar reactions [76].

7.5. The Active Species

7.5.1. The "18 Electron Rule"

It has been observed frequently that transition metal complexes are particularly stable if they have a total of 18 valence electrons, especially if π bonding ligands are present. By the total valence electrons we mean, for a neutral complex, the sum of the valence electrons of the metal, one electron from each covalently bonded ligand, two electrons from each coordinatively bonded ligand (*e.g.* lone pairs), and n electrons from each η^n-bonded ligand. For ionic complexes the overall charge must be taken into account also. The following stable species, for instance, all have 18 valence electrons:

$$Cr(CO)_6; \quad (\eta^5 - C_5H_5)_2Fe; \quad RhH(CO)[P(C_6H_5)_3]_3; \quad [Co(CN)_6]^{3-}; \quad [Co(NH_3)_6]^{3+}.$$

This experimental fact is often referred to as the "18 electron rule". It may be interpreted in terms of the MO theory of transition metal complexes (*cf.* Chapter 6). The metal possesses 9 valence orbitals [five nd, one $(n + 1)s$ and three $(n + 1)p$], from which up to nine bonding MO's may be derived provided the proper ligands are present. For an octahedral complex, six of the MO's have σ symmetry and three have π symmetry (*cf.* Figs. 6.14 and 6.15). The most effective stabilization will be obtained if π bonding lowers the energy of the t_{2g} MO's (Fig. 6.15), and if all 9 MO's are each occupied by two electrons.

It should be born in mind that we are speaking of a "rule" and not of a law, *i.e.* stability is certainly not rigorously restricted to 18 electron compounds. For octahedral complexes with only σ bonding ligands, for instance, there is no energy gain*[)] if there are between one and

---

* In some cases, the Jahn-Teller effect complicates this simple statement, as outlined in Section 5.2.

six electrons in the nonbonding t_{2g} level (*cf.* Fig. 6.14), and if the energy difference Δ between t_{2g} and e_g^* is small, compounds with even more than 18 electrons may be quite stable. Thus the hexamino and hexaquo complexes $[Cr(NH_3)_6]^{3+}$ (15 electrons: 3 in t_{2g}), $[Mn(H_2O)_6]^{2+}$ (17 electrons: 5 in t_{2g}), $[Fe(NH_3)_6]^{3+}$ (17 electrons: 3 in t_{2g}, 2 in e_g^*), $[Ni(H_2O)_6]^{2+}$ (20 electrons: 6 in t_{2g}, 2 in e_g^*), are all known stable compounds. In square-planar complexes, where the d_{z^2} orbital of the metal is necessarily nonbonding (d_{xz}, d_{yz}, and p_z can in principle participate in π bonding to the ligands), very often relatively stable compounds with a total of 16 valence electrons are found. Wilkinson's hydrogenation catalyst $RhCl[P(C_6H_5)_3]_3$ is a prominent example (*cf.* Section 7.4.4).

Nevertheless, transition metal centers show a remarkable tendency to optimize stabilization by accepting ligands which bring the total of valence electrons to 18. Much of transition metal coordination chemistry appears to be governed by this tendency. We saw already several examples of this in previous sections. Thus the oxidative addition of XY molecules to square-planar Ir(I) and Rh(I) complexes raises the total of valence electrons from 16 to 18 [Eqs. (7.8) and (7.12)]. The intramolecular oxidative addition represented in Eq. (7.13), as well as the formation of the dimeric Ru complex (Fig. 7.12) are further impressive examples. The series is augmented by the compound $Mo[P(CH_3)_2(C_6H_5)]_4$, first prepared by Chatt *et al.* [77]. This complex was considered originally as an astonishing example of a very stable, highly electron-deficient 14 electron species. X-ray structural analysis, however, has revealed that one of the phosphine ligands is bonded to the metal *via* the aromatic ring which, providing 6 electrons in its π MO's (η^6-bonding), brings the total of valence electrons to 18 [78].

$$L = P(CH_3)_2(C_6H_5)$$

The mechanism of ligand exchange depends upon the number of valence electrons in a characteristic (although not exclusive) way. Thus, an exchange involving an 18 electron complex will proceed generally *via* a dissociative process. For example an acetyl complex such as $CH_3COCo(CO)_4$ easily exchanges one CO ligand for a $P(C_6H_5)_3$ ligand [79], whereby the reaction rate is independent of the concentration of the phosphine [*cf.* Eqs. (7.15) and (7.16)]. The mechanism, with $R = CH_3$, is:

$$RCOCo(CO)_4 \xrightarrow{-CO} RCOCo(CO)_3 \xrightarrow{+P(C_6H_5)_3} RCOCo(CO)_3[P(C_6H_5)_3]$$

18 electrons   16 electrons   18 electrons

Exchanges involving complexes with 16 or less electrons, on the other hand, proceed preferentially *via* an associative process which increases the number of electrons in the transition state (provided a coordination site is available). The rate of ethylene exchange mentioned in conjunction with the complex $(acac)Rh(C_2H_4)_2$ (Section 7.4.4) is proportional to the ethylene pressure, indicating the validity of the rate law, Eq. (7.18). Presumably a third ethylene molecule is coordinated to the metal in the transition state:

$$(acac)Rh(C_2H_4)_2 \quad \underset{}{\overset{+ C_2D_4}{\rightleftharpoons}} \quad \left[\begin{array}{c} D_2C=CD_2 \\ O \\ \diagdown Rh \diagup \\ O \end{array} \right] \tag{7.20}$$

$$\xrightarrow{- C_2H_4} \quad (acac)Rh(C_2H_4)(C_2D_4)$$

Interestingly the very similar complex $(\eta^5\text{-}C_5H_5)Rh(C_2H_4)_2$ does not exchange with $D_2C=CD_2$, even at $100\,^\circ C$ [13]. In this case, the complex has 18 electrons, and the associative mechanism is not accessible. On the other hand, the relevant dissociation equilibrium [cf. Eq. (7.14)] lies too far to the left to provide a pathway for a dissociative process.

Complexes with 16 valence electrons are in principle open to either pathway. In the particular case of $Pt[P(C_6H_5)_3]_3$ exchange of a phosphine for a $CH_3C\equiv CC_6H_5$ ligand actually occurs simultaneously by the dissociative and the associative mechanism [80].

These tendencies of transition metal complexes concerning the preferred number of valence electrons evidently must have a bearing on catalytic processes. Very often, of course, it will be impossible to study such relationships in detail for a given catalytic cycle, because active species are generally too unstable to permit isolation and kinetic investigation. In favorable cases it may be possible to identify and study active intermediates by indirect methods such as NMR, infrared and optical [65], or EPR [81] spectroscopy. One may use the 18 electron rule as a rough guide line during such studies in the sense that, for instance, one would not expect an 18 electron complex to be an active species. If such a compound is employed as a catalyst it almost certainly loses a ligand prior to further reaction. Moreover, it appears from the bulk of known evidence that intermediates with more than 18 valence electrons are rare – if present at all – in catalytic cycles. These considerations may help in the establishment of reaction mechanisms.

7.5.2. Formation of the Active Species

For catalytic reactions involving only olefin insertion and β hydrogen transfer steps (such as isomerization, dimerization, oligomerization, and polymerization), the active site must possess one alkyl or hydride ligand, as well as an available coordination site [two for 1,4-*cis* insertion of conjugated diolefins; cf. Eqs. (7.1) and (7.3)]. Very frequently the active catalyst is prepared *in situ* from a transition metal compound not having the required active ligand, and an aluminum alkyl. Such a combination [in particular $TiCl_4/(C_2H_5)_3Al$] was first used by Ziegler as a polymerization catalyst for ethylene [82]. This constituted one of the greatest breakthroughs in catalysis during recent decades, and gained Ziegler the Nobel prize for Chemistry in 1964. Today, every combination of transition metal compound and an organometallic, containing a group III metal, either soluble or insoluble, is called a Ziegler catalyst, in honor of the great chemist. Ziegler systems may be derived from metals throughout the transition series.

Organometallic compounds, such as Grignard reagents or alkylaluminums, are powerful alkylation agents, exchanging an alkyl group for an anionic ligand of the transition metal. Alkylaluminums may have a second action on the metal compound in that they are effective complexing agents. They are themselves dimeric, bridged complexes in solution, in equilibrium with the monomer, *e.g.*:

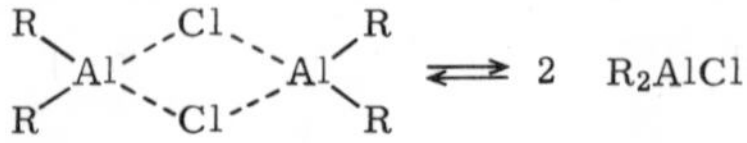

$$
\begin{array}{c}
R \diagdown \quad Cl \diagdown \quad R \\
Al \qquad Al \\
R \diagup \quad Cl \diagup \quad R
\end{array}
\rightleftharpoons 2 \quad R_2AlCl \tag{7.21}
$$

The two aluminum atoms are linked by two-electron three-center bonds. In the presence of a suitable transition metal compound, the monomer forms similar bridging complexes with the former.

As an example, $(\eta^5\text{-}C_5H_5)_2TiCl_2$ reacts with ethylmagnesium halide to give $(\eta^5\text{-}C_5H_5)_2Ti\text{-}(C_2H_5)Cl$, which may be isolated as a stable, pure, crystalline substance. Although possessing a metal-carbon σ bond, this compound is not a catalyst. As in the starting dichloride, the titanium is surrounded approximately tetrahedrally by the four ligands (see Fig. 7.13). In this symmetry all metal orbitals are involved in σ and π bonding to the ligands (*cf.* Section 6.3), and no site is available for coordination of an olefin molecule.

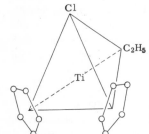

Fig. 7.13. Approximate structure of bis(cyclopentadienyl)ethyltitanium chloride.

In the presence of an excess of alkylaluminum, however, the complexing ability of the latter comes into action. Evidently, the tendency to form a complex with the transition metal compound is strong enough to force the titanium center into a different symmetry, see Fig. 7.14. (A similar change of symmetry, namely from square-planar to octahedral, occurs during oxidative addition, *cf.* Section 7.4.3.) Complex formation is essentially instantaneous.

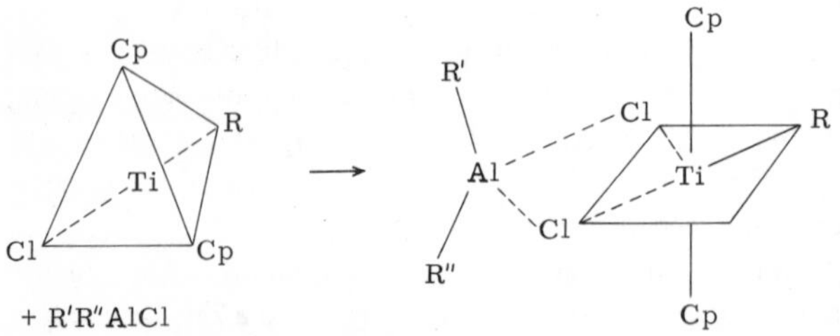

Fig. 7.14. Formation of an active catalyst species from Cp_2TiRCl and $R'R''AlCl$ [83]. ($Cp = \eta^5\text{-}C_5H_5$; R, R' = alkyl; R" = alkyl or chlorine.)

This complex is now an active catalyst for ethylene polymerization [83, 84]. The change of symmetry has made available an additional coordination site. The concept of an "octahedral structure" with one vacancy as drawn in Fig. 7.14 is, to a certain extent, a fiction (we return to this point in Section 7.5.3), but it simplifies the graphical representation and the discussion of the ligand environment, and often therefore it is used in the description of active sites.

The active catalyst represented in Fig. 7.14 could, of course, have been prepared *in situ*, from $(\eta^5\text{-}C_5H_5)_2TiCl_2$ and the alkylaluminum, without the isolation of the intermediate alkyltitanium [83]. Actually this direct procedure is frequently adopted if Ziegler catalysts are employed. In general, however, alkylation is a relatively slow equilibrium process. Induction periods in the catalytic reaction may result. Therefore, and particularly for kinetic work, it is often preferable to start from an alkylated transition metal species.

Frequently the alkylmetals obtained in the presence of an alkylaluminum (or other alkylating agent) are unstable and decompose by one of the pathways discussed in Section 7.3. If they do so by intramolecular β hydrogen transfer, an active metal hydride may result. However, if decomposition occurs by a bimolecular process, an inactive reduced metal species is often the consequence. This may be reactivated by further alkylation and, frequently, the alkylmetal containing the metal in a lower valency state is more stable. A typical example is the Ziegler system $Cr(acac)_3/C_2H_5AlCl_2$ where the active species formed *in situ* is an alkylchromium(II) or a hydridochromium(II) complex [85].

Catalytic reactions involving the oxidative addition of a small XY molecule, on the other hand, are found most frequently with square-planar group VIII complexes [122]. Here again, the concept of a "square-planar structure" is often fictitious because one of the ligands has to be expelled at some stage of the catalytic cycle, to make available a total of three coordination sites, two for the oxidatively added XY molecule (*e.g.* H_2), and one for the substrate molecule (*e.g.* ethylene). We mentioned already Wilkinson's catalyst which in a first (equilibrium) step loses a phosphine ligand L, according to Eq. (7.22), before the catalytic cycle [Eq. (7.9)] can start.

$$L_3RhCl \quad \rightleftarrows \quad L_2RhCl + L \tag{7.22}$$

Inhibition by an excess of free ligand L, as well as the occurrence of induction periods, are generally indications that the active species is formed by ligand expulsion.

7.5.3. The "Vacant Site" and the "Template Effect"

In the course of this chapter we have stressed repeatedly the necessity of a vacancy in the coordination sphere of the transition metal, where the substrate molecule may be coordinatively bonded prior to reaction. Do we really have to imagine, in solution, structures with a vacancy? From the example given in Fig. 7.14 it would appear barely plausible that a well-defined symmetry enlargement from tetrahedral to octahedral has taken place, and then one coordination site remains open. It has been suggested [84] that the pentacoordinate complex shown on the right in Fig. 7.14 might be represented better with the metal having a trigonal-bipyramidal environment, the cyclopentadienyl groups at the apices, the two chlorines together with the alkyl group in the equatorial plane. This would mean that the octahedral structure is adopted only if a sixth ligand (ethylene) approaches the catalyst center. In fact it has been calculated recently [48] that such a trigonal bipyramidal complex would be energetically more favorable than an "octahedral" one with a vacancy. Similar small changes in symmetry are, in principle, conceivable in most cases where a coordination site must become available. On the other hand, a higher symmetry (octahedral in Fig. 7.14) may be adopted from the very beginning with the aid of solvent molecules which are easily replaced by substrate molecules in the course of a catalytic cycle.

In the context of the problem of "free" coordination sites, the so-called "template effect" should be mentioned. This implies the bringing together of several substrate molecules at the same catalyst center, and hence requires even more than one vacancy. A well known example is the formation of cyclooctatetraene from acetylene reported by Reppe in 1948. This reaction proceeds homogeneously, at 80–95 °C and 20–30 atm, on a Ni(II) catalyst. Schrauzer investigated the mechanism of this reaction [86a], and concluded that the four acetylene molecules required are coordinated simultaneously to the Ni center (Fig. 7.15a). Blocking one coordination site by a phosphine ligand (P:Ni = 1; Fig. 7.15b) leads to the formation of benzene, whereas addition of the bidentate ligand *o*-phenanthroline inhibits reaction (Fig. 15c).

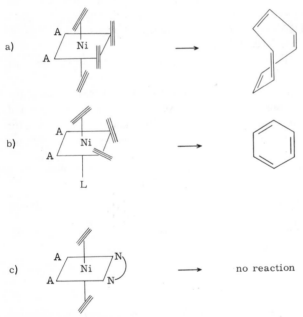

Fig. 7.15. Cyclooligomerization of acetylene, after Schrauzer [86a]. A: CN^- or acetylacetonate anion. a) Formation of cyclooctatetraene; b) L:P(C$_6$H$_5$)$_3$; formation of benzene; c) N−N:o-phenanthroline.

Again one is reluctant to believe that, in the absence of the substrate, the four coordination sites are essentially "vacant". Either there are solvent molecules within the complex, or several metal centers are associated in clusters, or both.

(For the related palladium(II) induced acetylene trimerization, on the other hand, a stepwise insertion mechanism has been suggested by Maitlis [86b], and an intermediate dienylpalladium complex has been isolated in the presence of a chelating ligand.)

7.6. Ligand Influences

7.6.1. General Aspects

In the previous sections of this chapter we have represented ligands which are not directly involved in the catalytic process, summarily as (L_n). We shall now discuss the influence of these ligands on the metal-carbon and metal-olefin (or CO) bond, and hence on catalytic activity. Depending upon their electron donating or accepting capacities, the ligands may change the electronic structure of the whole complex, i.e. they may provoke a decrease of the electron density in certain regions, an increase in others, and a consequent change in the bonding.

Let us first consider an undoubtedly greatly simplified electrostatic picture in which the central transition metal ion is positively charged[*], and all the bonds between the metal and the ligands, including R and olefin, must be regarded as more or less polar:

$$\overset{\delta+}{M}\text{------}\overset{\delta-}{L}$$

If a ligand L(1) is now replaced by another ligand L(2) which is a better electron donor (*i.e.* one which is more basic, or has a lower electron affinity, or however one may express it) than L(1), the positive charge on the metal is reduced, with the result that the bonding between the metal and all other ligands is weakened.

Of particular interest is the change of bond strength within a complex, if one ligand L is varied, thus changing systematically the electron donor properties of this ligand. A theoretical treatment of such a problem has been made by Zumdahl and Drago [87]. The authors used extended Hückel MO theory to calculate the overlap populations in the various bonds of the square-planar Pt(II) complex *trans*-$PtL(NH_3)Cl_2$, with varying L:

The overlap population is a measure of the electron density in the bonding region of two atomic orbitals; a summation over all involved orbitals can be used as an indication of relative bond strength. It turns out (*cf.* Table 7.3) that the more strongly bonding the ligand L (*i.e.* the higher the overlap population in the Pt-L bond), the lower the electron density in all other bonds, *i.e.* the weaker are these bonds. The effect is somewhat more pronounced in the *trans* position, but clearly noticeable also in the *cis* positions.

The ligands NH_3 and Cl do not have empty π orbitals available for electron back-donation, hence the electronic influence of L may be assumed to operate *via* the σ orbital system. Qualitatively one may understand the difference of the effect on the *trans* and *cis* positions (which in practical experiments is generally found to be much more pronounced; see Section 7.6.2) considering that all σ bonding metal orbitals ($d_{x^2-y^2}$, s, p_x or p_y) are shared by two ligands in *trans* positions, but only two of them ($d_{x^2-y^2}$ and s) are shared by *cis* ligands.

---

[*] This is valid even for complexes that have a formally zerovalent or negatively charged central atom, since the extra electrons are largely delocalized over the ligands.

Table 7.3. Overlap populations for *trans*-PtL(NH$_3$)Cl$_2$ complexes [87].

L	Pt—L	Pt—N	Pt—Cl
H$_2$O	0.241	0.377	0.391
NH$_3$	0.322	0.322	0.383
H$_2$S	0.429	0.324	0.374
PH$_3$	0.568	0.309	0.357
H	0.607	0.307	0.364

For ligands having orbitals of the appropriate symmetry available (antibonding π^* orbitals in olefins and CO, d orbitals of phosphorus in phosphines, *etc.*), the influence of electrons in π orbitals is superimposed on the action *via* the σ orbital system. A ligand that can act as a π acceptor withdraws electrons from the π bonds with other ligands, and the dπ orbitals of the metal act as a "conductor" for the electrons. This is illustrated schematically in Figure 7.16 for a metal complex with a CO and a phosphorus ligand. The possible effect on the metal-CO or metal-olefin bond in catalytic processes is obvious.

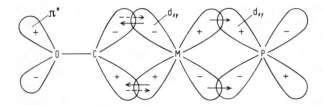

Fig. 7.16. Retrodative π bond between a metal and CO on the one hand and between a metal and a phosphine on the other. The π bond to CO is weakened by the presence of the phosphine (σ orbitals nor shown).

The ligand influences referred to are, in principle, experimentally accessible by vibrational spectroscopy (infrared), crystallographic bond length determination and other measurements on stable complexes (Section 7.6.2), although it is not always easy to separate σ and π effects. Concerning ligand influences in catalytic processes, on the other hand, it should again be born in mind that active species are very often too unstable to permit direct investigation. Furthermore, the catalytic processes are mostly complicated reaction sequences, in which the activation of different steps may place entirely different requirements on the ligands. To destabilize the M—R bond, the σ electron system must be influenced, while activation of the coordinated substrate (retrodative π bonding) may occur preferentially *via* the π electron system; a ligand L frequently must be dissociated in order to make a coordination site available, and the important factor then is strength of the M—L bond itself. There is, therefore, no reason to expect a simple relationship between electronic structure and activity. An attempt will be made here nevertheless to discuss the various factors separately. For an effective catalyst design it will be essential evidently to plan a positive ligand influence on the rate determining step(s) of a catalytic cycle.

7.6.2. "Trans Influence" and "Trans Effect"

These two terms have been introduced to designate different manifestations of the actions of a ligand on others, particularly on ligands in the *trans* position.

The *trans* influence as defined by Venanzi and coworkers [88] is a thermodynamic concept, describing the extent to which a ligand in a complex weakens the bond *trans* to itself, in the ground state of the complex. The *trans* effect is a kinetic phenomenon, namely the effect of a ligand upon the rate of exchange of the group *trans* to it [89]. Evidently bond weakening plays a role in the *trans* effect, but additional actions on the transition state in an exchange are the reason why *trans* influence and *trans* effects are not always parallel.

The experimental observation of the *trans* influence is made possible by vibrational spectroscopy, by X-ray crystallography, and by several other methods [90]. The bulk of available experimental evidence indicates that the *trans* influence of a ligand is, in fact, generally much greater than its *cis* influence. We consider here therefore only the former.

A group of widely investigated compounds is *trans* $Pt(II)XClL_2$, with L a tertiary phosphine and various X. The $Pt-Cl$ stretching vibration has been used as a measure of the $Pt-Cl$ bond strength. Some representative data are shown in Table 7.4. Where available, the $Pt-Cl$ bond distances for related complexes, as obtained from X-ray structural analyses, are also given in the Table. Both sets of data are perfectly consistent, showing increasing $Pt-Cl$ bond length, where a strong *trans* influence has lowered the bond strength and hence decreased the energy required to induce the $Pt-Cl$ stretching vibration.

Table 7.4. *Trans* influence of ligands X in *trans*-$PtXClL_2$, manifested as a decrease in the $Pt-Cl$ stretching frequency, and as an elongation of the $Pt-Cl$ bond [90] (L = phosphine).

X	ν_{Pt-Cl} (cm^{-1})	$Pt-Cl$ (Å)
Me_3Si	238	–
H	269	2.42
CH_3	274	–
$P(C_2H_5)_3$	295	2.37
$P(C_6H_5)_3$	298	–
$P(OC_6H_5)_3$	316	–
NH_3 [a]	321	2.33
pyridine	336	–
Cl	340	2.30
CO	344	2.28

[a] from *cis*-$PtCl_2(NH_3)_2$.

From similar experiments the following series of ligands concerning decreasing σ donor strength (decreasing *trans* influence) have been set up [91]:

σ *Donor Strength of Ligand L*

$C_2H_5O^- > CH_3O^- > CH_3CH_2^- > CH_3^- > CH_2 = CH^- > CH \equiv C^- > CN^- > OH^- > Cl^-$;
$(C_2H_5)_3P > (CH_3)_3P > (C_2H_5)_2P(C_6H_5) > (C_2H_5)P(C_6H_5)_2 > (C_6H_5)_3P$;
$H^- \approx CH_3^- > PR_3 > Cl^- > CO$;
$(CH_3OC_6H_4O)_3P > (C_6H_5O)_3P > (ClC_6H_4O)_3P$;
$R_3P > (RO)_3P$;
$R_3P > R_3As > R_3Sb$.

Some authors term good electron donor ligands "soft bases" and the less good donors "hard bases", with the following definitions [92]: soft bases (with donor atoms of low electronegativity) partially donate their valence electrons to the metal, reducing its positive charge and making it a softer Lewis acid; hard bases (electronegative donor atoms) retain their valence electrons, hence keeping the positive charge at the metal high and making it a hard Lewis acid. We may conclude that the *trans* influence is primarily a result of strong covalent bonding.

So far we have considered only experimental evidence for *trans* influences *via* the σ electron system. But the effects of π bonding ligands on the π bonds between the central metal and other ligands can also be determined experimentally (*cf.* Fig. 7.16); again this is done by infrared spectroscopy. In carbonyl complexes, for example, the CO stretching frequency ν_{CO} depends on how much electron density passes into the antibonding π* orbital of CO by retrodative π bonding from the metal to CO; the greater this quantity, the weaker is the bond between carbon and oxygen, and the lower is ν_{CO}. If electron density is withdrawn from π*(CO) by a π acceptor at another point in the complex, ν_{CO} is increased. Nitric oxide can be used in a similar manner as a "diagnostic ligand". By investigating complexes of the type $ML(CO)X_n$, or $ML(NO)X_n$ in which the metal M and the other ligands X_n are kept constant while L is varied, it is possible to arrange the ligands L in order of their π acceptor strength. The following series are obtained [91]:

π Acceptor Strength of Ligands L

$NO \approx CO > PF_3 > PCl_3 > PCl_2C_6H_5 > PCl(C_6H_5)_2 > P(C_6H_5)_3 > P(C_2H_5)_3;$
$PCl_2(OC_2H_5) > P(OC_6H_5)_3 > P(OC_2H_5)_3 \approx P(OCH_3)_3 > P(CH_3)_3 > P(C_2H_5)_3;$
$PR_3 \approx AsR_3 \approx SbR_3.$

It has been pointed out that there might be some ambiguity in these series, because the stretching frequency ν_{CO} is also influenced by the σ bonding and the contributions of the two effects to the measured result is not quite certain [93]. Nevertheless, the series may serve as a rough guide line for practical use.

Halide ions have no appreciable influence on ν_{CO}. They appear to act neither as π acceptors nor as π donors [93]. However, a change in a halide ligand (e.g. from Cl to Br) may have a considerable influence on a catalytic process; presumably the halide ligands act *via* the σ electron system.

The kinetic *trans* effect has also been investigated widely. The rate of exchange of ligand A has been measured as a function of ligand X in a great number of complexes of the general type *trans*-$MAX(L_n)$ [89]. Most work has been done with square planar Pt(II) complexes:

$$trans\text{-}PtAXLL' + B \quad \rightarrow \quad trans\text{-}PtBXLL' + A \qquad (7.23)$$

A characteristic of such exchange reactions is that substitution occurs without stereochemical change of the complex. The following X series lists several ligands in the order of decreasing *trans* effect [94]:

Trans Effect of Ligands X

$$C_2H_4 \approx CN \approx CO \approx NO > H^- \approx PR_3 > H_3C^- > I^- > Cl^- > Br^- > \text{pyridine} > NH_3.$$

The appearance of strong π acceptors at the top of this list is most remarkable. Chatt *et al.* [95] and Orgel [96] recognized that the *trans* effect operates not only by weakening the *trans* bond, but also by stabilizing the transition state. Their qualitative theory, recently confirmed by more elaborate quantitative MO calculations (Armstrong et al. [97]), may be summarized as follows: The activation energy of an exchange process may be lowered in one of two ways (see Fig. 7.17): a) by destabilization of the ground state of the initial complex (which is what a ligand with a strong *trans* influence such as H or CH_3 will do), or b) by stabilization of the transition state (which is what the strong π acceptors will do).

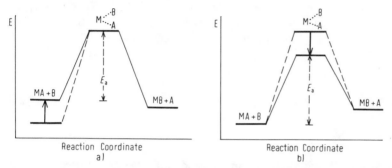

Fig. 7.17. The two ways of lowering the activation energy of a ligand exchange: a) increase in the ground state energy of the starting complex by *trans* influence; b) stabilization of the transition state.

The theory assumes that the ligand B approaches the complex as indicated in Fig. 7.18, for the particular case, where X [*cf.* Eq. (7.23)], is a strong π acceptor, *i.e.* an olefin. Because of the π Pt-olefin bond, the dπ orbital d_{xz} loses electron density, thus facilitating interaction of the electron(s) of the incoming ligand B with this orbital. (Note that such influence is only possible *trans* to the olefin ligand). As soon as A has left the complex, B will occupy its site forming then a σ bond with $d_{x^2-y^2}$. The stereochemistry of the complex is maintained.

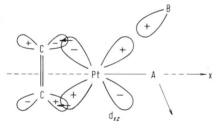

Fig. 7.18. Substitution of A by B in *trans*-Pt(olefin) ALL′; schematic representation of the *xz* plane. (Note that the *xy* plane is the molecular plane of the square-planar complex.)

Many otherwise puzzling phenomena of stable transition metal complex stereochemistry may be explained on the basis of *trans* influence and/or *trans* effect [92]. For example, dialkyl complexes tend to have *cis* structures; the high *trans* influence makes the *trans* position unfavorable for another strong donor. Evidently, it is more advantageous to have an ionic ligand (*e.g.* Cl⁻) *trans* to a strongly covalent ligand. Also, two hydride ligands, or H⁻ and PR_3, usually avoid being *trans* to each other, at least in square planar complexes. In octahedral complexes, the *trans* effect appears to parallel the *trans* influence, and strong π acceptors (*e.g.* CO) generally do not have a high *trans* effect.

7.6.3. Ligand Influences on the Stability of the Metal-Carbon Bond, and on Insertions

It is obvious that excessive stability of a metal-carbon bond is unfavorable for a catalytic process including, for instance, an insertion step [cf. Eq. (7.1)]. The bond must be stable enough to exist, but it must also be susceptible to the relevant concerted reaction, e.g. Eq. (7.4). Very fine adjustment of the reactivity of the $R-M$ bond (R = alkyl) is possible by using ligands with relatively small differences in electron donor properties. This is exemplified with the catalyst system represented in Fig. 7.14. This and similar systems have been widely used for mechanistic investigations [83, 84], although they are, from a technical point of view, relatively poor catalysts (for the polymerization of ethylene). They decompose thermally, in a bimolecular reaction [83], giving a Ti(III) complex, alkane and alkene. For the particular case where R is ethyl, the decomposition is described by Eq. (7.24):

$$2 \quad \begin{array}{c} R' \\ \diagdown \\ Al \\ \diagup \\ R'' \end{array} \begin{array}{c} Cl \\ \diagup \quad Cp \\ \diagup Ti(IV) \diagdown C_2H_5 \\ Cl \diagdown \\ Cp \end{array} \longrightarrow 2 \quad \begin{array}{c} R' \\ \diagdown \\ Al \\ \diagup \\ R'' \end{array} \begin{array}{c} Cl \diagdown \quad Cp \\ Ti(III) \\ Cl \diagup \quad Cp \end{array} + C_2H_4 + C_2H_6 \qquad (7.24)$$

Since Ti(IV) is diamagnetic, and Ti(III) is paramagnetic (3 d^1), the rate of decomposition can be determined by measuring the magnetic susceptibility of the reaction solution as a function of time. The magnetic moment of the reduced complex is $\mu_{eff} = 1.73$ B.M. [123] (cf. also Table 5.7); hence its concentration is easily calculated from the volume susceptibility of the reaction solution, using Eqs. (5.4) and (5.7).

Although the breaking of the $Ti-C$ bond [Eq. (7.24)] is almost certainly a concerted process (cf. Section 7.4), its rate depends – inter alia – upon the metal-carbon bond strength. Thus

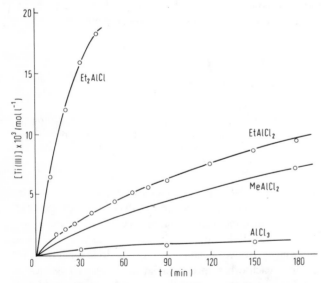

Fig. 7.19. Reduction Ti(IV)→Ti(III) in the catalyst system Cp$_2$TiEtCl/R'R"AlCl, with different aluminum compounds. $T = 15\,°C$, $[Ti(IV)]_0 = 20 \times 10^{-3}$ mol/l; Al/Ti = 2; solvent: toluene [84].

ligand influences on the strength of this bond may be monitored conveniently by measuring the rate of appearance of the paramagnetic Ti(III) complex. Varying the ligands R' and R" in the catalyst system (use of the appropriate alkylaluminums or AlCl₃), revealed that the rate of reduction depends on these ligands in the following way [84]:

$$R'=R''=Cl \ < \ R'=CH_3, R''=Cl \ < \ R'=C_2H_5, R''=Cl \ < \ R'=R''=C_2H_5.$$

This result is shown in Fig. 7.19. As expected from theoretical considerations (Section 7.6.1), the stronger the donor properties of R' and R", the stronger is the destabilization of the Ti−C bond. Amazingly, the "fine tuning" is effective even across the intermediate chlorine bridges.

Though direct measurement is not always quite so simple, the principle of adjusting the stability of the M−R bond by varying the ligands probably has general application.

So far we have studied the catalyst system in the absence of a substrate. Additional destabilization of the metal-carbon bond is brought about by the olefin itself. This is shown in Fig. 7.20, where the reduction Ti(IV) → Ti(III) was measured in the presence of various olefins. In order not to disturb the determination of the magnetic susceptibility by precipitating polymer, the work has been carried out with nonpolymerizable olefins. It turns out that the rate of reduction increases considerably in the series of additives:

$$trans\text{-octene-}2 \ < \ cis\text{-octene-}2 \ < \ \text{octene-}1.$$

This series also reflects the order of increasing coordination ability (cf. Section 7.2.1).

We conclude that the olefin acts predominantly as a donor towards the Ti(IV) compound, thus destabilizing the Ti−C bond. In this particular case the conclusion is immediately

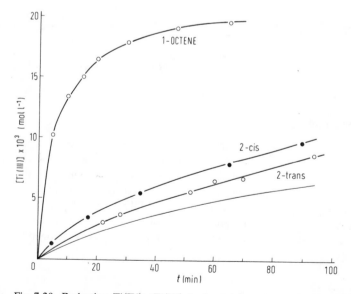

Fig. 7.20. Reduction Ti(IV)→Ti(III) in the catalyst system Cp₂TiEtCl/EtAlCl₂, in the presence of octenes; [Ti] = [Octene] = 20×10^{-3} mol/l; lowest curve: without octene [84].

plausible, since the metal has no electrons in $d\pi$ orbitals available for back-donation in a π bond. Also with more electron rich metal centers, where back-donation certainly plays a role, the donation of electrons from the olefin to the metal in the σ bond appears to predominate. Thus spectroscopic studies with iridium complexes [121] and dipole measurements on platinum complexes [98] have indicated that ethylene and other simple olefins exhibit basic character towards these metals*). To sum up, there is considerable evidence that coordinated olefinic substrate molecules take part in the destabilization of metal-carbon bonds.

It remains to be proved experimentally that destabilization of the $M-C$ bond actually favors a catalytic process. Fig. 7.21 shows initial rates of polymerization (ethylene uptake as a function of time) obtained with catalyst systems $Cp_2TiRCl/R'R''AlCl$, for three of the four aluminium compounds used in Fig. 7.19. The two sets of data (reduction and polymerization) are in fact parallel, indicating the importance of the activation of the $M-C$ bond by donor ligands. [With $(C_2H_5)_3Al$ the deactivation of the catalyst by reduction, Eq. (7.24), is too rapid; no measurable polymerization is observed.]

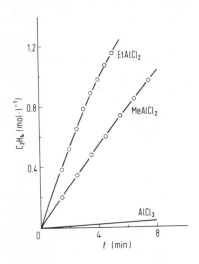

Fig. 7.21. Initial rate of polymerization of ethylene with catalyst systems, $Cp_2TiRCl/R'R''AlCl$. [Ti] $= 20 \times 10^{-3}$ mol/l; Al/Ti $= 2$; $T = 0\,°C$; ethylene pressure 9.46×10^4 Nm^{-2} ($= 710$ torr) [84].

7.6.4. Ligand Influences on β Hydrogen Transfer

In polymerizations and oligomerizations, the molecular weight of the product depends on the relative rates of chain propagation (olefin insertion) and β hydrogen transfer. The experimental evidence [99] supports the view that electron withdrawing ligands increase the relative frequency of the latter reaction, thus lowering the molecular weight. In polymerization studies with titanium based Ziegler catalysts it was found that, all other conditions being constant, the average molecular weight of the polyethylene product decreases, if strong donor ligands (C_2H_5O groups) at the titanium component are replaced be chlorine ligands. This tendency is shown in Table 7.5.

---

* This statement should not be generalized for all metal centers. In Ni(0) complexes, for example, electron back-donation plays the more important role (see Section 7.2.1).

Table 7.5. Molecular weight of polyethylene obtained with Ti/Al Ziegler systems [99]. Al component: $C_2H_5AlCl_2$; Al/Ti = 5; solvent: benzene; $T = 5\,°C$; $p_{C_2H_4} = 1$ atm.

Ti component	Oligomers[a] (%)	Polymers (%)
$(C_2H_5O)_4Ti$	31	69
$(C_2H_5O)_3TiCl$	56	44
$(C_2H_5O)TiCl_3$	77	23
$TiCl_4$	92	8

[a] The oligomer fraction is taken to be that part of product which is soluble in benzene at room temperature.

In order to rationalize this observation one may argue that an increase in positive charge at titanium (with chlorine ligands) facilitates the six-center transition state for β hydrogen transfer [cf. Eq. (7.7)], by stronger polarization of the adjacent bonds, including polarization of the C_β-H bond. Donor ligands, on the other hand, lower the positive charge at the metal, so that the four-center transition state, Eq. (7.4), is favored leading to insertion of the monomer*). In agreement with this hypothesis, it is observed that higher oxidation states favor β hydrogen transfer: Ti(IV) catalysts give, under comparable conditions, lower molecular weight polymers than do Ti(III) compounds [99, 100]; a diethyl platinum(IV) complex decomposes at 30 °C, whereas a comparable diethyl platinum(II) complex is stable at 100 °C [101].

7.6.5. Ligand Influences on Oxidative Addition

In Section 7.4.3 it was mentioned that oxidative addition is, in particular, a domain of square-planar d^8 complexes, and is favored by donor ligands. Table 7.6 shows some further examples of this behavior. The rate of oxidative addition of methyl iodide as well as of hydrogen to Vaska's complex, and to its rhodium analogue, clearly increases as the donor properties of the phosphine ligands increase. Variation of the halide ligand has the same tendency: the anion which reduces the electron density at the metal center less effectively (I^-), leads by far to the highest rate of oxidative addition.

With $RhCl[P(C_6H_5)_3]_3$ one phosphine ligand dissociates in solution. The complex $RhCl[P(C_6H_5)_3]_2$, undergoes oxidative addition of H_2 much more rapidly than the latter [cf. Eq. (7.9) and Section 8.2.2]. If such a dissociation equilibrium were to play a role also in the reactions mentioned in Table 7.6, the overall rates would depend not only on the oxidative addition process itself, but also on the ease of dissociation of the ligand PR_3. Since,

---

* The assumption of different transition states for the two reactions:

$$MCH_2CH_2R + CH_2 = CH_2 \xrightarrow{\text{insertion}} MCH_2CH_2CH_2CH_2R$$

$$MCH_2CH_2R + CH_2 = CH_2 \xrightarrow{\text{β hydrogen transfer}} MCH_2CH_3 + CH_2 = CHR$$

is implied in the experimental finding of different activation energies [E_A (insertion) $> E_A$ (β hydrogen transfer)]; cf. Section 8.4.3.

Table 7.6. Ligand influence in oxidative additions:
$trans$-$M(X)(CO)(PR_3)_2 + YZ \rightarrow M(X)(Y)(Z)(CO)(PR_3)_2$.

M	X	R	YZ	Rel. Rate	Ref.
Rh	Cl	$p-FC_6H_4$ C_6H_5 $p-CH_3OC_6H_4$	CH_3I	1.0 4.6 37.0	[102a]
Ir	Cl	$p-ClC_6H_4$ $p-CH_3C_6H_4$ $p-CH_3OC_6H_4$	CH_3I	1.0 40 95	[102b]
Ir	Cl	$p-ClC_6H_4$ $p-FC_6H_4$ $p-CH_3C_6H_4$ $p-CH_3OC_6H_4$	H_2	1.0 1.04 2.3 2.7	[102b]
Ir	Cl Br I	C_6H_5	H_2	1.0 15 $> 10^2$	[102c]

however, the metal-phosphine bond strength increases with the basicity of the phosphine [99], the positive influence of donor ligands on the rate of oxidative addition itself remains undisputed.

This experimental observation is readily understood if one takes into account that two electrons, originally localized at the metal, have to be brought into molecular orbitals representing more or less polarized covalent bonds (e.g. $M^{\delta+} - H^{\delta-}$). Hence, an electron flow from the metal to the neutral YZ molecule is required, and that is evidently favored by high electron density at the metal center.

7.6.6. "σ-π Equilibrium" of Allylic Ligands

The 1,4-insertion of a conjugated diolefin into a $M-R$ bond leads to an allylic ligand which tends to stabilize in the η^3-form [Eq. (7.3)]. Under the influence of basic ligands (*e.g.* phosphines) this configuration can change with formation of the localized σ bond between the metal and the carbon, whereas the "π complex" is favored in the presence of poor donors or of acceptor ligands.

$$
\begin{array}{c}
R \\
| \\
CH_2 \\
| \\
CH \\
M - \overset{\diagdown}{\underset{\diagup}{\text{CH}}} \\
CH_2
\end{array}
\underset{\text{Acceptor}}{\overset{\text{Donor}}{\rightleftarrows}}
MCH_2CH{=}CHCH_2R
\qquad (7.25)
$$

During the oligomerization and polymerization of butadiene (Chapter 9) the monomer itself, when coordinated to the metal center, exerts a donor function thus activating the growing chain and changing its configuration from π to σ.

A relevant example of ligand influence on the configuration of an allyl group was discussed in the context of Fig. 7.8, where a π-σ-π rearrangement is observed as a consequence of the *trans* influence of a donor ligand (phosphine).

7.6.7. Steric Effects

We have not considered until now the role of steric effects of the ligands. This facet is particularly important with tertiary phosphine ligands where considerable differences in bulkiness can easily be realized [compare, *e.g.* $P(CH_3)_3$ and $P(t-C_4H_9)_3$]. Most relevant properties of a transition metal complex containing one or more phosphine ligands parallel the electronic behavior of the phosphine only as long as the ligands have comparable steric hindrance. For instance, the rate of oxidative addition of small molecules to Rh(I) and Ir(I) complexes containing *para* substituted tertiary arylphosphines generally increases as the basicity of the phosphine increases (*cf.* Table 7.6). More bulky ligands can completely upset this ordering, partly because the active species may become too unstable and partly because the coordination of the substrate may become hindered. Tolman [103] has constructed cone-models of the MPR_3 arrangement in order to have some measure of the steric hindrance of a phosphine ligand. Consider a cone with its apex at the metal center, and enclosing the PR_3 ligand freely rotating around the $M-P$ axis. The cone angle, then, is a measure of the steric requirement of the ligand PR_3. The absolute values of the cone angles should not be taken too seriously (especially because a fixed $M-P$ distance was used for all phosphines), but they are still meaningful for a comparison of various ligands. Some relevant cone angles are: $P(CH_3)_3$, 118°; $P(C_2H_5)_3$, 132°; $P(C_6H_5)_3$, 145°; $P(i-C_3H_7)_3$, 160°; $P(t-C_4H_9)_3$, 182°; $P(o-CH_3C_6H_4)_3$, 194°.

Bulky ligands can, however, have an interesting influence on a catalytic process by forming a rigid frame (matrix) which predetermines the direction of coordination, and hence of incorporation, of the substrate molecule. A remarkable selectivity towards one of several possible reaction products may be achieved sometimes in this way. The dimerization of propylene with nickel or palladium based Ziegler type catalysts represents an interesting example. Whereas in most cases variable mixtures of the three isomeric structures for the dimer (linear hexenes, methylpentenes and dimethylbutenes) are found, a high selectivity yielding dimethylbutenes, or linear hexenes, can be obtained with bulky phosphines and otherwise sterically demanding ligands (*cf.* Section 8.4.2).

Matrix formation by ligands even can be used to synthesize optically active substances. One of the most remarkable recent examples is the synthesis of the important drug L-Dopa by Knowles *et al.* [104]. The authors succeeded in hydrogenating an α, β-unsaturated α-acyl-aminoacid stereospecifically, to yield the L form (80–90% optical purity) of the hydrogenation product which is an immediate precursor of L-Dopa. The catalyst is a Rh(I) complex, with a diolefinic ligand (*e.g.* 1,5-hexadiene), and with two chiral phosphine ligands. The authors were able to show that, apart from the steric matrix of the chiral phosphine ligands, a hydrogen bond between the hydrogen of the amide and the oxygen of the methoxy group was responsible for the stereoselective coordination and hydrogenation of the substrate.

(80 - 90% optical purity)

Catalyst: $\left[L_2Rh \overset{\| }{\underset{\| }{\bigcirc}} \right]^{+} BF_4^{-}$ L: $P-CH_3$ with cyclohexyl and OCH_3 substituents on benzene ring

Finally, the direction of a catalytic reaction can sometimes be changed by blocking coordination sites. The cyclooligomerization of acetylene mentioned in Section 7.5.3 (Fig. 7.15) represents a typical example of this steric effect.

7.6.8. Macromolecular Ligands

Homogeneous catalysts can exhibit problems associated with product contamination and catalyst loss, where the low molecular products are not readily separated from the low molecular weight catalyst. Anchoring homogeneous catalysts to insoluble polymers or inorganic supports has been used to "heterogenize" them [105]. Fixation of the catalytic center to soluble polymers has also been reported [106]. The most frequent procedure for anchoring the metal to a polymer is by chloromethylation of, *e.g.*, polystyrene and subsequent reaction with potassiumdiphenylphosphide. The metal species is then fixed to the polymeric phosphine ligand and the separation of the soluble macromolecular catalysts is effected by membrane filtration or by precipitation.

These methods are of recent interest and potentially combine the advantages of homogeneous catalysts (higher activity, milder reaction conditions, higher selectivity) with those of heterogeneous catalysts (easy catalyst recovery). However, several drawbacks have been reported. In swollen crosslinked polymers, diffusion of the substrate to the active sites may become rate limiting. With glass or ceramic surfaces, as yet undefined interactions of the transition metal species with the surface have lead to a deterioration in activity as compared with the homogeneous counterparts. In general, the activity appears to be lower and, moreover, decreases on recycling. If these shortcomings cannot be overcome, the recovery of the metal from the support in order to restore activity may run the risk of being more expensive than the present technique of recovery from homogeneous solutions.

7.6.9. Solvent Effects

The influence of solvent on homogeneous catalysis has not been studied very extensively. In many cases the choice of solvents is limited by the solubility of the catalyst and the substrate. It has, however, been reported that reactions with a highly polar transition state may be promoted by polar solvents [107]. Certain square-planar catalysts appear to become active only when a strongly coordinating solvent molecule occupies one of the axial positions (*cf.* Sections 8.2.3 and 11.2). The formation of the active center in Ziegler type catalysts is also solvent dependent, in the sense that methylene chloride and aromatic solvents are better than acyclic hydrocarbons [108, 109]. Finally, solvent molecules may form part of apparently coordinatively unsaturated (*e.g.* 14 or 16 electrons) intermediates in catalytic cycles, such as those depicted in Eq. (7.9) (*cf.* also Section 7.5.3).

7.7. Metal Catalysis of Symmetry Forbidden Reactions

In the mid nineteen sixties several authors, nearly simultaneously, extended simple molecular orbital theory into the area of reaction chemistry [110–113]. In particular, rules have been set up for concerted reactions (simultaneous bond breaking and bond making), in order to predict whether or not a certain reaction of this type would be, either thermally or photochemically, "allowed" (*i.e.* whether or not it would occur with a reasonable activation energy), or "forbidden" (excessively high activation energy). These rules are based on symmetry arguments concerning the involved molecular orbitals of reaction partners and product. They have found broad acceptance in organic chemistry and, in their most widespread form, are known as the "Woodward-Hoffmann Rules".

Symmetry forbidden reactions are rarely observed in organic chemistry. In our context, however, it is noteworthy that transition metal complexes may catalyze effectively certain forbidden reactions. The interpretation of this phenomenon is somewhat controversial [114–117, 122], and the problem does not yet appear to be quite settled. Although a profound discussion of different views is beyond the scope of this book we shall present, so to speak as an introduction, and to stimulate further literature study, some of Pearson's ideas [116], in very abbreviated form.

As in all other treatments of symmetry rules, MO theory is the basis; a chemical reaction is considered as a perturbation on the reactant system. Using the quantum mechanical method of second order perturbation theory, as well as group theory, symmetry rules are stated in a rigorous way. Fortunately for our purpose, the conclusions can be stated in a simple, pictorial manner.

In the course of a concerted process, certain molecular orbitals must be vacated (to break bonds) and others must be filled (to create the new bonding situation). The most important of these changes involves a flow of electrons from the highest occupied molecular orbital (HOMO) of one reaction partner to the lowest unoccupied molecular orbital (LUMO) of the other. Electron movement between two orbitals cannot occur unless the orbitals meet the symmetry requirement. For a bimolecular reaction the requirement is simply that the two orbitals should have a net overlap (*cf.* Section 6.1.1). A further requirement may be added based upon chemical knowledge rather than quantum mechanical arguments: the HOMO must represent a bond that is broken, and the LUMO one that is formed during the reaction. To illustrate these principles consider the reaction of hydrogen with ethylene, in which the two molecules collide broadside, giving rise to a four-center transition state:

$$
\begin{array}{c}
\text{H} \text{------} \text{H} \\
\vdots \qquad \vdots \\
\text{H--C} \text{=====} \text{C--H} \\
\quad | \qquad\qquad | \\
\quad \text{H} \qquad\quad \text{H}
\end{array}
\quad \xrightarrow{\;\;/\!\!\!/\;\;} \quad
\begin{array}{c}
\text{H} \;\; \text{H} \\
\;\; | \;\;\; | \\
\text{H--C--C--H} \\
\;\; | \;\;\; | \\
\text{H} \;\; \text{H}
\end{array}
\qquad\qquad (7.26)
$$

The HOMO's, representing bonds to be broken, are the bonding σ of H_2 and the bonding π of ethylene; the LUMO's are σ^* and π^*. But neither σ and π^*, nor π and σ^* have any net overlap (Fig. 7.22, *cf.* Fig. 6.3). Hence the reaction is "symmetry-forbidden".

One may easily deduce that four-center concerted reactions of diatomic or pseudo-diatomic molecules (H_2, O_2, N_2, CO, C_2H_4 *etc.*) are generally forbidden. Nevertheless, many impor-

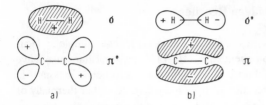

Fig. 7.22. HOMO's and LUMO's of ethylene and dihydrogen. Electron flow from H_2 to C_2H_4 (a), or from C_2H_4 to H_2 (b), is forbidden by symmetry.

tant examples in catalysis involve just such molecules. The role of the catalyst is then to circumvent the symmetry restrictions.

The key role is played probably by the coordinative metal-ligand bond. Consider the particular case of the metal-olefin bond (*cf.* Fig. 7.1). If electron back-donation transfers electron density into the π^* orbital of the olefin, this orbital becomes now (partially) a HOMO; evidently it has the right symmetry to interact with the LUMO of the H_2 molecule (see Fig. 7.23). A similar statement may be made concerning the bonding π orbital of the olefin which loses electron density in the metal-olefin σ bond and thus might become susceptible to a (partial) electron flow from σ (H_2) to π (olefin). One may express these actions of the transition metal center as exchanging electrons of the incorrect, for others of the correct, symmetry *via* its d orbital system.

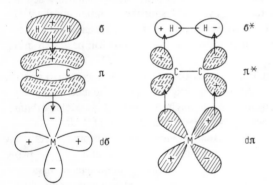

Fig. 7.23. Concerted addition of H_2 to olefin may become allowed by metal influence. ///: occupied; ///: partly occupied.

The four-center process, Eq. (7.26), could thus become symmetry-allowed under the influence of a transition metal catalyst. Nevertheless, the catalytic hydrogenation of olefins most probably is not as simple as this and appears to proceed in a series of concerted steps, where each step is, of course, subject to symmetry restrictions. In the particular case of the hydrogenation cycle shown in Eq. (7.9) the successive steps are oxidative addition of the hydrogen molecule, insertion of a coordinated olefin into one of the metal-hydrogen bonds and, finally, reductive elimination of the alkane. Similar stepwise mechanisms have been postulated for several other catalyzed symmetry-forbidden reactions [115, 122].

References

[1] J. J. Berzelius, Jahresber. Chem., *15*, 237 (1836).   [2] O. Roelen, Angew. Chem., *60*, 62 (1948).
[3] M. J. S. Dewar, Bull. Soc. Chim. Fr. *18*, C17 (1951).   [4] J. Chatt and L. A. Duncanson, J. Chem.

Soc., *1953*, 2939. [5] L. S. Bartell, E. A. Roth, C. D. Hollowell, K. Kuchitsu and J. E. Young, J. Chem. Phys., *42*, 2683 (1965). [6] K. W. Muir and J. A. Ibers, J. Organometal. Chem., *18*, 175 (1969). [7] D. A. Bekoe and K. N. Trueblood, Z. Kristallogr., *113*, 1, (1960). [8] R. A. Love, T. F. Koetzle, G. J. B. Williams, L. C. Andrews and R. Bau, Inorg. Chem., *14*, 2653 (1975). [9] P. T. Cheng, C. D. Cook, C. H. Koo, S. C. Nyburg and M. T. Shiomi, Acta Crystallogr., *B27*, 1904 (1971). [10] G. Bombieri, E. Forsellini, C. Panattoni, R. Graziani and G. Bandoli, J. Chem. Soc. (A) *1970*, 1313. [11] J. A. McGinnety and J. A. Ibers, J. Chem. Soc. Chem. Commun., *1968*, 235. [12] H. C. Allen and E. K. Plyler, J. Chem. Phys., *31*, 1062 (1959). [13] R. Cramer, J. Amer. Chem. Soc., *86*, 217 (1964). [14] B. F. G. Johnson and J. A. Segal, J. Chem. Soc. Chem. Commun., *1972*, 1312. [15] R. Cramer, J. B. Kline and J. D. Roberts, J. Amer. Chem. Soc., *91*, 2519 (1969); J. Ashley-Smith, Z. Douek, B. F. G. Johnson and J. Louis, J. Chem. Soc. Dalton Trans., *1974*, 128. [16] P. M. Henry, J. Amer. Chem. Soc., *88*, 1595 (1966). [17] C. A. Tolman, J. Amer. Chem. Soc., *96*, 2780 (1974). [18] R. Cramer, J. Amer. Chem. Soc., *89*, 4621 (1967). [19] K. Vrieze, *Fluxional Allyl Complexes*, in: L. M. Jackman and F. A. Cotton (Eds.), *Dynamic Nuclear Magnetic Resonance Spectroscopy*, Academ. Press 1975 (and references therein). [20] H. L. Clarke, J. Organometal. Chem., *80*, 155, 369 (1974) (and references therein).
[21] F. A. Cotton, J. Amer. Chem. Soc., *90*, 6230 (1968). [22] P. W. N. M. van Leeuven and A. P. Praat, J. Organometal. Chem., *21*, 501 (1970). [23] D. A. Brown and A. Owens, Inorg. Chim. Acta, *5*, 675 (1971). [24] W. R. McClellan, H. H. Hoehn, H. N. Cripps, E. L. Muetterties and B. W. Howk, J. Amer. Chem. Soc., *83*, 1601 (1961). [25] F. A. Cotton, J. W. Faller and A. Musco, Inorg. Chem., *6*, 179 (1967). [26] R. P. Hughes and J. Powell, J. Amer. Chem. Soc., *94*, 7723 (1972). [27] M. Oslinger and J. Powell, Can. J. Chem., *51*, 274 (1973), and references therein. [28] V. I. Telnoi, I. B. Rabinovich, V. D. Tikhanov, V. I. Latyaeva, L. I. Vysginskaya and G. A. Razuvaev, Dokl. Akad. Nauk SSSR, *174*, 1374 (1967). [29] D. Lalage, S. Brown, J. A. Connor and H. A. Skinner, J. Organometal. Chem., *81*, 403 (1974). [30] S. J. Ashcroft and C. T. Mortimer, J. Chem. Soc., A, *1967*, 930.
[31] K. W. Egger, J. Organometal. Chem., *24*, 501 (1970). [32] G. M. Whitesides, E. R. Stedronski, C. P. Casey and J. San Fillipo, J. Amer. Chem. Soc., *92*, 1426 (1970). [33] P. S. Braterman and R. J. Cross, Chem. Soc. Rev., *2*, 271 (1973). [34] M. C. Baird, J. Organometal. Chem., *64*, 289 (1974). [35] M. R. Collier, M. F. Lappert and M. M. Truelock, J. Organometal. Chem., *25*, C36 (1970), M. R. Collier, M. F. Lappert and R. Pearce, J. Chem. Soc. Dalton Trans., *1973*, 445. [36] G. Yagupsky, W. Mowat, A. Shortland and G. Wilkinson, J. Chem. Soc. Chem. Commun., *1970*, 1369; W. Mowat, A. Shortland, G. Yagupsky, N. J. Hill, M. Yagupsky and G. Wilkinson, J. Chem. Soc. Dalton Trans., *1972*, 533. [37] F. Calderazzo, Pure Appl. Chem., *33*, 453 (1973), and references therein. [38] K. Thomas, J. A. Osborn, A. R. Powell and G. Wilkinson, J. Chem. Soc. (A), *1968*, 1801. [39] E. O. Fischer and A. Maasböl, Angew. Chem. Internat. Edit., *3*, 580 (1964). [40] For recent reviews of transition metal-carbene complexes see: a) F. A. Cotton and C. M. Lukehart, Progr. Inorg. Chem., *16*, 487 (1972); b) D. J. Cardin, B. Cetinkaya and F. M. Lappert, Chem. Rev., *72*, 545 (1972).
[41] H. E. O'Neal and S. W. Benson, J. Phys. Chem., *71*, 2903 (1967). [42] R. Cramer, J. Amer. Chem. Soc., *87*, 4717 (1965). [43] R. Cramer, J. Amer. Chem. Soc., *94*, 5681 (1972). [44] F. Calderazzo and F. A. Cotton, Inorg. Chem., *1*, 30 (1962). [45] K. Noack and F. Calderazzo, J. Organometal. Chem., *10*, 101 (1967). [46] P. Cossee, J. Catal., *3*, 80 (1964). [47] P. Cossee, Rec. Trav. Chim. Pays-Bas, *85*, 1151 (1966). [48] D. R. Armstrong, P. G. Perkins and J. J. P. Stewart, J. Chem. Soc. Dalton Trans., *1972*, 1972. [49] V. W. Markownikoff, Compt. Rend., *81*, 668 (1875). [50] See, e. g. P. Longi, G. Mazzanti, A. Roggero and A. M. Lachi, Makromol. Chem., *61*, 63 (1963). [51] J. Evans, J. Schwartz and P. W. Urquhart, J. Organometal. Chem., *81*, C37 (1974). [52] F. H. Westheimer, Chem. Rev., *61*, 265 (1961). [53] K. W. Egger, J. Chem. Kinet., *1*, 459 (1969). [54] G. Henrici-Olivé and S. Olivé, J. Polym. Sci., Polym. Lett. Edit. *12*, 39 (1974). [55] O. Novaro, S. Chow and P. Magnouat, J. Polym. Sci., Polym. Lett. Edit., *13*, 761 (1975); J. Catal. *41*, 91 (1976). [56] G. M. Whitesides, J. F. Gaasch and E. R. Stedronsky, J. Amer. Chem. Soc., *94*, 5258 (1972). [57] J. K. Kochi, Accounts Chem. Res., *7*, 351 (1974). [58] L. Vaska and J. W. DiLuzio, J. Amer. Chem. Soc., *84*, 679 (1962). [59] M. Kubota and D. M. Blake, J. Amer. Chem. Soc., *93*, 1368 (1971). [60] D. M. Blake, S. Shields and L. Wyman, Inorg. Chem., *13*, 1595 (1974).
[61] R. G. Pearson and W. R. Muir, J. Amer. Chem. Soc., *92*, 5519 (1970). [62] E. M. Miller and B. L. Shaw, J. Chem. Soc. Dalton Trans. *1974*, 480. [63] M. P. Brown, R. J. Puddephatt and C. E. E. Upton, J. Chem. Soc. Dalton Trans., *1974*, 2457. [64] A. V. Kramer, J. A. Labinger, J. S. Bradley and J. A. Osborn, J. Amer. Chem. Soc., *96*, 7145 (1974). [65] C. A. Tolman, P. Z. Meakin, D. L. Lindner

and J. P. Jesson, J. Amer. Chem. Soc., *96*, 2762 (1974).   [66] D. Forster, J. Amer. Chem. Soc., *97*, 951 (1975).   [67] M. A. Bennett and D. L. Milner, J. Amer. Chem. Soc., *91*, 6983 (1969).   [68] J. Chatt and J. M. Davidson, J. Chem. Soc., *1965*, 843.   [69] F. A. Cotton, B. A. Frenz and D. L. Hunter, J. Chem. Soc. Chem. Commun., *1974*, 755.   [70] J. Evans, J. Schwartz and P. W. Urquhart, J. Organometal. Chem., *81*, C37 (1974).

[71] F. H. Jardine, J. A. Osborn and G. Wilkinson, J. Chem. Soc. A, *1967*, 1574.   [72] M. A. Bennett and D. L. Milner, J. Amer. Chem. Soc., *91*, 6983 (1969).   [73] M. W. Adlard and G. Socrates, J. Chem. Soc. Chem. Commun., *1972*, 17.   [74] T. W. Swaddle, Coord. Chem. Rev., *14*, 217 (1974).   [75] C. H. Langford and H. B. Gray, *Ligand Substitution Dynamics*, W. A. Benjamin, New York, 1965.   [76] J. K. Stille and D. E. James, J. Amer. Chem. Soc., *97*, 674 (1975); J. K. Stille, D. E. James and L. H. Hines, J. Amer. Chem. Soc., *95*, 5062 (1973).   [77] J. Chatt and A. G. Wedd, J. Organometal. Chem., *27*, C15 (1971).   [78] R. Mason, K. M. Thomas and G. A. Heath, J. Organometal. Chem., *90*, 195 (1975). [79] R. F. Heck, J. Amer. Chem. Soc., *85*, 657 (1963).   [80] J. Halpern and T. A. Weil, J. Chem. Soc. Chem. Commun., *1973*, 631.

[81] I. H. Elson and J. K. Kochi, J. Amer. Chem. Soc., *97*, 1264 (1975).   [82] K. Ziegler, E. Holzkamp, H. Breil and H. Martin, Angew. Chem., *67*, 541 (1955).   [83] G. Henrici-Olivé and S. Olivé, Angew. Chem. Internat. Edit., *6*, 790 (1967), and refs. therein.   [84] G. Henrici-Olivé and S. Olivé, Advan. Polym. Sci., *6*, 421 (1969), and refs. therein.   [85] G. Henrici-Olivé and S. Olivé, Angew. Chem. Internat. Edit., *10*, 776 (1971).   [86] a) G. N. Schrauzer and S. Eichler, Chem. Ber., *95*, 550 (1962); b) B. E. Mann, P. M. Bailey and P. M. Maitlis, J. Amer. Chem. Soc., *97*, 1275 (1975).   [87] S. S. Zumdahl and R. S. Drago, J. Amer. Chem. Soc., *90*, 6669 (1968).   [88] A. Pidcock, R. E. Richards and L. M. Venanzi, J. Chem. Soc. A, *1966*, 1707; L. M. Venanzi, Chem. Brit., *1968*, 162.   [89] F. Basolo and R. G. Pearson, Progr. Inorg. Chem., *4*, 381 (1962) and refs. therein.   [90] T. G. Appleton, H. C. Clark and L. E. Manzer, Coord. Chem. Rev., *10*, 335 (1973).

[91] G. Henrici-Olivé and S. Olivé, Angew. Chem. Internat. Edit., *10*, 105 (1971); and refs. therein. [92] R. G. Pearson, Inorg. Chem., *12*, 712 (1973).   [93] C. A. Tolman, J. Amer. Chem. Soc., *92*, 2953 (1970) and refs. therein.   [94] L. G. Vanquickenborne, J. Vranckx and C. Görller-Walrand, J. Amer. Chem. Soc., *96*, 4121 (1974).   [95] J. Chatt, L. A. Duncanson and L. M. Venanzi, J. Chem. Soc., *4*, 4456 (1955).   [96] L. E. Orgel, J. Inorg. Nucl. Chem., *2*, 137 (1956).   [97] D. R. Armstrong, R. Fortune and P. G. Perkins, Inorg. Chim. Acta, *9*, 9 (1974).   [98] M. A. Muhs and F. T. Weiss, J. Amer. Chem. Soc., *84*, 4697 (1962).   [99] G. Henrici-Olivé and S. Olivé, Advan. Polym. Sci., *15*, 1 (1974).   [100] A. Schindler, J. Polym. Sci., Polym. Lett. Edit., *4*, 193 (1966).

[101] M. P. Brown, R. J. Puddephatt, C. E. E. Upton and S. W. Lawington, J. Chem. Soc. Dalton Trans., *1974*, 1613.   [102] a) I. C. Douek and G. Wilkinson, J. Chem. Soc. A. *1969*, 2604; b) R. Ugo, A. Pasini, A. Fusi and S. Cenini, J. Amer. Chem. Soc., *94*, 7364 (1972); c) J. Halpern and P. B. Chock, J. Amer. Chem. Soc., *88*, 3511 (1966).   [103] C. A. Tolman, J. Amer. Chem. Soc., *92*, 2956 (1970). [104] W. S. Knowles, M. J. Sabacky and B. D. Vineyard, J. Chem. Soc. Chem. Commun., *1972*, 10; Chem. Tech., *2*, 590 (1972).   [105] C. U. Pittman, L. R. Smith and R. M. Hanes, J. Amer. Chem. Soc., *97*, 1742 (1975), and refs. therein.   [106] E. Bayer and V. Schurig, Angew. Chem., Internat. Edit., *14*, 493 (1975).   [107] E. M. Hyde and B. L. Shaw, J. Chem. Soc. Dalton Trans., *1975*, 765.   [108] H. Bestian and K. Clauss, Angew. Chem., Internat. Edit., *2*, 704 (1963).   [109] G. Henrici-Olivé and S. Olivé, J. Organometal. Chem., *35*, 381 (1972).   [110] R. Hoffmann and R. B. Woodward, J. Amer. Chem. Soc., *87*, 395 (1965), R. B. Woodward and R. Hoffmann, Angew. Chem. Internat. Edit., *8*, 781 (1969).

[111] H. C. Longuet-Higgins and E. W. Abrahamson, J. Amer. Chem. Soc., *87*, 2045 (1965).   [112] K. Fukui, Tetrahedron Lett., *1965*, 2009. K. Fukui and S. Inagaki, J. Amer. Chem. Soc., *97*, 4445 (1975). [113] M. J. S. Dewar, Tetrahedron, Suppl. *8*, Part 1, 75 (1966); The Molecular Orbital Theory of Organic Chemistry, McGraw-Hill, New York 1969.   [114] R. Pettit, H. Sugahara, J. Wristers and W. Merk, Disc. Faraday Soc., *47*, 71 (1969).   [115] T. J. Katz and S. A. Cerefice, J. Amer. Chem. Soc., *91*, 6519 (1969).   [116] R. G. Pearson, Accounts Chem. Res., *4*, 152 (1970).   [117] F. D. Mango, Topics in Current Chem., *45*, 39 (1974).

Suggested Additional Reading:

[118] R. B. King, *Transition Metal Compounds*, in: J. E. Eisch and R. B. King, Eds. *Organometallic Syntheses* Vol. 1, Academic Press, 1965.   [119] P. J. Davidson, M. F. Lappert and R. Pearce, *Metal σ-Hydrocarbyls*, Chem. Rev., *76*, 219 (1976).   [120] J. P. Colman, *Patterns of Organometallic Reac-*

tions Related to Homogeneous Catalysis, Accounts Chem. Res., *1*, 136 (1968). [121] L. Vaska, *Reversible Activation of Covalent Molecules by Transition Metal Complexes*, Accounts Chem. Res., *1*, 335 (1968). [122] J. Halpern, *Oxidative-Addition Reactions of Transition Metal Complexes*, Accounts Chem. Res., *3*, 386 (1970). [123] G. Henrici-Olivé and S. Olivé, *The Active Species in Homogeneous Ziegler-Natta Catalysts for the Polymerization of Ethylene*, Angew. Chem. Internat. Edit., *6*, 790 (1967). [124] G. Henrici-Olivé and S. Olivé, *Olefin Insertion in Transition Metal Catalysis*, Topics Curr. Chem. *67*, 1 (1976).

8. Reactions of Olefins

In this chapter we initiate discussion of particular reactions catalyzed by soluble transition metal complexes. Although, in general, we have in view primarily the metal center itself, and the mechanistic aspects of the processes occurring in its coordination sphere, we have chosen to divide the material to be treated according to substrates rather than to metal centers in order to minimize overlap.

The examples given are necessarily limited. They have been selected to illustrate special mechanistic aspects of the catalyzed reaction under consideration in each particular section. However, this is not as restrictive as it may appear, because if a reaction pattern is found suitable for a given transition metal catalyst, frequently complexes of other metals can be found (although perhaps with different ligand environment and/or valency) which fit the same pattern. For broader coverage the reader is referred to many excellent review articles and monographs which are cited at the end of each chapter, and of which we have made liberal use.

8.1. Isomerization

The isomerization of olefins (double bond migration as well as *cis-trans* isomerization) is catalyzed by nearly all relevant transition metal compounds, and hence is almost omnipresent in the processes discussed in this chapter. Very often the success of a catalytic reaction depends on the prevention of an undesirable isomerization, but the opposite may be valid also: an inert inner olefin may become reactive in a certain catalytic process as a result of migration of the double bond to the α position. Some examples illustrate the situation.

Ethylene can be dimerized with a rhodium catalyst [101]; the primary product is butene-1. Isomerization is much faster than dimerization, so butene-1 would be expected to isomerize to butene-2 as it is formed. An equilibrated system of linear butenes (at room temperature) would contain only *ca.* 4% of butene-1, nevertheless as much as 40% has been obtained when dimerization is stopped after 50% of conversion. In this case the effective principle is a thermodynamic one: the coordination of ethylene is favored by a factor of 1000 over that of butene-1, hence selective olefin coordination controls the course of the reaction. Certain nickel catalysts dimerize propylene rapidly to 4-methylpentene-2 (*cf.* Section 8.4.2). Attempts to dimerize butene-1 under the same conditions failed. Isomerization prevails, and the inner olefin does not dimerize [1].

On the other hand, the polymerization of butene-2 by a titanium based catalyst gives a product having the same structure as that obtained by polymerization of butene-1 under the same conditions [2]. It follows that the *conditio sine qua non* for the polymerization of butene-2 is its isomerization to butene-1 which is carried out by the same catalyst.

Stepwise as well as multiple isomerizations have been observed. Evidently, the relative rates of isomerization and olefin dissociation are the decisive factors. An interesting example is the isomerization of 4-methylpentene-1 with an iron catalyst which gives 2-methylpentene-1 as the major product, under conditions where the isomerization of 4-methylpentene-2 is rather slow [3]:

$$CH_2 = CHCH_2CH(CH_3)_2 \searrow$$
$$CH_3CH = CHCH(CH_3)_2 \nrightarrow \quad CH_3CH_2CH_2C(CH_3) = CH_2$$

Multiple isomerization occurs before the olefin leaves the complex, whereas the isomerization of the 4-methylpentene-2 is hampered by the poor coordinating ability of this inner olefin. Catalysts with bulky ligands, such as $RuCl_2[P(C_6H_5)_3]_3$, appear to be good candidates for selective stepwise olefin isomerization [4].

Two mechanistically different pathways to double bond migration are supported by experiment, one including alkylmetal, the other π-allylmetal intermediates.

8.1.1. Double Bond Migration with Alkylmetal Intermediates

This mechanism requires a metal hydride catalyst. The olefin inserts into the $M-H$ bond; β hydrogen transfer from a different carbon atom to the metal gives the isomerized product:

$$M-H + CH_2 = CHCH_2R \quad \rightarrow \quad MCH(CH_3)CH_2R$$
$$\rightarrow \quad M-H + CH_3CH = CHR \qquad (8.1)$$

(For the sake of clarity, the other ligands of the complex are omitted.) The metal does not change its valence state. Note that isomerization of a terminal olefin necessarily implies insertion of the olefin according to the anti-Markownikoff mode (*cf.* Section 7.4.1). The Markownikoff mode regenerates the starting material after β hydrogen transfer.

The metal hydride catalyst may be introduced as such (*e.g.* $HRh(CO)[P(C_6H_5)_3]_3$), or it may be formed *in situ*. A variety of hydride generating reactions is available, *e.g.* the oxidative addition of H_2 or HCl to a low valent metal species, and the heterolytic splitting of H_2. Eqs. (8.2–4) exemplify these reactions for rhodium centers [5]:

$$Rh(I) \quad + H_2 \quad \rightarrow \quad Rh(III)H(H) \qquad (8.2)$$

$$Rh(I) \quad + HCl \quad \rightarrow \quad Rh(III)H(Cl) \qquad (8.3)$$

$$Rh(III)Cl + H_2 \quad \rightarrow \quad Rh(III)H + H^+ + Cl^- \qquad (8.4)$$

Good evidence for the mechanism depicted in Eq. (8.1) has been obtained by Asinger *et al.* [6] using a palladium catalyst (palladium chloride + HCl). Tritium labeled octene-1, $CH_2 = CHCHT(CH_2)_4CH_3$, was isomerized in the presence of unlabeled hexene-1. After almost complete isomerization of both compounds, 49% exchange of tritium with the hexene was observed.

Although anti-Markownikoff insertion of the olefin is required for effective isomerization of α olefins [Eq. (8.1)], Markownikoff addition is expected to occur as well, to an extent depending on the metal itself and on the ligands [*cf.* Section 7.4.1]. Cramer [101] investigated this aspect during the isomerization of butene-1 with a rhodium catalyst formed *in situ* from a Rh(I) complex and DCl (actually HCl in solvent CH_3OD was used). The observation of deuterated, nonisomerized butene-1 revealed the presence of a reaction sequence including Markownikoff addition, according to:

$$\begin{array}{ccc}
\text{Cl} & & \text{Cl} \\
| \quad \text{CH}_2 & & | \quad \text{CH}_2 \\
\text{Rh}\cdots\parallel & \rightarrow \text{ClRhCH}_2\text{CHDC}_2\text{H}_5 \rightarrow & \text{Rh}\cdots\parallel \\
| \quad \text{CHC}_2\text{H}_5 & & | \quad \text{CDC}_2\text{H}_5 \\
\text{D} & & \text{H}
\end{array} \qquad (8.5)$$

The experimental results also required the assumption of a rapid HCl/DCl exchange (with H or D on the metal) indicating that Eq. (8.1) describes the total process only superficially.

8.1.2. Double Bond Migration with π Allylmetal Intermediates

Many transition metal complexes catalyze double bond migration under conditions where the only source of hydrogen is the olefin itself. It is generally assumed that in these cases the isomerization occurs *via* a 1,3-hydrogen shift, with a π allylmetal intermediate:

$$\begin{array}{ccccc}
& \text{HCR} & & \text{H}\quad\text{R} & & \text{H}_2\text{CR} \\
& \parallel & & \diagup\diagdown & & | \\
\text{M}\cdots\parallel\text{CH} & \rightleftarrows & \text{H-M}-\text{C}\diagdown\text{CH} & \rightleftarrows & \text{M}\cdots\parallel\text{CH} \\
& | & & \diagdown & & \text{HCR}' \\
& \text{H}_2\text{CR}' & & \text{H}\quad\text{R}' & &
\end{array} \qquad (8.6)$$

This mechanism includes an oxidative addition/reductive elimination cycle, experimental support for which comes from the detection, by ^{1}H NMR spectroscopy, of the related equilibrium in the case of a nickel complex (Bönnemann [7]):

$$\begin{array}{ccc}
\text{F}_3\text{P}\quad\text{H}\quad\text{H} & & \\
\diagdown\quad\text{C} & -40\,^\circ\text{C} & \text{CH}_2 \\
\text{Ni}-\text{C}\diagdown\text{CH} \rightleftarrows & & \text{F}_3\text{P-Ni}\cdots\parallel \\
\text{H}\quad\diagup\quad\text{C} & -50\,^\circ\text{C} & \text{CHCH}_3 \\
\text{H}\quad\text{H} & &
\end{array} \qquad (8.7)$$

The π allylNi(II)hydride was prepared and isolated at $-130\,^\circ$C. It is stable in solution up to $-50\,^\circ$C and shows the characteristic NMR signals of a π allyl group (*cf.* Section 7.2.2). After warming to $-40\,^\circ$C during several hours, three new signals appear which can be attributed to coordinatively bonded propylene. The process is reversible.

In transition metal hydride complexes containing a symmetric π allyl group, such as in Eq. (8.7), migration of the hydrogen from the metal to either of the two relevant carbon atoms can, of course, yield only one and the same olefin. Effective isomerization can be expected only if the hydrogen atom can move to both ends of an asymmetric π allyl group [*cf.* Eq. (8.6)]. Some illustrative experiments have been reported by Nixon and Wilkins [8]. HCl was added to several asymmetric π allylrhodium(I) complexes. At $-75\,^\circ$C, NMR evidence for the intermediate formation of π allylRh(III)H(Cl) was observed. At room temperature the hydrogen ligand migrated to the allyl group giving the olefin. With 1-methylallyl [R' = H, R = CH$_3$; Eq. (8.6)], and with 1-methyl-3-ethylallyl (R' = CH$_3$, R = C$_2$H$_5$) complexes, an almost 1:1 mixture of the expected isomeric olefins is found. For sterically hindered π allyl groups such as 1,1-dimethylallyl, hydrogen migration to the least hindered carbon atom is preferred.

Further evidence for isomerization *via* a π allylmetal hydride mechanism was found by Casey and Cyr [9], during isomerization of 3-ethylpentene-1 with $Fe_3(CO)_{12}$. Experiments with deuterated isomers showed that only intramolecular 1,3-hydrogen shifts take place. Since no primary isotope effect was observed, the splitting of a C−H bond to give the π allylmetal hydride can be discarded as the rate determining step. It was suggested that the slowest reaction is either the formation of the monometallic olefin complex [Fe(CO)₄olefin], or the dissociation of one CO ligand, resulting in the tetracoordinate[Fe(CO)₃olefin]. Upon formation of the π allylmetal hydride, the metal changes its oxidation number from zero to two.

The positional isomerization, 1,4-dichlorobutene-2 $\rightleftarrows$ 3,4-dichlorobutene-1, which is catalyzed strongly by soluble iron complexes, is accompanied by a 1,3-chlorine shift, indicating intermediate formation of a π allylmetal chloride [10]:

$$\underset{\text{1,4-DCB}}{ClCH_2CH\!=\!CHCH_2Cl} \underset{\text{Cat}}{\rightleftarrows} \underset{\underset{ClCH_2}{\overset{\displaystyle CH}{|}}}{ClFe\!-\!\!\!\overset{\displaystyle CH_2}{\underset{\displaystyle CH}{\diagup}}\!\!\!CH} \underset{\text{Cat}}{\rightleftarrows} \underset{\text{3,4-DCB}}{CH_2\!=\!CHCHClCH_2Cl} \tag{8.8}$$

An investigation of kinetics of this isomerization, using $[(\eta^5\text{-}C_5H_5)Fe(CO)_2]_2$ as catalyst, has been reported. At 120 °C after *ca.* 3 hours the reaction attains equilibrium which may be approached from either side, as shown in Fig. 8.1. (For abbreviations, see Eq. (8.8); Cat = catalyst.)

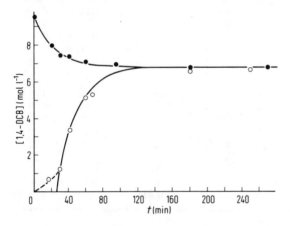

Fig. 8.1. The isomerization equilibrium 1,4-DCB $\rightleftarrows$ 3,4-DCB, starting from 1,4-DCB (upper curve) or from 3,4-DCB (lower curve). Solid lines: calculated with $k_1 = 0.1$ l mol⁻¹ s⁻¹ and $k_2 = 0.26$ l mol⁻¹ s⁻¹ [Eqs. (8.11) and (8.12) resp.]. [Cat] = 2×10^{-3} mol l⁻¹; $T = 120$ °C.

The rate for positional isomerization is accomodated by the rate law ([Cat] = const.):

$$-\frac{d[\text{1,4-DCB}]}{dt} = k_1[\text{Cat}][\text{1,4-DCB}] - k_2[\text{Cat}][\text{3,4-DCB}] \tag{8.9}$$

This equation can be integrated, if the concentration of the 3,4-isomer is replaced by $[\text{3,4-DCB}]_t = [\text{1,4-DCB}]_0 - [\text{1,4-DCB}]_t$. Making use of the equilibrium constant:

$$K = \frac{k_1}{k_2} = \frac{[\text{3,4-DCB}]_{\text{eq.}}}{[\text{1,4-DCB}]_{\text{eq.}}} \tag{8.10}$$

one obtains:

$$k_1 = -\frac{2.3\,K}{(K+1)[Cat]\,t} \cdot \log_{10}\left\{\frac{(K+1)[1,4\text{-DCB}]_t}{K[1,4\text{-DCB}]_0} - \frac{1}{K}\right\} \tag{8.11}$$

If the equilibrium is approached from the 3,4-DCB side, the rate constant k_2 can be obtained:

$$k_2 = -\frac{2.3\,K}{(K+1)[Cat]\,t}\;\log_{10}\left\{\frac{(K+1)[3,4\text{-DCB}]_t}{[3,4\text{-DCB}]_0} - K\right\} \tag{8.12}$$

Values of k_1 and k_2, calculated according to Eqs. (8.10) and (8.11), (if starting from 1,4-DCB), or Eqs. (8.10) and (8.12) (if starting from 3,4-DCB) are given in Table 8.1. From the consistency of the rate constants at each temperature the rate expression would appear satisfactory.

Table 8.1. Kinetic data for the isomerization 1,4-DCB $\rightleftharpoons$ 3,4-DCB. No solvent.

T ($^\circ$C)	K Eq. (8.10)	$[Cat] \times 10^3$ (mol l^{-1})	Starting isomer	t (min)	$\dfrac{[1,4\text{-DCB}]_t}{[1,4\text{-DCB}]_0}$	$\dfrac{[3,4\text{-DCB}]_t}{[3,4\text{-DCB}]_0}$	k_1	k_2
90	0.34	2.0	1,4-DCB	60	0.953		0.0071	0.021
				120	0.932		0.0050	0.016
				180	0.912		0.0050	0.015
				300	0.858		0.0058	0.017
				600	0.808		0.0050	0.015
				900	0.775		0.0051	0.015
		6.0		30	0.943		0.0059	0.017
120	0.39	1.0	1,4-DCB	60	0.773		0.128	0.33
				120	0.762		0.073	0.19
		2.0		20	0.837		0.101	0.26
				30	0.783		0.112	0.29
				40	0.781		0.087	0.22
				60	0.749		0.090	0.23
		2.0	3,4-DCB	14		0.642	0.102	0.26
				32		0.457	0.103	0.26
				40		0.444	0.087	0.22

Taking average values $k_1 = 5.5 \times 10^{-3}$ l mol^{-1}s^{-1}, $k_2 = 1.6 \times 10^{-2}$ l mol^{-1}s^{-1} at 90 $^\circ$C, and $k_1 = 0.11$ l mol^{-1}s^{-1}, $k_2 = 0.26$ l mol^{-1}s^{-1} at 120 $^\circ$C, from Table 8.1, one obtains

$$\log_{10}k_1 = 14.3 - (27{,}500/2.303\,RT)$$
$$\log_{10}k_2 = 14.4 - (26{,}900/2.303\,RT)$$

The difference between the activation energies of the forward and backward reaction, $E_1 - E_2 = 2.7$ kJ $\cdot$ mol^{-1}, is of the correct order of magnitude and sign expected from consideration of the difference between the bond dissociation energies of a carbon-chlorine bond involving primary and secondary carbon atoms [11].

The kinetic results are in agreement with either of the following two mechanisms:

1. Essentially all the iron is present as the π allyl complex. The equilibration occurs *via* ligand displacement, according to:

$$1,4\text{-DCB} + \text{Complex} \underset{k_2}{\overset{k_1}{\rightleftharpoons}} 3,4\text{-DCB} + \text{Complex} \tag{8.13}$$

From this mechanism the rate law, Eq. (8.9), follows straightforwardly.

2. Either of the two dichlorobutenes is in equilibrium with the active complex:

$$1,4\text{-DCB} + \text{Fe} \underset{k_{-a}}{\overset{k_a}{\rightleftharpoons}} \text{Complex} \underset{k_{-b}}{\overset{k_b}{\rightleftharpoons}} 3,4\text{-DCB} + \text{Fe}$$

If the concentration of the complex is small compared with $[\text{Fe}]_0$, so that $[\text{Fe}] \simeq [\text{Fe}]_0$, and if stationary state conditions are assumed for the complex, the same formal rate law follows, with

$$k_1 = \frac{k_a k_{-b}}{k_{-a} + k_{-b}} \qquad\qquad k_2 = \frac{k_b k_{-a}}{k_{-a} + k_{-b}}$$

$$K = \frac{k_a k_{-b}}{k_b k_{-a}} = \frac{[3,4\text{-DCB}]_{eq.}}{[1,4\text{-DCB}]_{eq.}}$$

8.1.3. cis-trans Isomerization

The positional isomerization of 1-alkenes gives preferentially the *cis*-2-alkene at first, and only when the system has had time to establish equilibrium, the thermodynamically expected *cis*: *trans* ratio is found. This behavior has been observed with catalysts based on a variety of transition metals [5, 12, 13].

The *cis-trans* isomerization of internal olefins is generally slower than the positional isomerization of terminal olefins (with the same catalyst), which appears plausible in view of the superior coordinating ability of the latter. Nevertheless an extremely rapid *cis-trans* isomerization was observed [5] in some cases, where the active metal hydride catalyst was prepared *in situ* with a strong acid [*cf.* Eq. (8.3)]. It was, however, suggested that this result may reflect acid rather than metal catalysis [14].

With metal hydride catalysts, the *cis-trans* isomerization most likely proceeds *via* olefin insertion and β hydrogen elimination [15]:

$$\tag{8.13}$$

In systems with π allylmetal intermediates, the isomerization goes evidently *via* the *syn-anti* equilibrium which is characteristic for such species (*cf.* Eq. (8.6) and Fig. 7.8).

8.2. Hydrogenation

8.2.1. General Aspects

A wide variety of soluble transition metal complexes now are known to function as hydrogenation catalysts under mild conditions (25 °C, 1 atm). For a broad coverage, as well as for details of catalyst preparation and other experimental aspects, the reader is referred to a recent review book [102].

Most of the hydrogenation catalysts are derived from electron rich, low valent group VIII metal ions. Very often the metal is surrounded by several high field ligands (CO, CN⁻, phosphines). Under these conditions the splitting of the d electrons under the influence of the ligand field may be assumed to be large, so that low spin complexes are expected to be the rule (*cf.* Section 5.1.2). If the number of d electrons exceeds six, the energetic advantage of octahedral symmetry (optimal ligand field stabilization energy, Section 5.1.4) diminishes, and complexes with lower coordination number (4 or 5) become favored. The availability of potential coordination sites, on the other hand, is essential evidently for the activation of the hydrogen molecule, and the subsequent transfer of this hydrogen to a coordinatively bonded substrate (*e.g.* olefin) molecule, and in general for catalytic activity.

Fig. 8.2 shows the relative energy ordering of the d orbitals of a d^8 metal ion in different environments (*cf.* Section 5.1.3).

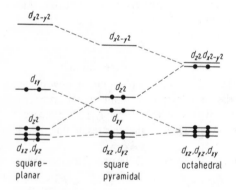

Fig. 8.2. Energy levels of the d orbitals of a d^8 system in different symmetries.

Consider, as an example, a square planar complex such as $Rh(I)Cl[P(C_6H_5)_3]_3$, and the orbital interaction between this complex and an approaching hydrogen molecule. Evidently the approach must proceed in the z direction, since the equatorial xy plane is occupied by ligands. The metal d orbitals which extend above and below the xy plane (d_{z^2}, d_{xz}, d_{yz}) are all occupied in this symmetry. Hence the interaction with an approaching H_2 molecule may be assumed to be *via* the empty antibonding σ* orbital of the H_2 molecule (*cf.* Section 6.2.1). The approach may be "end-on", with a σ bond-like interaction between σ* and d_{z^2}, or "edge-on" involving d_{xz} or d_{yz} in a π bond-like interaction (see Fig. 8.3). In either case, electron density is transferred into the σ* orbital of the H_2 molecule, weakening the H−H bond and hence activating the molecule. There might be also some interaction between the bonding σ orbital of the hydrogen molecule and an empty $(n+1)$s or $(n+1)$p valence orbital of the metal, but the enhancing effect of donor ligands towards the oxidative addition of H_2 (cf. Section 7.6.5) appears to indicate that this is less important.

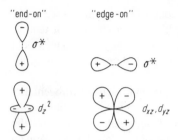

"end-on"   "edge-on"

Fig. 8.3. Modes of approach of the hydrogen molecule to a d^8 transition metal ion in a square planar complex.

Which of the two modes of approach actually takes place is not known, but the final product of this activation step is known: the two hydrogens will occupy two cis positions of the metal center, each linked to the metal by a σ bond (oxidative addition, see Section 7.4.3). The "edge-on" attachment appears to be more suitable for this rearrangement.

Some of the most investigated hydrogenation catalysts are summarized in Table 8.2, together with the rates of hydrogenation of heptene-1.

The complex $RuH(Cl)[P(C_6H_5)_3]_3$ has been described as the most active catalyst known for the hydrogenation of terminal monosubstituted olefins. Moreover the catalysis is highly selective; internal and cyclic alkenes as well as vinylidene compounds are reduced at rates which are lower by a factor of 10^3 to 10^4 than those of 1-alkenes [18].

Table 8.2. Rates of reduction of heptene-1 with several hydrogenation catalysts. $[Cat] = 2 \times 10^{-3}$ mol l^{-1}; [heptene-1] = 0.8 mol l^{-1}; $p_{H_2} = 9.67 \times 10^4\ Nm^{-2}$ (= 725 torr); solvent: toluene.

Catalyst[a]	T (°C)	Hydrogenation rate $\times 10^3$ (mol $l^{-1} \cdot s^{-1}$)	Ref.
$RuH(Cl)L_3^{[b]}$	25	34	[16]
$RhClL_3$	25	5.7	[16]
$IrClL_3$	25	0.6	[17]
$RhH(CO)L_3$	25	2.8	[16]
$IrH(CO)L_3$	25	0.3	[16]
$IrCl(CO)L_2$	80	8.9	[16]

[a] $L = P(C_6H_5)_3$.
[b] $[Cat] = 8.4 \times 10^{-4}$ mol l^{-1}; [heptene-1] = 1.2 mol l^{-1}; solvent: benzene.

The complex $RhCl[P(C_6H_5)_3]_3$, known as Wilkinson's catalyst, is also fairly active. In contrast to the ruthenium catalyst mentioned above, it hydrogenates also inner and cyclic olefins, although at varying rates. In the following series relative rates under comparable conditions are given in parentheses [19]: cyclohexene (100), hexene-1 (91), 2-methyl-pentene-1 (84), cis-pentene-2 (72), cis-4-methylpentene-2 (31), trans-4-methylpentene-2 (6). Alkynes also are reduced with $RhClL_3$ under mild conditions; in benzene the rate of reduction of straight chain C_6-C_{12} terminal acetylenes is *ca.* 0.85 that of C_6-C_{12} linear terminal olefins [102]. Conjugated olefins are hydrogenated with more difficulty; butadiene, for instance, requires a hydrogen pressure of 60 atm for effective hydrogenation [19]. Presumably the butadiene molecule acts as a strong chelating ligand, blocking the coordination site required for hydrogen activation.

Wilkinson's catalyst permits the selective hydrogenation of double bonds in the presence of a variety of potentially reducible functional groups, as long as these groups do not tend to form very stable complexes with the metal; *e.g.* α, β unsaturated carbonyl compounds yield saturated carbonyl compounds; the same is true for α, β unsaturated nitro compounds, nitriles, ketones and aldehydes [20].

From Table 8.2 it follows that, generally, rhodium complexes are more active hydrogenation catalysts than are their iridium analogs. In most cases one or more phosphine ligands must dissociate from the metal center in order to provide coordination sites for H_2 and the substrate (*cf.* Section 8.2.2). It has been observed that these ligands are much more strongly bonded in iridium complexes than in the corresponding rhodium complexes [16]. This appears to be an important reason for the difference in activity.

Vaska's complex, $IrCl(CO)L_2$, the first for which the oxidative addition of H_2 was reported (*cf.* Section 7.4.3), is a poor hydrogenation catalyst at 25 °C; it can, however, be used at higher temperatures (Table 8.2).

Some other types of hydrogenation catalysts deserve mention. Complexes of Pd(II) and Ni(II) with square-planar chelating Schiff base ligands, such as Pd(salen) [N,N'-ethylenebis-(salicylideneiminato)Pd(II)] have been suggested as models for natural hydrogenating systems (hydrogenases) because of mechanistic similarity [21].

The water soluble anionic cobalt complex $[Co(CN)_5]^{3-}$ can be used to hydrogenate conjugated and other activated double bonds; isolated olefinic double bonds are unaffected [102]. The very stable hydridometallocarboranes, particularly of rhodium, have recently been reported as catalysts for the hydrogenation of olefinic double bonds [22]. Ziegler catalysts [*cf.* Section 7.5.2], which are formed *in situ* from transition metal salts, usually acetylacetonates, with organoaluminum compounds, have been used to hydrogenate unsaturated compounds under relatively mild conditions [102]. Mössbauer spectroscopy demonstrates the homogeneous character of a catalyst derived from iron(III)acetylacetonate and triethylaluminum, and in particular the exclusion of very small metallic particles [23].

Soluble hydrogenation catalysts of the lower transition metal groups have been reported, but are less important. Occasionally dimeric carbonyls such as $Re_2(CO)_{10}$ [102] or $Mn_2(CO)_{10}$ [24] have been used; they are active only under severe conditions of temperature and pressure, where reaction or decomposition products might be the true catalyst.

In addition to the reduction of double and triple bonds in aliphatic hydrocarbons reduction of aromatics has been achieved, first with Ziegler systems under high pressures and temperatures [102], and more recently with a well-defined, benzene soluble cobalt complex, η^3-$C_3H_5Co[P(OCH_3)_3]_3$, under mild conditions [25]. Finally, the common heteronuclear unsaturated bonds, such as carbonyl, nitrile, imine *etc.*, have all been effectively reduced, although in most cases severe conditions of temperature and hydrogen pressure are necessary [102].

In many examples of hydrogenation with soluble transition metal catalysts, isomerization of the unsaturated substrate also takes place to a smaller or greater extent. During hydrogenation of 1-alkenes, the 2-*cis* and 2-*trans* isomers are formed and generally react more slowly than the 1-alkenes, hence the rate of hydrogenation tends to slow down with increasing conversion. For a series of iridium and rhodium complexes the activity for hydrogenation parallels isomerization [26]. For the highly active catalyst $RuH(Cl)[P(C_6H_5)_3]_3$ (Table 8.2), however, no isomerization accompanies 1-alkene hydrogenation [18]. With Wilkinson's catalyst, isomerization appears also to be unimportant in benzene and other hydrocarbon solvents, whereas considerable isomerization is observed in benzene ethanol mixtures [27]. We shall return to this point when discussing the reaction mechanism of hydrogenation with Wilkinson's catalyst, in Section 8.2.2.

A fairly detailed mechanistic picture of the catalytic hydrogenation process has been obtained for some systems (*cf.* Sections 8.2.2 and 8.2.3). Apart from the aim to learn how catalysis works (as a chemical reaction), these mechanistic investigations always have the making of more active and/or more selective catalysts as the ultimate goal.

We shall discuss separately mechanisms involving homolytic and heterolytic splitting of the hydrogen molecule. Homolytic splitting will result in an increase in oxidation number of the metal; in other words: a hydrogen atom requires an electron from the valence shell of the metal to form a σ bond with the latter. Thus, for instance:

$$M(I) + H_2 \longrightarrow M(III)H(H) \tag{8.14}$$
$$2\,M(II) + H_2 \longrightarrow 2\,M(III)H \tag{8.15}$$

Reaction (8.14), the oxidative addition of a H_2 molecule to a metal center, is discussed in Section 7.4.3; it is characteristic for square-planar d^8 complexes. Reaction (8.15) is observed with certain pentacoordinate d^7 complexes such as $[Co(CN)_5]^{3-}$ [28]. (See also Section 9.2.)

Heterolytic splitting of H_2, on the other hand, results in no change of the formal oxidation number of the metal. Generally, an anionic ligand A^- is replaced by the hydride ion:

$$MA + H_2 \longrightarrow MH + H^+ + A^- \tag{8.16}$$

It is believed generally that both the hydrogen molecule and substrate (*e.g.* olefin) must be present simultaneously in the coordination sphere of the metal for effective hydrogenation. Concerning the order of entry of the two reaction partners into the coordination sphere, two possibilities have been suggested, and experimental evidence has been found for either. The "hydride route" includes activation of the H_2 molecule at the catalyst center, Cat, as the primary step, with subsequent coordination and hydrogenation of the substrate molecule, S:

$$Cat + H_2 \;\rightleftharpoons\; Cat(H)H \xrightarrow{\;S\;} SH_2 + Cat \tag{8.17}$$

In the "olefin route", the olefinic substrate molecule is first coordinatively bonded to the catalyst:

$$Cat + S \;\rightleftharpoons\; CatS \xrightarrow{\;H_2\;} SH_2 + Cat \tag{8.18}$$

In most cases attempts have been made to deduce the mechanism through kinetic studies of the overall process. The information obtained in this way may, however, be incomplete in view of the complexity of the multistep processes [*cf.*, *e.g.*, Eq. (7.9) in Chapter 7]. In favorable cases, one or more of the component steps may be examined separately [29, 30]. Useful supplementary mechanistic information may also be provided by the identification of intermediates, by exchange experiments using dideuterium, or deuterium labelled solvents, by kinetic isotope effects, and by the study of ligand influences.

8.2.2. Mechanisms involving Homolytic Splitting of Dihydrogen

The hydrogenation of cyclohexene (or styrene) with Wilkinson's catalyst $RhClL_3$ (L = aromatic tertiary phosphine) is probably the most thoroughly investigated hydrogenation in the mechanistic sense. Wilkinson and coworkers carried out in 1966 the first kinetic measurements on the overall hydrogenation [31]. Although their suggested reaction scheme later turned out to be oversimplified, several of the significant mechanistic features were already recognized at that time. The importance of dissociation of one phosphine ligand has been discussed, although overestimated at the beginning [31, 32]. Inhibition terms in the overall rate law have been interpreted as consequences of the equilibria expressed in Eqs. (8.17) and (8.18), and the "hydride route" has been postulated as the more important one.

More recently, Halpern [29], Tolman [30] and their respective coworkers have undertaken very detailed studies of the many individual steps involved in the complicated multistep process. We shall give a somewhat broad review on the present state of knowledge, because the work done with this system is a beautiful – and possibly to date unique – example of the elucidation of all important individual steps of a complicated catalytic cycle. Where not stated otherwise, the following presentation is based mainly on the extensive kinetic work of Halpern *et al.*

First consider the complex $RhClL_3$ in an inert solvent, say benzene, in the absence of the reactants, hydrogen and olefin. Dissociation of one phosphine ligand is revealed by spectrophotometric studies, in the following way: At low concentrations ($< 10^{-3}$ mol l^{-1}) a departure from Beer's law is observed, the apparent extinction coefficient in the range 350 to 500 nm decreasing with increasing dilution. At the same time the absorption maximum is somewhat shifted. These observations indicate a partial conversion into a compound having a smaller absorption than $RhClL_3$ in this range. Addition of an excess of phosphine ligand to

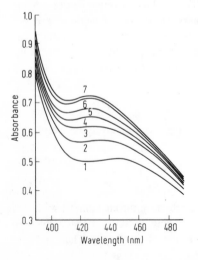

Fig. 8.4. Effect of added L on the spectrum of $RhClL_3$ in benzene (Halpern *et al.* [29]). L = $P(C_6H_5)_3$; [Rh] $= 5.1 \times 10^{-4}$ mol l^{-1}; added L, mol l^{-1}: 1, 0.0; 2, 2.5 $\times$ 10^{-4}; 3, 5.0×10^{-4}; 4, 7.5×10^{-4}; 5, 12.5×10^{-4}; 6, 25×10^{-4}; 7, $>5 \times 10^{-3}$ ($= A_\infty$).

the diluted solution restores the absorbance characteristic of concentrated ($>10^{-3}$ mol l^{-1}) solutions of RhClL$_3$. The effect of L addition to a diluted solution is shown in Fig. 8.4.

Thus, a dissociation equilibrium, giving the coordinatively unsaturated species RhClL$_2$, is indicated. But RhClL$_2$ dimerizes very rapidly, see Eq. (8.19):

$$2 \quad \overset{L \diagup \quad \diagdown L}{\underset{L \diagdown \quad \diagup Cl}{\boxed{Rh}}} \quad \underset{+2\,L}{\overset{-2\,L}{\rightleftarrows}} \quad 2 \quad \overset{L \diagup \quad \diagdown \square}{\underset{L \diagdown \quad \diagup Cl}{\boxed{Rh}}} \quad \rightleftarrows \quad \overset{L \diagup \quad Cl \quad \diagdown L}{\underset{L \diagdown \quad Cl \quad \diagup L}{\boxed{Rh \diagup Rh}}} \tag{8.19}$$

The intermediate RhClL$_2$ cannot be detected spectroscopically, but its presence in very low concentration can be inferred from kinetic evidence as discussed below. Thus, although the two separate equilibria, Eqs. (8.20) and (8.21), are operative, the spectral data given in Fig. 8.4 represent the overall equilibrium, Eq. (8.22):

$$\text{RhClL}_3 \quad \underset{\leqslant 10^{-5}}{\overset{K_1}{\rightleftarrows}} \quad \text{RhClL}_2 + \text{L} \tag{8.20}$$

$$2\,\text{RhClL}_2 \quad \underset{\geqslant 10^6}{\overset{K_1'}{\rightleftarrows}} \quad \text{Rh}_2\text{Cl}_2\text{L}_4 \tag{8.21}$$

$$2\,\text{RhClL}_3 \quad \underset{10^{-4}}{\overset{K_1''}{\rightleftarrows}} \quad \text{Rh}_2\text{Cl}_2\text{L}_4 + 2\,\text{L} \tag{8.22}$$

Hence, the equilibrium constant K_1'' can be determined from the spectral data (see Appendix 8.1); at room temperature it's value is *ca.* 10^{-4} mol l^{-1}. On the other hand, the dimer Rh$_2$Cl$_2$L$_4$ can be prepared pure [31]. Even in dilute solutions of this compound, no dissociation to RhClL$_2$ can be observed (NMR evidence [30]). Based on this fact, a lower limit of 10^6 l mol^{-1} was set for the equilibrium constant K_1', and from there it follows directly that $K_1 \leqslant 10^{-5}$ mol l^{-1} [$K_1 = K_1''/K_1')^{0.5}$]. From these numbers we can deduce that most of the rhodium is present in the undissociated form, with three phosphine ligands per rhodium.

Next we shall review the reaction of the catalyst with hydrogen, which leads to the following octahedral complex:

The solution configuration of this complex has been determined from its proton decoupled ^{31}P NMR spectrum [30]. The absorption spectra of RhClL$_3$ and RhClL$_3$H$_2$ in the 400 to 550 nm range are shown in Fig. 8.5. The rate of the reaction of RhClL$_3$ with H$_2$ can be determined spectrophotometrically by monitoring the absorbance change, *e.g.* at 470 nm.

The kinetic measurements on this reaction have been performed in the presence of an excess of L, in order to avoid the accumulation of the rhodium dimer according to Eq. (8.22). In the

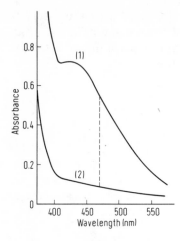

Fig. 8.5. Absorbance of RhClL$_3$ (1) and of RhClL$_3$H$_2$ (2) in benzene; [Rh] = 5.6 × 10^{-4} mol l^{-1} [29 d].

presence of an excess of H$_2$ the hydrogenation of RhClL$_3$ goes to completion under pseudo first order conditions, following the rate law:

$$-\frac{d\,[\text{RhClL}_3]}{dt} = k_{\text{obs}}\,[\text{RhClL}_3] \tag{8.23}$$

However, the observable, apparent rate constant k_{obs} depends in a characteristic way on the hydrogen pressure and on the concentration of added phosphine ligand, L. The experimental data are accomodated quantitatively by a mechanistic scheme which includes, in addition to the direct hydrogenation of RhClL$_3$ [Eq. (8.24)], a contribution from the dissociative pathway described by Eqs. (8.20), (8.25) and (8.26):

$$\text{RhClL}_3 + \text{H}_2 \underset{k_{-2}}{\overset{k_2}{\rightleftharpoons}} \text{RhClL}_3\text{H}_2 \tag{8.24}$$

$$\text{RhClL}_3 \underset{k_{-1}}{\overset{k_1}{\rightleftharpoons}} \text{RhClL}_2 + \text{L} \tag{8.20}$$

$$\text{RhClL}_2 + \text{H}_2 \underset{k_{-3}}{\overset{k_3}{\rightleftharpoons}} \text{RhClL}_2\text{H}_2 \tag{8.25}$$

$$\text{RhClL}_2\text{H}_2 + \text{L} \underset{k_{-4}}{\overset{k_4}{\rightleftharpoons}} \text{RhClL}_3\text{H}_2 \tag{8.26}$$

According to this scheme, the overall rate of the reaction of the catalyst with H$_2$ is given by:

$$-\frac{d\,[\text{RhClL}_3]}{dt} = k_2\,[\text{RhClL}_3]\,[\text{H}_2] + k_3\,[\text{RhClL}_2]\,[\text{H}_2]$$

Application of the stationary state approximation to the shortlived intermediate $RhClL_2$ gives [cf. Eqs. (8.20) and (8.25)]:

$$[RhClL_2] = \frac{k_1 [RhClL_3]}{k_{-1}[L] + k_3[H_2]}$$

and hence:

$$-\frac{d[RhClL_3]}{dt} = \left(k_2 + \frac{k_1 k_3}{k_{-1}[L] + k_3[H_2]} \right) [RhClL_3][H_2] \tag{8.27}$$

Comparison of (8.23) with (8.27) shows that:

$$\frac{1}{k_{obs}} = \frac{1}{k_2[H_2]} + \frac{k_{-1}[L]}{k_1 k_3[H_2]} + \frac{1}{k_1} \tag{8.28}$$

From a series of determinations of k_{obs}^{-1} as a function of $[H_2]^{-1}$, at constant [L], k_1^{-1} can be obtained as the intercept of a straight line. On the other hand, if $[H_2]$ is maintained constant, and k_{obs}^{-1} is plotted as a function of [L], k_2 and the ratio k_3/k_{-1} become available from the intercept and slope respectively of a straight line. Since the upper limit of $K_1 = k_1/k_{-1}$ is also known [Eq. (8.20)], all relevant rate constants, or at least limiting values, can be determined. Numerical values, obtained at 25 °C in benzene solution are given in Fig. 8.6.

It turns out that $RhClL_2$ is at least 10^4 times more reactive towards H_2 than is $RhClL_3$ ($k_3 \gg k_2$). Direct support is thus provided for the original suggestion of Wilkinson *et al.* [31] concerning the high reactivity of the dissociated species and its probable importance (despite its low concentration in solution) in $RhClL_3$ catalyzed hydrogenations of olefinic substrates. The formation of the dimer $[Rh_2Cl_2L_4]$, Eq. (8.19), appears to be suppressed completely in the presence of an excess of L and of H_2. Thus, evidently the species $RhClL_3H_2$ is the most stable one in the substrate free system.

Finally, the kinetics of substrate hydrogenation remains to be discussed. Over a wide range of conditions, the hydrogenation of $RhClL_3$ to $RhClL_3H_2$, according to the mechanism described above, is fast compared to the subsequent steps in olefin hydrogenation; one of the latter is, accordingly, rate determining. The relevant rate and equilibrium constants for this last phase of the catalytic cycle have been determined in the reaction of $RhClL_3H_2$ with an olefin (*e.g.* cyclohexene), in the absence of an excess of hydrogen (*i.e.* under noncatalytic conditions), but again in the presence of an excess of phosphine. One of the phosphine ligands of the coordinatively saturated octahedral rhodium dihydride complex must be exchanged for the olefin (formal equilibrium constant K_5, although the actual pathway is presumably *via* K_4 and K_7, cf. Fig. 8.6). The olefin then inserts into one of the Rh—H bonds (rate constant k_6). The subsequent reductive elimination of alkane, with reentry of the rhodium into the catalytic cycle, is assumed to be very rapid, because no optical or NMR evidence could be detected for a significant concentration of an alkylrhodium species. The rate of hydrogenation of the substrate S, r_H, is then given by Eq. 8.29:

$$r_H = -\frac{d[S]}{dt} = -\frac{d[RhClL_3H_2]}{dt} = k_6[RhClL_2H_2S] \tag{8.29}$$

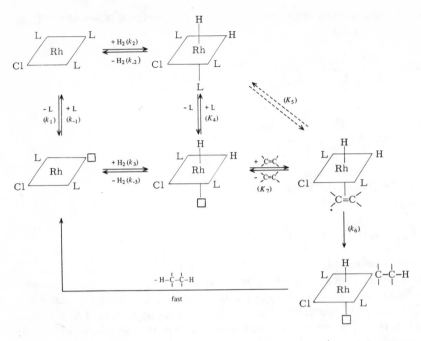

$k_1 = 0.68 \text{ s}^{-1}$

$k_{-1} \geq 7 \times 10^4 \text{ l mol}^{-1} \text{ s}^{-1}$

$K_1 \leq 10^{-5} \text{ mol l}^{-1}$

$k_{-1}/k_4 \cong 1$

$K_5 = 3 \times 10^{-4}$

$k_2 = 4.8 \text{ l mol}^{-1} \text{ s}^{-1}$

$k_{-2} = 2.8 \times 10^{-4} \text{ s}^{-1}$

$K_2 = 1.7 \times 10^4 \text{ l mol}^{-1}$

$k_3 \geq 7 \times 10^4 \text{ l mol}^{-1} \text{ s}^{-1}$

$k_6 = 0.22 \text{ s}^{-1}$

Fig. 8.6. Mechanism of the hydrogenation of olefin with Wilkinson's catalyst, $RhClL_3$ (Halpern *et al.* [29]). The constants K_5 and k_6 refer to the hydrogenation of cyclohexene [29d]. □ = vacant coordination site.

where $[RhClL_3H_2]$ and $[RhClL_2H_2S]$ are related by the equilibrium:

$$RhClL_3H_2 + S \underset{}{\overset{K_5}{\rightleftarrows}} RhClL_2H_2S + L \qquad (8.30)$$

Furthermore:

$$[RhClL_3H_2] + [RhClL_2H_2S] \simeq [Rh]_{total} \qquad (8.31)$$

(Under the noncatalytic conditions described, relationship (8.31) is valid only for initial rates; under catalytic conditions however, it is a useful approximation throughout the whole process, because of the fast hydrogenation of the rhodium species $RhClL_2$ and $RhClL_3$.)
From Eq. (8.30) it follows that:

$$\frac{[RhClL_2H_2S][L]}{K_5[S]} = [RhClL_3H_2] \qquad (8.32)$$

After adding $[RhClL_2H_2S]$ to either side of Eq. (8.32) and applying Eq. (8.31), Eq. (8.32) is easily rearranged to give:

$$[RhClL_2H_2S] = \frac{K_5[S][Rh]_{total}}{K_5[S]+[L]} \tag{8.33}$$

Introducing (8.33) into (8.29), one obtains an expression for the rate of hydrogenation of the substrate, r_H, which depends only on rate and equilibrium constants, and on measurable concentrations of $[Rh]_{total}$, $[S]$ and $[L]^{*)}$:

$$r_H = \frac{k_6 K_5[S][Rh]_{total}}{K_5[S]+[L]} \tag{8.34}$$

Experimentally, this rate can be determined by spectrophotometry, making use of the absorbance of $RhClL_3H_2$ and $RhClL_3$ (see Fig. 8.5; note that under the conditions of this investigation, *i.e.* in the absence of free hydrogen, and in the presence of an excess of phosphine, all rhodium is finally transformed into $RhClL_3$). A series of initial rate determinations, at varying substrate concentrations, allows one to evaluate the constants K_5 and k_6, from the intercept and slope of a straight line, $r_H^{-1} = f([S]^{-1})$, according to:

$$\frac{1}{r_H} = \frac{1}{k_6[Rh]_{total}} + \frac{[L]}{k_6 K_5[Rh]_{total}[S]} \tag{8.35}$$

All these data are included in Fig. 8.6, which gives a comprehensive view of all important individual steps in the hydrogenation cycle, at 25 °C in benzene solution. This mechanism evidently corresponds to what has been defined earlier as the "hydride route" Eq. (8.17). The "olefin route", *i.e.* the primary reaction of an olefin with $RhClL_3$ (or $RhClL_2$), is thermodynamically unfavorable, as born out by the equilibrium constant of the following reaction [30]:

$$RhClL_3 + C_2H_4 \;\xrightleftharpoons{\;K_8\;}\; Rh(C_2H_4]L_2 + L$$

with a value of $K_8 = 0.4$ (as compared with $K_2 = k_2/k_{-2} = 1.7 \times 10^4\,l\,mol^{-1}$).

The reaction scheme outlined in Fig. 8.6 has stood the test also under catalytic conditions, with cyclohexene as substrate (see Table 8.3). For a relatively broad range of substrate concentrations, the measured hydrogenation rate is in good agreement with the rate calculated from Eq. (8.34).$^{**)}$ The predicted independence on hydrogen pressure is also reasonably well obeyed in the experimental range; the observed deviation is, of course, a consequence of the approximate character of Eq. (8.31), which becomes less reliable at low hydrogen pressures where nonhydrogenated rhodium species start to play a role.

---

* If the concentration of the excess of ligand, $[L]_{excess} \gg [Rh]_{total}$, the concentration of free ligand in the solution is *ca.* $[L]_{excess}$.

** With styrene ($k_6 = 0.11$ s^{-1}, $K_5 = 1.7 \times 10^{-3}$) the hydrogenation proceeds faster than calculated according to Eq. (8.34) [29d]. This was interpreted as the consequence of an additional reaction path including a species in which two L ligands have been replaced by styrene molecules.

Table 8.3. Hydrogenation of cyclohexene with $RhClL_3$ in benzene at $25\,°C$ [29 d]. $L = P(C_6H_5)_3$; $[Rh]_{total} = 7.6 \times 10^{-5}\,mol\,l^{-1}$; $[L]_{excess} = 7.8 \times 10^{-4}\,mol\,l^{-1}$; $k_6 = 0.22$; $K_5 = 3 \times 10^{-4}$.

Cyclohexene ($mol\,l^{-1}$)	H_2 (atm)	Rate $\times 10^6$ ($mol\,l^{-1} \cdot s^{-1}$) Observed	Calc. Eq. (8.34)
0.37	0.79	2.2	2.1
0.62	0.79	3.3	3.2
0.83	0.79	4.1	4.1
1.24	0.79	5.6	5.4
0.62	0.81	3.3	3.2
0.62	0.54	2.9	3.2
0.62	0.36	2.5	3.2

As expected, the overall hydrogenation rate increases with increasing substrate concentration. Eq. (8.34), however, indicates that there is a limiting rate, $k_6 \times [Rh]_{total}$, which would be attained if $K_5[S] \gg [L]$. This points to the insertion of the olefin into the $Rh-H$ bond as the actual rate-determining step in the catalytic cycle.

In Section 7.6.3 it was shown that donor ligands tend to accelerate insertions. Indeed, an increase in donor capacity of the phosphine ligands in Wilkinson's catalyst brings about a considerable increase of the hydrogenation rate which, presumably, is attributable primarily to the ligand influence on the rate-determining insertion. The following data [33] illustrate this point. For $L = R_3P$, the activity of the catalyst increases as R varies in the order $p-ClC_6H_4(1.8) < C_6H_5(41) < p-CH_3C_6H_4(86) < p-CH_3OC_6H_4(100)$. The figures in parentheses are the relative rates of hydrogenation of cyclohexene. The basicity of the phosphines increases in the same order. A similar trend is shown by the observation that the rate is reduced greatly on replacement of $(C_6H_5)_3P$ by $(C_6H_5)_3As$, and even more with $(C_6H_5)_3$-Sb. As mentioned in Section 7.6.2 the basicity (σ donor strength) decreases in the same order. This picture is also in agreement with the discovery that when chlorine is replaced by other halides, the rate increases in the order $Cl < Br < I$.

The last question to be considered in the context of Wilkinson's catalyst is its isomerization ability. Under hydrogenation conditions in inert nonprotic solvents, no isomerization of the olefinic substrates is observed [31]. This means that the insertion step is essentially irreversible, presumably because of the very rapid reductive elimination of the alkane from the complex. In protic solvents (e.g. benzene ethanol mixtures [27]), isomerization and hydrogenation take place simultaneously. A different mechanism as yet undefined appears to operate.

For catalysts containing one metal-hydride bond, such as $RhH(CO)L_3$ or $RuH(Cl)L_3$ $[L = P(C_6H_5)_3]$, the "olefin route" [Eq. (8.18)] is generally assumed. The first step is again the dissociation of one of the phosphine ligands. In benzene solution, dissociation gives a square-planar species, with trans phosphines, which on dilution can lose a second phosphine:

$$RhH(CO)L_3 \underset{+L}{\overset{-L}{\rightleftharpoons}} RhH(CO)L_2 \underset{+L}{\overset{-L}{\rightleftharpoons}} RhH(CO)L \qquad (8.36)$$

At a concentration of ca. $10^{-3}\,mol\,l^{-1}$, the bisphosphine complex predominates [34].

The assumption that the "olefin route" prevails is corroborated by the following observations [34, 102]:

a. There is no hydrogen absorption by solutions of the catalyst in the absence of substrate, hence the equilibrium (8.37) must lie far to the left:

$$RhH(CO)L_2 + H_2 \ \rightleftarrows \ RhH_3(CO)L_2 \qquad\qquad (8.37)$$

b. The rhodium catalyst reacts with tetrafluoroethylene (at 5 atm) to give the stable, square-planar *trans*-Rh(CF$_2$CF$_2$H)(CO)L$_2$; this complex reacts with hydrogen (50 °C, 70 atm), in the presence of an excess of phosphine, to give C$_2$F$_4$H$_2$, with regeneration of the catalyst:

$$(8.38)$$

c. In contrast with RhClL$_3$, the catalyst RhH(CO)L$_3$ isomerizes 1-olefins at rates comparable to the hydrogenation rates, indicating that the olefin insertion [step a, reaction (8.38)] is reversible.

d. The hydrogenation rate is directly proportional to the hydrogen pressure, over a range of $(3–8) \times 10^4$ N m^{-2}, clearly indicating a different mechanism to that found with Wilkinson's catalyst. This hydrogen pressure dependence points to oxidative addition of H$_2$ [step b, reaction (8.38)] as the rate determining step.

A detailed study of all important individual steps, as shown above for RhClL$_3$, has yet to be reported, but the overall kinetics are in agreement with the scheme indicated by Eqs. (8.36–38) [34]. As with RuH(Cl)L$_3$ (*cf.* Section 8.2.1), the RhH(CO)L$_3$ catalyst is highly selective for terminal olefins; relative rates of hydrogenation of different olefins are [16]: heptene-1 (100) $\gg$ *cis*-heptene-2 (6.4) > *trans*-heptene-2 (4.9) $\simeq$ *trans*-heptene-3 (4.7).

8.2.3. Mechanisms involving Heterolytic Splitting of Dihydrogen

Heterolytic splitting of the hydrogen molecule to give a metalmonohydride and a proton has been demonstrated or invoked in several catalytic hydrogenation systems, particularly in polar solvents. The overall reaction may be depicted as:

$$M + H_2 \ \longrightarrow \ MH + H^+ \ \xrightarrow{+S} \ M + SH_2$$

Frequently the hydride ion replaces an ionic ligand of the metal, *e.g.*:

$$L_nMCl + H_2 \ \longrightarrow \ L_nMH + H^+ + Cl^-$$

(Several examples are to be found in James' hydrogenation review [102].)
An easy exchange between H$_2$ gas and deuterium from deuterated, protic solvents (alcohol, water) is taken generally as a criterion for heterolytic splitting:

$$M + H_2 \longrightarrow MH + H^+ \tag{8.39}$$

$$H^+ + D_2O \longrightarrow D^+ + HDO \tag{8.40}$$

$$MH + D^+ \longrightarrow M + HD \tag{8.41}$$

Frequently some D_2 is also found in such exchange experiments. This fact has sometimes been interpreted in terms of the exchange ability of hydrogen in a metal hydride:

$$MH + D_2O \longrightarrow MD + HDO$$
$$MD + D^+ \longrightarrow M + D_2$$

Taking into account the polarity of $M^{\delta+} - H^{\delta-}$ bonds (*cf.* Section 7.4.1), this interpretation appears less plausible. The following more probable and easy path leading to D_2 formation has been suggested [21]. The HD molecule formed according to Eqs. (8.39–41) may, without leaving the coordination sphere of the metal, be split again, but this time into MD and H^+:

$$\begin{array}{c} H-D \\ \vdots \\ MD + H^+ \; \rightleftarrows \; M \; \rightleftarrows \; MH + D^+ \end{array} \tag{8.42}$$

Natural hydrogen activating enzymes (hydrogenases) appear to operate *via* heterolytic splitting of the hydrogen molecule at a transition metal center [102]. It is generally assumed that the proton released in the first step [*cf.* Eq. (8.39)] is accepted by a nonidentified "basic site" B within the enzyme entity:

$$\begin{array}{cc} & H^- \;\; H^+ \\ M \quad B + H_2 \; \longrightarrow \; M \quad B \\ \rule{1.5em}{0.4pt} & \rule{1.5em}{0.4pt} \end{array}$$

This formulation readily accounts for the observed H^+/D_2O exchange. It also explains another characteristic of hydrogenases, namely a pH activity dependence, with a maximum at *ca.* pH = 8 [35]. In acidic solutions the heterolytic splitting of H_2 would be retarded, since at low pH's most of the basic site B would be neutralized and hence could not accept the proton. In basic solution, however, the slow step would be the reformation of H_2 since there would now be very little [H^+B] present. A maximum activity would therefore be expected at intermediate pH values. Obviously it would be difficult to explain a pH maximum if simple homolytic cleavage occurred.

The nature of the active site in natural hydrogenases is far from clear, and its investigation is complicated by the macromolecular protein component. It is known, however, that Nature very frequently uses transition metal catalysts with square-planar, chelating ligands such as porphyrins, phthalocyanines, *etc.* A synthetic palladium(II) complex with a square-planar, chelating Schiff base ligand has been suggested recently as a model for hydrogenase [21]. This complex, N,N'-ethylenebis(salicylideneiminato)Pd(II), Pd(salen), shows a remarkable mechanistic similarity to natural hydrogenase, with respect to H/D exchange with solvent, kinetics of catalytic hydrogenation of substrates, and pH activity dependence.

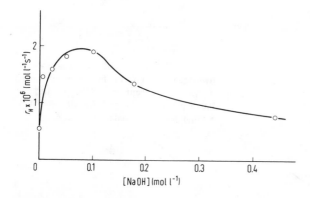

$$\text{Pd(salen)}$$

Pd(salen) is a yellow microcrystalline substance, very poorly soluble in alcohols, more soluble in dimethylformamide (DMF) or pyridine. Suspensions of Pd(salen) in EtOD activate H/D exchange with gaseous H_2 (Table 8.4). Alcoholic suspensions as well as solutions in DMF catalyze the hydrogenation of olefinic substrates. All other conditions being equal, the ratio of the hydrogenation rates for hexene-1: *cis*- and *trans*-hexene-2: cyclohexene is $1:0.85:0.7$. Evidently, the catalyst does not show any great selectivity; it also isomerizes the olefin, at a rate comparable to that of hydrogenation. It is strongly pH dependent: Work in acidic solutions leads to complete inhibition of the catalytic activity; on addition of alkali, on the other hand, the rate passes through a maximum (see Fig. 8.7: [NaOH] is given, since the pH in alcoholic solution is not defined).

Table 8.4. H/D Exchange. Pd(salen) in EtOD; H_2 atmosphere. $T = 20\,°C$.

t (h.)	H_2 (%)	HD (%)	D_2 (%)
3	98.68	1.25	0.07
16	92.88	6.90	0.22
3*)	100.00	–	–
16*)	99.66	0.30	0.04

* Control runs in absence of Pd(salen).

Fig. 8.7. Dependence of the initial rate of hydrogenation of hexene-1 on the NaOH concentration. $T = 20\,°C$; $p_{H_2} = 9.6 \times 10^4$ N m^{-2} ($= 720$ torr); heterogeneous reaction in EtOH.

The kinetics of the hydrogenation of hexene-1 have been measured in DMF (homogeneous solution). Figs. 8.8–10 show the initial rates at varying substrate concentrations, hydrogen pressures, and catalyst concentrations, respectively.
From these figures it follows that:

$$1/r_H = a + \frac{b}{[S]} \tag{8.43}$$

$$1/r_H = a' + \frac{b'}{P_{H_2}} \tag{8.44}$$

$$r_H \sim [Cat]_0 \tag{8.45}$$

where r_H = rate of hydrogenation, S = substrate (olefin), and a, a', b, and b' are constants. The same formal relationships have been reported for the hydrogenation of substrates with hydrogenase extracts from *Clostridium pasteurianum* [36].

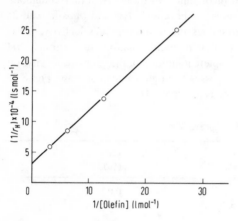

Fig. 8.8. Dependence of the initial rate of hydrogenation on the olefin concentration. $T = 20\,°C$. Solvent = DMF. $p_{H_2} = 6.7 \times 10^4$ N m^{-2} ($= 500$ torr). [Cat]$= 5.4 \times 10^{-3}$ mol l^{-1}. Straight line calculated from Eq. (8.49) with the rate and equilibrium constants as given in the text.

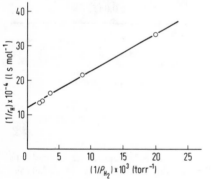

Fig. 8.9. Dependence of the initial rate of hydrogenation on the hydrogen pressure. $T = 20\,°C$. Solvent = DMF. [Olefin] = 0.079 mol l^{-1}. [Cat] = 5.4×10^{-3} mol l^{-1}. Straight line calculated as in Fig. 8.8.

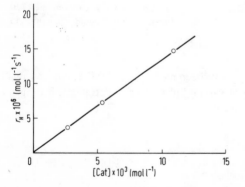

Fig. 8.10. Dependence of the initial rate of hydrogenation on the catalyst concentration. $T = 20\,°C$. Solvent = DMF. $p_{H_2} = 6.7 \times 10^4$ N m^{-2} ($= 500$ torr). [Olefin] = 0.079 mol l^{-1}. Straight line calculated as in Fig. 8.8.

Before fitting the kinetic results to a reaction scheme we must consider possible mechanisms. In the solid state (suspension in EtOH), the metal is surrounded by oxygen and nitrogen atoms of the chelating ligand arranged in the equatorial plane, with the next palladium unit being situated below this plane. In solution at least one of the axial positions can be assumed to be occupied by the strongly complexing solvent DMF. In neither case are sufficient co-ordination sites available in suitable *cis* positions for the simultaneous coordination of hydrogen and substrate. We are forced to conclude that a ligand (or ligands) are removed. In view of the astonishing similarity with natural hydrogenase, and particularly because of the pH dependence of the hydrogenation, heterolytic splitting of H_2 may be assumed to occur simultaneously with the opening of one of the equatorial metal-ligand bonds. One phenoxy ligand is then transformed into a phenol group, and the electroneutrality is maintained by the formation of a metal hydride. We may call such a process, which is supposed to proceed *via* a four-center transition state, a "ligand assisted heterolytic splitting" of the H_2 molecule.

(O---N---N---O stands for the salen ligand, L for Pd or solvent.)

A free site is now available and can be used for coordination of the olefin. Subsequently, the olefin inserts between the hydrogen and the metal with formation of an alkyl group, which then reacts intramolecularly with the phenol group giving alkane and the original catalyst.

This mechanistic picture suggests that the "basic site" in the enzyme is an anionic ligand which leaves the metal on accepting the proton, but remains inherently fixed within the complex. As a consequence, alkane formation involves a first order decomposition of the alkyl intermediate, *i.e.* formation does not involve any of the hydrogen ions of the medium, but only the proton attached to the ligand.

Regarding the limited space which would be available for the coordination of the olefin in this model (about 3.5 Å, even if the aromatic ring were to be tilted by 90°), one may argue that the vicinal neutral nitrogen ligand is simultaneously removed from the coordination sphere of the metal. This would lead to a reorientation of one half of the salen ligand by

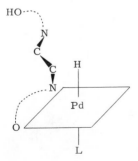

Fig. 8.11. Assumed structure of the Pd(salen) catalyst, after heterolytic splitting of a H_2 molecule, and reorientation of one-half of the salen ligand by rotation about the N−C−C−N moiety of the ligand.

rotation about the central $C-C$ and $N-C$ σ bonds, thus leaving ample space available for the coordination of the olefin (see Fig. 8.11). This configuration would also account for the lack of catalyst selectivity. A recent crystal structure determination on $(\eta^3$-methylallylPd)$_2$-salen [37] has indicated that such a rotation is feasible energetically, which supports this suggestion.

A reaction scheme which takes into account these considerations is given by Eqs. (8.46–48). The following abbreviations are used:

Cat$_f$ = free catalyst;
Cat(H)(OH) = catalyst which has reacted with H_2, *i.e.* possessing one hydrogen ion linked to the metal, one phenolic hydroxyl group in the salen ligand, and a free co-ordination site;
Cat(al)(OH) = catalyst which has had an olefin molecule inserted into the metal-H bond, and hence now possesses one alkyl group linked to the metal and a phenolic hydroxyl group in the salen ligand.

$$\text{Cat}_f \qquad + H_2 \quad \underset{\longleftarrow}{\overset{K_1}{\rightleftharpoons}} \quad \text{Cat(H)(OH)} \tag{8.46}$$

$$\text{Cat(H)(OH)} + S \quad \underset{\longleftarrow}{\overset{K_2}{\rightleftharpoons}} \quad \text{Cat(al)(OH)} \tag{8.47}$$

$$\text{Cat(al)(OH)} \quad \overset{k}{\longrightarrow} \quad \text{Cat}_f + \text{alkane} \tag{8.48}$$

Equilibrium (8.46) describes the heterolytic splitting of the hydrogen; equilibrium (8.47) takes care of the observed isomerization of the olefin; reaction (8.48) leads to the formation of the product alkane, as well as to the regeneration of the catalyst, with the quadridentate salen ligand again in the equatorial plane of the complex.

The hydrogenation rate, r_H, is given by

$$r_H = d\,[\text{alkane}]/dt = k\,[\text{Cat(al)(OH)}]$$

If the partition of the overall initial catalyst concentration, $[\text{Cat}]_0$, is taken into account:

$$[\text{Cat}]_0 = [\text{Cat}]_f + [\text{Cat(H)(OH)}] + [\text{Cat(al)(OH)}]$$

the following expressions follow readily:

$$[\text{Cat(al)(OH)}] = K_2[\text{Cat(H)(OH)}]\,[S]$$

$$[\text{Cat(H)(OH)}] = K_1[\text{Cat}_f]\,[H_2]$$

$$[\text{Cat}_f] \qquad = [\text{Cat}]_0/(1 + K_1[H_2] + K_1 K_2[H_2]\,[S])$$

In the initial stages when $[S] \approx [S]_0$, it follows that:

$$r_H = \frac{k K_1 K_2 [H_2]\,[S]_0\,[\text{Cat}]_0}{1 + K_1[H_2] + K_1 K_2[H_2]\,[S]_0} \tag{8.49}$$

Eq. (8.49), in fact, is in perfect agreement with the experimental findings shown in Figs. 8.8–10. The individual rate and equilibrium constants k, K_1, and K_2 may easily be evaluated from the slopes and intercepts of the experimental curves, giving the values (for $T = 20\,°C$): $k = 6.2 \times 10^{-3}\,s^{-1}$; $K_1 = 3.6 \times 10^3\,l\,mol^{-1}$; $K_2 = 4.2\,l\,mol^{-1}$. The straight lines depicted in Figs. 8.8–10 were actually calculated from Eq. (8.49) employing these constants, thus demonstrating the high self-consistency of the three sets of experimental data, and hence corroborating the suggested reaction scheme.

Eq. (8.47) is, of course, a simplified version of the insertion whose individual steps cannot be isolated by this kinetic study. Thus, for instance, it is possible that the olefin can enter and leave the complex many times before insertion actually occurs, which would mean that the formal constant K_2 would, in fact, be a product of two equilibrium constants. From the relatively low value determined for k it would appear that the last step in the process is probably rate-determining.

Alltogether, Pd(salen) appears to be, in many respects, very similar in behaviour to the enzyme hydrogenase. It has been suggested [21], that other enzymes containing a transition metal ion may also function by a similar mechanism, whereby an anion of a chelating ligand, e.g. S^-, O^-, CO_2^-, or N^- may be displaced from the metal, acting as a "basic acceptor site". The activation of small molecules such as H_2, N_2, O_2, and possibly of $C-N$, $C-H$, and $C-C$ bonds, then may occur by a ligand assisted heterolytic splitting as suggested for H_2 with Pd(salen).

8.2.4. Selectivity and Stereospecificity

The high selectivity of $RuH(Cl)L_3$ and $RhH(CO)L_3$ [$L = P(C_6H_5)_3$] for 1-alkenes (cf. Sections 8.2.1 and 8.2.2) generally is ascribed to steric factors. It is assumed [34] that the formation of the intermediate, square-planar alkylmetal species [cf. Eq. (8.38), step a], in which the alkyl group is mutually cis to two trans phosphine ligands, is seriously hindered; in other words, the insertion through a four-center transition state, as shown in formula (8.50), cannot occur readily, except when $R' = H$.

(8.50)

The same conditions appear to apply in the formation of the pentacoordinate alkyl inter-mediate during hydrogenation with $RhClL_3$ (cf. Fig. 8.6). Nevertheless this complex also ef-fectively hydrogenates internal and cyclic olefins. Differences in $M-P$ bond lengths in the octahedral olefinrhodium complex (Fig. 8.6) may be important; alternatively this complex may have a somewhat different configuration from that given in the figure, namely with two phosphine ligands in the *cis* position, and hence providing less of an obstacle to insertion.

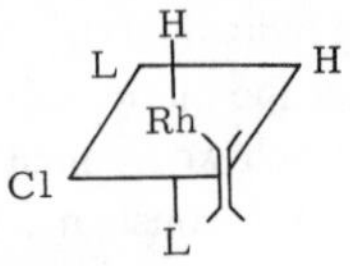

Numerous other homogeneous catalysts have been reported to selectively hydrogenate certain multiple bonds in the presence of others, *e.g.* conjugated double bonds or acetylenic bonds in the presence of isolated double bonds. This feature of many hydrogenations has been reviewed by Lyons *et al.* [37].

Another interesting aspect deals with stereospecificity, and in particular with asymmetric hy-drogenation. In 1968, Horner *et al.* [38a] suggested the use of optically active phosphine ligands, and Knowles and Sabacky [39a] reported, also in the same year, the first example of a homogeneous, catalytic asymmetric hydrogenation of an optically inactive substrate. The latter authors used a Wilkinson type catalyst with a "Horner phosphine", namely $L = (C_6H_5)CH_3(C_3H_7)P^*$, to add hydrogen to α-phenylacrylic acid. The resulting product was optically active.

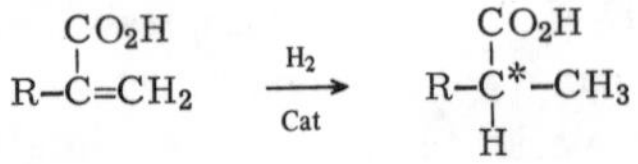

At that time, 5–15% enantiomeric excess (= "optical yield") was considered a remarkable result [38b, 39a]. But since then this type of asymmetric synthesis has been improved greatly. Knowles *et al.* [39b, c] used *o*-anisylcyclohexylmethylphosphine (I) or a related bisphosphine (II) as chiral ligands, and achieved asymmetric hydrogenation of α-acetamido-acrylic acid with 80–96% enantiomeric excess (*cf.* Section 7.6.7).

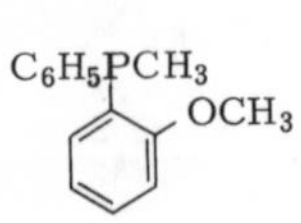

(I) 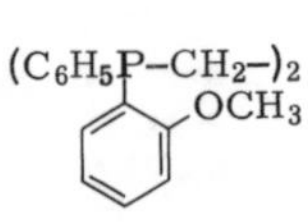   (II)

Another chiral diphosphine ligand frequently used in asymmetric hydrogenation with Wilkinson type rhodium catalysts is 2,2-dimethyl-4,5-bis(diphenylphosphinomethyl)-1,3-dioxolane (DIOP), introduced by Kagan and coworkers [40]:

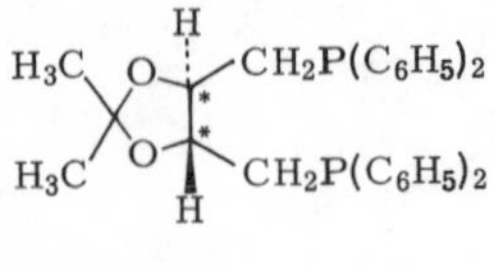

The achievement of asymmetric syntheses with optical yields $> 90\%$ is a breakthrough in a field hitherto reserved for enzymes.

8.3. Oxidation

8.3.1. General Aspects

The catalytic oxidation of ethylene to acetaldehyde (and higher olefins to ketones and aldehydes) in homogeneous aqueous solution, with a palladium/copper catalyst at 20–60 °C, is known as the Wacker process, after the company where it was developed by J. Smidt and coworkers in the late nineteen fifties [41]. Together with hydroformylation (cf. Chapter 10) it is one of the early examples of commercially attractive homogeneous catalysis. In the years since Smidt's discovery, considerable attention has been paid to the reaction mechanism and to the possibilities of influencing the product composition. A recent account by Maitlis [103] contains references to most of the important work in this field.

The actual oxidation process is described by Eq. (8.51). This stoichiometric reaction in which palladium(II) is reduced to atomic palladium, was reported first in 1894 [41a]. However, Smidt and coworkers discovered that palladium(0) can be reoxidized in situ by cupric chloride [Eq. (8.52)]. This fact, combined with the ready reoxidation of cuprous chloride by oxygen or air [Eq. (8.53)], turned the reaction into an industrially important process. The net reaction is air oxidation of the olefin, as shown in Eq. (8.54).

$$PdCl_2 + C_2H_4 + H_2O \longrightarrow CH_3CHO + Pd^\circ + 2\ HCl \tag{8.51}$$

$$Pd^\circ + 2\ CuCl_2 \longrightarrow PdCl_2 + 2\ CuCl \tag{8.52}$$

$$2\ CuCl + 2\ HCl + {}^1/_2\ O_2 \longrightarrow 2\ CuCl_2 + H_2O \tag{8.53}$$

$$C_2H_4 + {}^1/_2\ O_2 \longrightarrow CH_3CHO \tag{8.54}$$

The oxidation of Pd° by Cu^{2+} is favored by the presence of an excess of Cl^- ions. The oxidation potential $Pd^\circ \rightarrow Pd^{2+}$ is lowered considerably by the formation of chloro complexes, e.g. $[PdCl_4]^{2-}$, thus making possible the dissolution of metallic palladium [41a]. The technical process operates at a Cl^-/Pd ratio of ca. 200 to 400 [41b]. Other oxidizing agents (e.g. Fe^{3+}, K_2CrO_7) are applicable in principle, but Cu^{2+} is preferred generally because of its ready reaction with oxygen. For mechanistic work, p-benzoquinone is frequently used as an oxidizing agent.

The Wacker process is applicable to most olefins. In general, α olefins give methylketones (with some aldehydes):

$$RCH = CH_2 \longrightarrow RCOCH_3 \tag{8.55}$$

Substituted olefins with electron-withdrawing groups (e.g. $Y = CN$, NO_2) linked directly to the double bond, add oxygen at the carbon remote from this group, e.g.:

$$YCH = CH_2 \longrightarrow YCH_2CHO \tag{8.56}$$

As we shall see in the next section, the most important step in the catalytic sequence of the Wacker process is the insertion of the olefin into a polar $Pd^{\delta+} - OH^{\delta-}$ bond. Hence, the behavior expressed in Eqs. (8.55) and (8.56) is that predicted by Markownikoff's rule (*cf.* Section 7.4.1).

Olefins with a vinylic halogen or a carboxylic acid group lose this substituent during reaction, and methylketones result, *e.g.*:

$$RCH = CHCO_2H \longrightarrow RCOCH_3$$

$$RCCl = CH_2 \longrightarrow RCOCH_3$$

Cyclic olefins up to cycloheptene give cyclic ketones:

Dienes react with double bond migration:

$$CH_2 = CHCH = CH_2 \longrightarrow CH_3CH = CHCHO$$

$$CH_2 = CHCH_2CH = CH_2 \longrightarrow CH_3CH_2CH = CHCHO$$

Higher olefins require higher temperatures [41a]. They tend to give a number of ketones, due to double bond isomerization. It has been reported that isomerization can be minimized in water dimethylformamide mixtures. Thus dodecanone-2 has been made, at $60\,°C$, in 87% yield and with 96% selectivity, from dodecene-1 [42].

If ethylene is oxidized in acetic acid solvent, vinylacetate is formed, although together with a number of byproducts [103]. There is evidence that in this solvent the oxidation proceeds through insertion of ethylene into an acetoxy-palladium bond, with subsequent β elimination of hydrogen:

$$PdOAc + C_2H_4 \longrightarrow PdCH_2CH_2OAc \longrightarrow PdH + CH_2 = CHOAc$$

8.3.2. The Mechanism of the Wacker Process

Eq. (8.51) describes the overall oxidation of ethylene, neglecting the individual steps taking place at the metal center. Several groups, in particular those of Smidt, Moiseev, Henry, and their colaborators [41, 43, 44] have carried out mechanistic studies in order to elucidate the detailed reaction scheme. Some of their most important conclusions, as well as certain unclarified points are summarized below.

The overall experimental reaction rate is described approximately by Eq. (8.57), showing proportionality to concentrations of catalyst and substrate, and inhibition by H^+ and Cl^-.

$$r_{0x} = -\frac{d\,[C_2H_4]}{dt} = \frac{k\,[PdCl_4^{2-}]\,[C_2H_4]}{[H^+]\,[Cl^-]^2} \tag{8.57}$$

In the initial stage, a very rapid ethylene uptake is observed, followed by slower ethylene absorption. The volume of ethylene initially taken up exceeds that required to saturate the reaction solution with ethylene (as determined in the absence of palladium salt). This points clearly to the formation of an ethylenepalladium complex. The volume taken up in excess of solubility requirements decreases as the Cl^- concentration in the solution increases, but is unaffected by the acid concentration. A reasonable conclusion is that the initial reaction can be represented by the ligand exchange equilibrium, Eq. (8.58):

$$[PdCl_4]^{2-} + C_2H_4 \; \overset{K_1}{\rightleftharpoons} \; [PdCl_3(C_2H_4)]^- + Cl^- \qquad (8.58)$$

Evidently, an excess of chlorine drives this equilibrium to the left. The subsequent slower ethylene uptake follows approximately the rate law Eq. (8.57). If the square-planar complex $[PdCl_3(C_2H_4)]^-$ is assumed to be a key intermediate in the total process, then the additional inhibition by Cl^- {square term in $[Cl^-]$, Eq. (8.57)} and by H^+ can be accounted for by the following two equilibria:

$$[PdCl_3(C_2H_4)]^- + H_2O \; \overset{K_2}{\rightleftharpoons} \; [PdCl_2(H_2O)(C_2H_4)] + Cl^- \qquad (8.59)$$

$$[PdCl_2(H_2O)(C_2H_4)] \; \overset{K_3}{\rightleftharpoons} \; [PdCl_2(OH)(C_2H_4)]^- + H^+ \qquad (8.60)$$

The next and rate-determining step is assumed to be the insertion of the hitherto coordinatively bonded olefin into the $Rd-OH$ bond to give a σ organopalladium intermediate:

$$[PdCl_2(OH)(C_2H_4)]^- \; \overset{k_4}{\underset{slow}{\longrightarrow}} \; [Cl_2Pd-CH_2-CH_2-OH]^- \qquad (8.61)$$

(This last step is sometimes called "hydroxopalladation" of the olefin.) Thereafter, rapid rearrangement and decomposition gives the aldehyde and palladium metal.

$$[Cl_2PdCH_2CH_2OH]^- \; \overset{rapid}{\longrightarrow} \; CH_3CHO + Pd^{\circ} + HCl + Cl^- \qquad (8.62)$$

The details of this rearrangement are still subject to speculation. There is evidence (*vide infra*) that hydrogen transfer from the β-carbon to the α-carbon attached to palladium is involved; and it is assumed that some type of metal hydrogen interaction is entailed in this process:

$$\left[Cl_2PdCH_2\overset{..}{C}H_2OH \right]^- \; \rightleftharpoons \; \left[\begin{array}{c} H_2C=CHOH \\ \vdots \\ Cl_2PdH \end{array} \right]^- \qquad (8.63)$$

$$\longrightarrow \left[Cl_2PdCH(CH_3)OH \right]^- \longrightarrow CH_3CHO + Pd^{\circ} + HCl + Cl^-$$

Some aspects of the final rearrangement, as well as the rate-determining character of the reaction (8.61), are inferred from work with deuterated ethylene. The oxidation of C_2D_4 gives only CD_3CDO, indicating that the rearrangement, Eq. (8.63), goes *via* a hydride shift. This hydrogen transfer would be expected to exhibit a kinetic isotope effect, but the overall reaction rate is essentially the same for C_2H_4 and C_2D_4, showing that no $M-H$ bond breaking is involved in the rate determining step, and that the hydride shift occurs after the slow step of the reaction. These observations point to olefin insertion, Eq. (8.61), as the slow step. Evidently, the isotope effect of β hydrogen transfer cannot be determined by kinetic measurements under these conditions, but it has become accessible by a competitive method, *viz.* by oxidizing CHD = CHD and determining the ratio of CH_2DCDO to CHD_2CHO in the product:

An isotope effect of $k_H/k_D = 1.7$ was measured; this relatively small value is in accord with a concerted reaction [*cf.* Section 7.4.2].

If, based on this evidence, we accept the reaction scheme as given by Eqs. (8.58–62), the overall rate of oxidation may be written:

$$r_{0x} = -\frac{d[C_2H_4]}{dt} = k_4[PdCl_2(OH)(C_2H_4)]^- \qquad (8.64)$$

because all subsequent steps are fast. Substituting the concentration of the olefin(hydroxy)-palladium complex in Eq. (8.64) with the aid of the equilibria (8.58–60), we obtain:

$$r_{0x} = k_4 K_1 K_2 K_3 [H_2O] \frac{[PdCl_4^{2-}]_{eq}[C_2H_4]_{eq}}{[H^+][Cl^-]^2} \qquad (8.65)$$

This equation is formally in agreement with the experimental rate law, Eq. (8.57), with the exception that (8.65) contains equilibrium concentrations of $[PdCl_4]^{2-}$ and C_2H_4. If the rate of oxidation is determined experimentally at constant ethylene pressure, $[C_2H_4]_{eq}$ is given by the solubility of ethylene in the reaction medium and can be determined in the absence of palladium salt. The term $[PdCl_4^{2-}]_{eq}$, however, is a complicated function of $[PdCl_4^{2-}]_0$, the ethylene pressure, as well as of $[H^+]$ and $[Cl^-]$. This function is available from the equilibria (8.58–60), but it yields Eq. (8.65) in an unwieldy form which is useless for experimental verification.

A simplification would have been expected in the extreme case where $[C_2H_4] \ll [Pd]$, and hence $[PdCl_4^{2-}]_{eq} \simeq [PdCl_4^{2-}]_0$. But under these conditions deviation from the relatively simple rate law (8.57) was found [44]. The overall rate is given by a two-term equation:

$$r_{ox} = \frac{k_1[PdCl_4^{2-}][C_2H_4]}{[H^+][Cl^-]^2} + \frac{k_2[PdCl_4^{2-}]^2[C_2H_4]}{[H^+][Cl^-]^3}$$

It was suggested that Eq. (8.57) is a limiting expression valid only for low palladium concentrations. The second term, quadratic in $[PdCl_4^{2-}]$, was interpreted as a consequence of an additional route to acetaldehyde, *via* a dimeric palladium complex.

One further point requires comment. The displacement of Cl^- by HO^- [Eqs. (8.59) and (8.60)], would be expected to involve the Cl^- ligand *trans* to the ethylene ligand because of the strong *trans* effect of coordinated olefins (*cf.* Chapter 7.6.2). On the other hand, the insertion, Eq. (8.61) requires *cis* positioning of olefin and HO^- ligand, according to present knowledge.*) Presumably, a kinetically significant concentration of the *cis* species is provided by a reversible isomerization, which may go *via* a pentacoordinate intermediate, in the course of equilibrium (8.60):

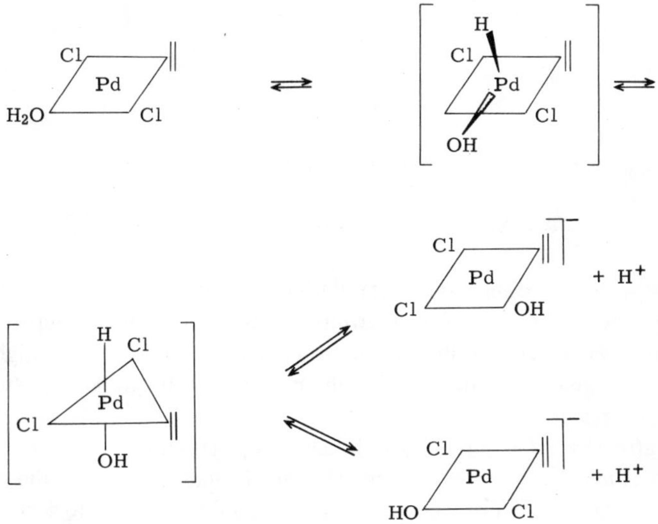

The ability of π back-donating ligands such as C_2H_4 to stabilize pentacoordinate d^8 intermediates is well recognized (*cf.* Section 7.6.2).

The mechanism as given by Eqs. (8.58–62) involves the reaction of coordinated ^-OH with coordinated olefin. An external nucleophilic attack of the coordinated olefin by free ^-OH from the solution has been suggested instead (see Section 7.4.5 and references given there). However, as pointed out by Henry [44], free ^-OH could not exist to any appreciable extent in solution under the acidic conditions of the Wacker reaction. Hence the coordinative mechanism appears more probable.

---

* Recently it was reported [45 a], that a very slow ethylene insertion takes place, if [*trans* PtH(acetone)-L−L]$^+$ [BF$_4$]$^-$ suspended in acetone is treated with ethylene (1.2 atm) during 12 hours; L−L is a rigid, bidentate ligand which, for steric reasons, can occupy only *trans* positions in a square-planar platinum-(II) complex. Hence, this insertion takes place under conditions where hydride and olefin ligand cannot be in *cis* positions in a square-planar complex. Presumably, the olefin coordinates axially prior to insertion. (See also 45 b.)

8.4. Dimerization, Oligomerization and Polymerization

8.4.1. General Aspects

The catalytic cycle in these reactions consists of only two types of individual steps, namely olefin insertion into a $M-R$ bond (*cf.* Section 7.4.1):

$$\overset{\diagdown}{\underset{\diagup}{C}} = \overset{\diagup}{\underset{\diagdown}{C}}$$

$$M-R \quad \xrightarrow{k_p} \quad M-\overset{|}{\underset{|}{C}}-\overset{|}{\underset{|}{C}}-R \qquad\qquad (8.66)$$

where R may be hydrogen or alkyl, and β hydrogen transfer (*cf.* Section 7.4.2):

$$MCH_2CH_2R \quad \xrightarrow{k_{tr}} \quad MH + CH_2 = CHR \qquad\qquad (8.67)$$

The relative rates of these two reactions determine the molecular weight of the product. If $r_p \gg r_{tr}$, a great number of insertion (polymerization, chain propagation) steps will occur, before the growth of the molecule is terminated by a β hydrogen transfer, and high molecular weight polymer is obtained. Where $r_{tr} \gg r_p$, the product will be predominantly dimeric, and for $r_{tr} \simeq r_p$ oligomers are to be expected.

Several features affecting the relative rates of chain propagation and transfer are known. Thus, donor ligands tend to increase, whereas acceptor ligands to decrease the r_p/r_{tr} ratio and hence the molecular weight. For the same metal, a higher oxidation state decreases r_p/r_{tr}. A tentative explanation of these effects has been given in Section 7.6.4. However, the influence of the most important factor, the transition metal itself, has been treated in a more or less empirical way to date. In general, group VIII transition metal centers tend to favor β hydrogen transfer, and most of the dimerization catalysts are from this group. On the other hand, olefin polymerization catalysts are usually derived from group IV to VI transition metals. Some exceptions are given in the following sections.

Numerous catalysts which will dimerize, oligomerize or polymerize olefins are known. Apart from a free site for coordination of the olefin molecule, the catalyst must possess an active ligand R bonded to the metal [*cf.* Eq. (8.66)]. Several procedures exist to favor such an active $M-R$ σ bond:

a. Activation of a stable $M-R$ bond by complex formation with a Lewis acid, in particular with an organoaluminum halide. (For an example, see Fig. 7.14).

b. *In situ* formation (and simultaneous activation) by a transition metal salt and an organometallic (*cf.* Section 7.5.2).

c. Metal hydride formation by oxidative addition of HCl or an organic acid: this procedure appears to be restricted to group VIII d^8 and d^{10} species, *e.g.* Rh(I) [101] or Ni(0) [46].

As may be expected from their relative coordinating abilities (Section 7.2.1), only ethylene and monosubstituted terminal olefins will undergo readily the reactions discussed in the following sections. Cyclic olefins such as cyclohexene or cyclooctene may react slowly, whereas internal or disubstituted olefins do not react, unless they are particularly strained. For unsymmetrical olefins (*e.g.* propylene), monomer insertion according to Markownikoff's rule or to the "anti-Markownikoff" mode (*cf.* Section 7.4.1) gives different products. The introduction of bulky ligands, especially of tertiary phosphines, sometimes permits highly selective formation of a certain isomer. With optically active ligands, even asymmetric synthesis by dimerization of nonoptically active monomers can be achieved.

8.4.2. Dimerization of Olefins

The dimerization of ethylene, propylene and of higher α olefins, as well as their codimerization has been achieved with an amazingly large number of catalytic systems. Some of them are typical Ziegler systems (*cf.* Section 7.5.2), for instance, cobaltous acetylacetonate/ triethylaluminum [47]; in others the active species is formed by oxidative addition of HCl (*e.g.* to Rh(I) [101]). Very often the exact nature of the active species has not been elucidated in the various "recipes" for catalytic systems. The reader is referred to a recent review [104].

By far the most active dimerization catalysts for ethylene and propylene are based on nickel compounds, and in particular on π allylnickel halides. These catalysts first reported and thoroughly investigated by Wilke, Bogdanović, and their coworkers [48], will be discussed in further detail. The catalysts are generally composed of a π allylnickel halide and a Lewis acid (aluminum halide or alkylaluminum halide). The π allylnickel halides are already slightly active on their own, at least at higher temperatures and pressures. However, the addition of the Lewis acid leads to extremely active catalysts which permit the dimerization of ethylene and propylene even at -30 to $-40\,^\circ$C and at subatmospheric pressures, with high reaction rate.

The π allylnickel halides are bridge-bonded dimers. For the Lewis acid activated systems it is assumed that, after opening of the bridge bonds, one electron pair of the halide ion is shared with the aluminum:

$$\left\langle\!\!-Ni\!\!\begin{array}{c}{}^{\cdot\cdot Cl\cdot\cdot}\\{}_{\cdot\cdot Cl\cdot}\end{array}\!\!Ni\!-\!\right\rangle + 2\,AlX_3 \longrightarrow 2 \left\langle\!\!-Ni\!\!\begin{array}{c}Cl\cdot AlX_3\end{array}\right. \tag{8.68}$$

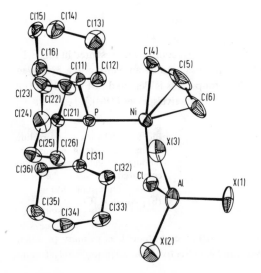

Fig. 8.12. Molecular structure of the complex comprising η^3-allylnickel chloride, tricyclohexylphosphine and methylaluminum dichloride [48c].

This creates free coordination sites around the nickel. If these sites are occupied by electron acceptors (*e.g.* CO, in a 2:1 CO/Ni ratio), the system becomes inactive. If, however, only one site is occupied by a phosphine molecule, the system remains active, and by changing the phosphine, the course of the catalytic reaction may be controlled (*vide infra*). A three di-mensional X-ray crystallographic structure of a complex of this type is shown in Fig. 8.12. The phosphine molecule is coordinatively bound to the nickel center; the Lewis acid mole-cule interacts with the halogen atom of the π allylnickel halide.

It has been shown [48 b] that the first step after addition of olefin to the catalyst system is the displacement of the π allyl group, presumably according to the reaction sequence depicted in Eq. (8.69):

$$L_nNi-\text{⟩} \; + \; \text{C=C} \; \longrightarrow \; L_nNi-C-C=C \; \longrightarrow \tag{8.69}$$

$$L_nNi-C-C-C-C=C \; \longrightarrow \; L_nNiH \; + \; C=C-C-C=C$$

(For the π−σ conversion of the allyl ligand, see Section 7.2.2.) This reaction sequence precedes the actual catalytic cycle [Eq. (8.70)], where the first step is the insertion of a monomer (ethylene or propylene) into a metal-H bond; β hydrogen transfer at this stage regenerates the monomeric olefin and the metal hydride, *i.e.* the first step is reversible and may occur many times before a second growth step takes place. But β hydrogen transfer after the second growth step gives the dimeric product, plus the metal hydride which contin-ues the catalytic cycle.

$$L_nNi-H \; \underset{\pm \; \text{C=C}}{\rightleftharpoons} \; L_nNi-\overset{|}{C}-\overset{|}{C}-H \tag{8.70}$$

The active metal center is assumed to differ from the π allyl complex represented in Fig. 8.12 only in that in the former two coordination sites are occupied by an olefin and a hydrogen atom (or an alkyl group) instead of by a π allyl group. This is depicted in Fig. 8.13. (The positions of hydrogen and olefin might be reversed in the active species [48 d].)

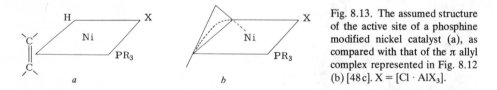

Fig. 8.13. The assumed structure of the active site of a phosphine modified nickel catalyst (a), as compared with that of the π allyl complex represented in Fig. 8.12 (b) [48 c]. X = [Cl · AlX₃].

The dimerization of ethylene according to Eq. (8.70) gives butene-1 as primary product. However, essentially all dimerization catalysts are also effective isomerization catalysts, and in most cases the equilibrated mixture of butenes*) is obtained. It has been stated by Cramer [101] that a $M-H$ bond inserts olefin considerably faster than does $M-R$. Although this has been observed in the particular case of ethylene dimerization with a rhodium catalyst, the rate determining character of the second monomer insertion is probably general in olefin dimerizations.

For dimerization of propylene at a transition metal center M, four reaction modes are conceivable, depending upon the orientation of the monomer (the "regiospecificity") in each of the two insertion steps. They are shown in Fig. 8.14.

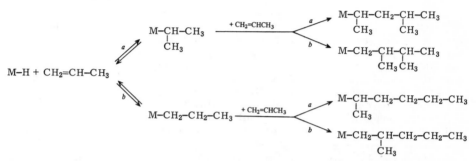

Fig. 8.14. Reaction scheme for the dimerization of propylene at a transition metal center, M.

From Fig. 8.14 it follows that, for instance, the *bb* route gives, on β hydrogen abstraction, 2-methylpentene-1 (2MP1) which may isomerize to give 2-methylpentene-2 (2MP2). The route *ab* gives the two isomeric 2,3-dimethylbutenes (2,3 DMB); *etc.*, hence Fig. 8.14 may be summarized as follows:

route *aa:*   4MP2 (4MP1) $\longrightarrow$ 2MP2
route *ab:*   2,3 DMB
route *ba:*   linear hexenes
route *bb:*   2MP1 $\longrightarrow$ 2MP2

Consequently, the isomers 4MP2, 2,3 DMB, linear C_6 and 2MP1 are each characteristic for one particular route, whereas 2MP2 reflects the tendency of a catalytic system to isomerize but does not distinguish between routes *aa* and *bb*.

The work of Wilke *et al.* with propylene and the catalyst cited above [*cf.* Eqs. (8.68–70)], as well as other work with related nickel catalysts [50] has indicated that, in the absence of special steric effects, the route *aa*, giving 4-methylpentenes, far predominates. Evidently, the *a* mode of monomer incorporation is electronically favored. This mode has been defined in Section 7.4.1 as the "anti-Markownikoff" mode, and a possible explanation for its predominance in certain cases has been discussed there: electron-rich low-valent group VIII metal

---

* The equilibrium concentrations of the three linear butenes have been measured by Benson *et al.* [49] in the range 87–332 °C. At 100 °C, 8% butene-1, 26% *cis*-butene-2, and 66% *trans*-butene-2 are present. Extrapolation to room temperature gives the following approximate values: 4% butene-1, 22% *cis*-butene-2, and 74% *trans*-butene-2. The equilibrium composition at −20 °C has been measured by Bogdanović and Karmann [48 a]: 1.7% butene-1, 19.5% *cis*-butene-2 and 78.8% *trans*-butene-2.

centers appear to reverse the polarization of the olefinic double bond, mainly by electron back-donation into the antibonding olefin π^* orbital. Table 8.5 supports this view in several respects. (See also Ref. [57d].)

Table 8.5. Olefin insertion into metal-R bonds; R = H, alkyl, or aryl. *a:* anti-Markownikoff mode; *b:* Markownikoff mode.

Metal-R	Substrate	T (°C)	Product	Insertion mode	Ref.
Ti(III)−R	propylene	<80	polymer	...*bbb*...	[51]
Ti(IV)−R	butene-1	−70	2-ethylhexene-1 (88%)*)	*bb*	[52]
Ti(IV)−R	pentene-1	−70	2-(*n*-propyl)heptene-1 (90%)*)	*bb*	[52]
Ti(IV)−R	hexene-1	−70	2-(*n*-butyl)octene-1 (90%)*)	*bb*	[52]
Ni(II)−H	propylene	30–40	NiCH(CH$_3$)$_2$ (70–80%)	*a*	[48b]
Co(I)−H	propylene	26	CoCH(CH$_3$)$_2$ (70%)	*a*	[53]
Pd−R	propylene	30	PdCH(CH$_3$)CH$_2$R (84%)	*a*	[54]
Rh(III?)−R	propylene	40	4-methylpentene-2 (60%)	*aa*	[55, 104]
Ir(III?)−R	propylene	40	4-methylpentene-2 (60%)	*aa*	[55]

* Refers to the dimeric fraction of the product.

Insertion into Ti−R bonds (one or no d electrons, no back-donation) occurs according to the Markownikoff rule (*b* mode), whereas for a number of low valent group VIII metals the olefin is incorporated in the "anti-Markownikoff" sense. It should, however, be noted that this tendency is not without exceptions. Thus, the insertion of α olefins into Pd−OH bonds in aqueous solution during the Wacker process (Section 8.3.1) generally follows Markownikoff's rule. Presumably the complete absence of donor ligands impedes electron back-donation to the olefin in this case [*cf.* Eq. (8.61)]. Markownikoff insertion of propylene into the Pt−H bond of the complex *trans*-PtHCl[P(C$_2$H$_5$)$_3$]$_2$ also has been reported [56]; however in this case possible steric hindrance arising from the two phosphine ligands could inhibit formation of any isopropyl groups [57d].

Quite an important steric ligand influence of this type has been revealed by Wilke *et al.* during the dimerization of propylene with their π allylnickel-based catalyst system [Eqs. (8.68–70)]. These authors have shown that very bulky phosphine ligands such as C$_2$H$_5$P-(*t*-C$_4$H$_9$)$_2$ or (*i*-C$_3$H$_3$)$_3$P can force the reaction to adopt the *ab* route, giving predominantly (>70%) 2,3-dimethylbutene [48b]. With (*i*-C$_3$H$_7$)$_2$P(*t*-C$_4$H$_9$), even 96% selectivity in favor of this dimer was attained, although at a lower temperature (−60°C) [48d]. To explain this unusual reaction sequence, it is assumed that, after the first monomer insertion into a Ni−H bond according to the electronically favored *a* mode, the combined bulk of the newly formed isopropyl group plus the sterically demanding phosphine ligand provides a matrix which permits the second monomer molecule to become incorporated in one way only, namely according to the *b* mode.

The 2,3-dimethylbutenes resulting from the *ab* reaction sequence are interesting intermediates; they can be hydrogenated to 2,2-dimethylbutane, which has good antiknock properties, or dehydrogenated to 2,3-dimethylbutadiene, which can be used for polymerizations, and for Diels-Alder reactions, *etc.* [48b].

Very high selectivity ($>90\%$) *via* the other unusual reaction sequence, namely *ba*, has been found for certain palladium catalysts [55, 57]. As an example, Pd(II)acetylacetonate, in the presence of an alkylaluminum and a phosphine gives a catalyst system which dimerizes propylene predominantly to linear hexenes, although at a relatively low rate [57c]. Again, the particular structure of the phosphorus containing ligand is critical, as shown in Table 8.6.

Table 8.6. Dimerization of propylene with Pd/Al catalysts in the presence of phosphorus-containing additives. $[Pd(acac)_2] = 2 \times 10^{-3}$ mol l^{-1}; $[C_2H_5AlCl_2] = 22 \times 10^{-3}$ mol l^{-1}; $[P]/[Pd] = 2$; [propylene] = 1.3 mol l^{-1}; $T = 20\,°C$; $t = 3$ h. [57c].

Phosphorus additive	Conversion [%]	Selectivity [%]		
		linear hexenes	methyl- pentenes	dimethyl- butenes
$(n\text{-}C_4H_9)_3P$	4	95.1	4.9	–
$(C_6H_5)_3P$	4	81.8	18.2	–
$(C_6H_5O)_3P$	12	23.6	68.2	8.2

Evidently the formation of linear hexenes by the *ba* sequence is also the consequence of a steric phenomenon, because electronic tailoring of the catalyst center would influence both steps in the same manner. It is assumed that, also in this system, the first step occurs predominantly according to the electronically favored *a* mode, but that the combined bulk of the resulting isopropyl group plus the surrounding ligands impedes the second step. Because of the reversibility of the first step the less favorable (slower) *b* mode can thus gain importance. The incorporation of the first monomer gives the less bulky n-propyl group which permits the second step to proceed according to electronic preferences, *viz.* by route *a* (*cf.* Fig. 8.14). Comparison of the data obtained with tributyl- and triphenylphosphine (Table 8.6) underlines the steric character of the phosphine influence. Differing considerably in the electron donor ability, the two phosphines have comparable bulkiness. Reaction rate and isomer distribution are similar. The low reaction rate implies that the catalyst spends most of its time in the inactive form attached to the isopropyl group. The less bulky phosphite ligand, on the other hand, is not able to build up an efficient barrier.

The P/Pd ratio is also critical; best results are obtained with P/Pd = 2; at lower ratios inactive, black colloidal Pd^0 is formed even at low alkylaluminum concentration. At P/Pd ≥ 5 no catalytic activity is observed, presumably due to the blocking of free sites.

Relatively high Al/Pd ratios are required, indicating that the catalytic species is formed in an equilibrium. Activity is found for $5 \leq$ Al/Pd ≤ 25, with an optimum at *ca.* 20. In this region, an absorption peak at 25,600 cm^{-1} is observed, suggesting the presence of an alkylaluminum in the complex.

Based on these data, and by analogy with other Ziegler type catalyst systems (see Fig. 7.14), the following structure has been suggested for the active site:

A series of steps must be postulated in order to account for the formation of this species, but analogous reaction sequences are common during the *in situ* formation if Ziegler type catalysts (*cf.* Section 7.5.2): the acetylacetonate ligands are displaced by chlorine ligands (from the ethylaluminum chloride); subsequent exchange of one chlorine with an ethyl group forms an alkylpalladium derivative which, on β hydrogen abstraction, produces the hydride. The other chlorine is used in the aluminum bridge. The phosphines keep the complex in solution [57c].

Unfortunately this linear dimerization cannot be applied to other olefins (*e.g.* butene-1 → linear octenes) because the palladium catalyst very effectively isomerizes the terminal to internal olefin; slow reaction and branched dimers are the necessary consequence.

An interesting recent development in the field of catalytic olefin dimerization should be noted, namely the asymmetric synthesis involving C−C bond formation (as opposed to asymmetric hydrogenation, Section 8.2.4) by codimerization of norbornene or norbornadiene with ethylene, as reported by Wilke, Bogdanovic *et al.* [48c]. The nickel catalyst system [Eq. (8.68)] was modified with phosphines containing chiral carbon atoms [*e.g.* (−)dimenthyl(isopropyl)phosphine]. It was found that norbornene and ethylene codimerize in chlorobenzene to give optically active *exo*-(+)-2-vinylnorbornane as the predominant primary product. Ethylene added only from the sterically less hindered side of the norbornene C = C double bond:

Part of the product isomerizes to 2-ethylidenenorbornane. In this process the chiral center at C^2 is lost, but those at C^1 and C^4 remain, thereby retaining the total chirality of the molecule. Due to the fact that the norbornene double bond is activated by ring strain, this codimerization can be effected at very low temperature. The optical yield increases as the temperature is lowered; at −97 °C an optical yield of 80.6% has been obtained.

At present, the choice of optically active ligands is empirical, but there is optimism that with time more experimental data may enable correlations to be made between structure, catalyst configuration and the optically active reaction product. With the help of such knowledge it might become possible in the future to prepare "tailor-made" asymmetric catalysts [48c].

8.4.3. Oligomerization and Polymerization of Olefins

Catalysts for the oligomerization and polymerization of olefins are derived generally from group IV−VI transition metals, in particular from Ti, V, Cr and Zr. Most significant soluble catalysts are Ziegler type systems (*cf.* Section 7.5.2). We shall discuss in this Section several Ti/Al systems made up from Ti(IV) compounds and alkylaluminum halides. As a working hypothesis, we shall adopt the following general structure for the catalysts under consideration:

with R = alkyl or a growing chain, L = variable ligands, and □ = vacant coordination site (*cf.* Sections 7.5.2 and 7.5.3). The three-center bridge bonds between titanium and aluminum have been proved (by X-ray crystallography [58]) only for certain Ti(III)Al complexes derived from catalysts by thermal decomposition, as described in Section 7.6.3, Eq. (7.24). Nevertheless, such bridge bonds are usually assumed to be present also in the catalytically active Ti(IV) species which are not stable enough to be investigated by X-ray structural analysis.

The relative rates of chain growth and chain transfer, r_p/r_{tr}, and hence the molecular weight of the product, may be varied considerably according to the ligands used. Table 8.7 shows the catalytic reaction of ethylene with a series of Ziegler catalysts, with varying ligands bonded to titanium as well as to the aluminum component. A rough fractionation between oligomer (soluble in benzene) and polymer (precipitating from the reaction medium) is reported.

Table 8.7. Oligomerization and polymerization of ethylene with Ti/Al systems in benzene. $T = 5\,°C$; ethylene pressure, 1 atm [105].

Titanium compound	Aluminum compound	Oligomer (%)	Solid polymer (%)
$TiCl_4$	$C_2H_5AlCl_2$	92	8
$(C_2H_5O)TiCl_3$	$C_2H_5AlCl_2$	77	23
$(C_2H_5O)_3TiCl$	$C_2H_5AlCl_2$	56	44
$(C_2H_5O)_4Ti$	$C_2H_5AlCl_2$	31	69
$TiCl_4$	$(C_2H_5)_2AlCl$	30	70
$(C_2H_5O)TiCl_3$	$(C_2H_5)_2AlCl$	10	90

The average molecular weight of the product clearly increases with the number of donor group ligands in the system. Interestingly, substitution of one chlorine by ethyl at aluminum has the same effect on the molecular weight as has the substitution of four chlorines by ethoxy groups at the titanium. (See Section 7.6.2 for the relative donor capacities of C_2H_5O, C_2H_5 and Cl.)

Commercially, the most attractive result arising from ethylene oligomerization is evidently the production of linear low molecular weight α olefins which have become very important industrial intermediates during recent decades. This is due partly to recent developments in homogeneous catalysis, such as the Wacker process (*cf.* Section 8.3) or the Oxo process (Chapter 10), and partly to increased demand for biodegradable detergents, for plasticizers and for rubber or gasoline additives. However, the goal of making linear products is seriously handicapped: the α olefins formed during ethylene oligomerization are able to compete with ethylene for the coordination sites at the transition metal center. Although the coordinating ability falls with increasing molecular weight, the butene-1 accumulating in the reaction solution with higher conversion becomes a serious competitor for ethylene, especially at higher temperatures, where the solubility of ethylene in the reaction medium is relatively low.

The incorporation of α olefins into the growing chain leads to branched molecules; in most cases it limits the useful conversion to some 10–15 moles of ethylene per liter of reaction solution. Moreover, if β hydrogen transfer takes place after such a copolymerization step, unwanted vinylidene end groups form [Eq. (8.71), formulated for the incorporation of butene-1]:

$$TiR + CH_2=CHCH_2CH_3 \longrightarrow TiCH_2CHRCH_2CH_3$$
$$\longrightarrow TiH + CH_2=CRCH_2CH_3 \tag{8.71}$$

Apart from more trivial ways such as keeping the conversion low and the ethylene concentration high, a potential method for preventing incorporation of α olefins is to take advantage of the higher coordination ability of ethylene, increasing the electron density at the metal center by varying systematically the ligands to the extent that, hopefully, only ethylene is able to coordinate.

Thus, donor ligands would tend to maintain the branching at a low level but, as shown in Table 8.7, they also tend to increase the molecular weight. Fortunately, temperature is a variable acting upon the molecular weight without changing essentially the electronic conditions at the metal center. Since the activation energy of β hydrogen transfer is lower than that of chain propagation [105], the molecular weight decreases upon lowering the temperature. Table 8.8 shows an attempt to optimize the variables temperature ($-20\,°C$), ethylene pressure, and ligands, thus leading to low molecular weight linear α olefins.

Table 8.8. Oligomerization of ethylene [59]. $[Ti] = 20 \times 10^{-3}$ mol l^{-1}; $[C_2H_5AlCl_2]/[Ti] = 5$; $T = -20\,°C$; solvent: toluene.

Ti Compound	Ethylene pressure (atm)	Conversion mol l^{-1}	Oligomer (%)	α-Olefins[*] (%)
TiCl$_4$	6	10.8	100	34
(C$_2$H$_5$O)$_3$TiCl	3	13.5	96	50
(C$_2$H$_5$O)$_3$TiCl	6	10.6	99	85
(C$_2$H$_5$O)$_3$TiCl	12	13.0	99	93

* referred to oligomer.

Another example of olefin oligomerization, namely that of propylene, with Ziegler type catalysts may serve to emphasize the possibilities of "catalyst tailoring". Usually, oily or waxy propylene oligomers are prepared by cationic mechanisms [60]. The very reactive growing chain end (carbonium ion) tends to react not only with monomer, but also with tertiary hydrogen atoms of the polymer chain (hydride transfer). The consequence is a highly branched product. Table 8.9 shows a series of experiments made with the aim of preparing unbranched low molecular weight polypropylene [61].

Table 8.9. Ligand influence on the linearity of low molecular weight polypropylene [60]. $T = 5\,°C$; solvent: benzene.

Titanium component	Aluminum component	Molecular weight	linear growth (%)
$TiCl_4$	$C_2H_5AlCl_2$	516	28
$(C_2H_5O)_3TiCl$	$C_2H_5AlCl_2$	340	49
$(C_2H_5O)_4Ti$	$C_2H_5AlCl_2$	432	87
$TiCl_4$	$(C_2H_5)_2AlCl$	557	70
$(C_2H_5O)TiCl_3$	$(C_2H_5)_2AlCl$	1100	94

The extraordinary tendency of propylene to undergo cationic polymerization is well known. As a consequence, catalytic systems which have a certain acidic character, as $TiCl_4/EtAlCl_2$, yield oligomers of irregular structure, comparable to those obtained with, for instance, the purely cationic catalyst $AlCl_3$. However, stepwise replacement of chlorine by donor ligands gradually improves the linear growth. (Again donor ligands attached to aluminum have a stronger effect than those joined to titanium; cf. Table 8.7.) One may interpret this finding in the following manner. In the relatively "acid" system $TiCl_4/EtAlCl_2$, polymerization occurs by a more or less free cationic mechanism, i.e., the chain end is positively charged, and has a negative counterion. Decreasing the electron affinity of the Ti(IV) compound, i.e., diminishing its "acidity", cationic chain initiation is gradually depressed in favor of monomer insertion into a metal-alkyl or metal-hydride bond. As a consequence, the site of chain growth changes from the solution (in free cationic catalysis) to the coordination sphere of titanium.

A kinetic study of the detailed mechanism of olefin oligomerization, and in particular of the β hydrogen transfer step, has been carried out with ethylene and the catalyst system $(C_2H_5O)_3TiCl/C_2H_5AlCl_2$ [105]. This system has been shown (Table 8.8) to give, under certain restricted reaction conditions, essentially (or predominantly) linear α olefins. In this case, and in the absence of chain termination (constant overall rate), only two reactions have to be considered in the scheme, namely chain propagation and chain transfer:

$$\begin{array}{c} H_2C{=}CH_2 \\ \vdots \\ MCH_2CH_2R \end{array} \longrightarrow MCH_2CH_2CH_2CH_2R \qquad (8.72)$$

$$MCH_2CH_2R \longrightarrow MH + CH_2{=}CHR$$
$$\big\downarrow {\scriptstyle +\,C_2H_4}$$
$$MCH_2CH_3 \qquad (8.73)$$

The rate of chain propagation r_p may be formulated as:

$$r_p = k_p[P^*]\,[M] \qquad (8.74)$$

where [P*] and [M] are the concentrations of active sites and of monomer, respectively. This rate law is generally assumed to apply to soluble Ziegler-type catalysts.

If the β hydrogen transfer were to be a monomolecular reaction involving only the metal center with the growing chain attached, the rate would be given by:

$$r_{tr} = k_{tr} [P^*]$$

and, from standard kinetics [105] it would follow that the number average degree of polymerization, P_n, would be:

$$P_n = r_p/r_{tr} = k_p [M]/k_{tr}$$

Experimentally, however, it has been found that P_n does not depend upon the monomer concentration, but only on the temperature. The conclusion is that monomer is involved in the rate determining step of the chain transfer [cf. Eq. (8.73)]:

$$r_{tr} = k_{tr} [P^*] [M] \tag{8.75}$$

This means that monomer is used up not only by chain propagation but also in the transfer step. Hence, the overall rate of monomer consumption is [cf. Eqs. (8.74) and (8.75)]:

$$-d [M]/dt = r_p + r_{tr} = (k_p + k_{tr}) [P^*] [M]$$

The number-average degree of polymerization is, under these conditions, given by Eq. (8.76), in accordance with the experimental findings.

$$P_n = \frac{r_p + r_{tr}}{r_{tr}} = \frac{k_p}{k_{tr}} + 1 \tag{8.76}$$

Furthermore, very low values have been reported for Arrhenius activation energies and steric factors of the β hydrogen transfer, as compared with similar β hydrogen abstraction (see Table 8.10), indicating a mechanism different from those leading to HCl or AlH.

Presumably, the alkyl group (growing chain) is highly polarized even in the ground state under the influence of the transition metal. It is suggested that a six-center, cyclic, highly polar transition state including the monomer could account for the unusual activation parameters as well as for the rate law Eq. (8.75):

This transition state has been discussed in more detail in Sections 7.4.2 and 7.6.4 [see Eq. (7.7)].

Table 8.10. Arrhenius parameters of several β hydrogen abstraction reactions.

Reaction	E_A (kJ/mol)	log A	Ref.
$\begin{array}{cc} Ti & H \\ \mid & \mid \\ -C-C- \\ \mid & \mid \end{array} \longrightarrow TiH + \begin{array}{c} \diagdown \diagup \\ C=C \\ \diagup \diagdown \end{array}$	≈25	≈3	[105]
$\begin{array}{cc} Al & H \\ \mid & \mid \\ -C-C- \\ \mid & \mid \end{array} \longrightarrow AlH + \begin{array}{c} \diagdown \diagup \\ C=C \\ \diagup \diagdown \end{array}$	84–125	10–12	[65]
$\begin{array}{cc} Cl & H \\ \mid & \mid \\ -C-C- \\ \mid & \mid \end{array} \longrightarrow ClH + \begin{array}{c} \diagdown \diagup \\ C=C \\ \diagup \diagdown \end{array}$	209–222	13–14	[66]

So far in this section we have treated oligomerization rather than polymerization. Actually, the technically important catalysts for the production of high molecular weight polymers from simple monoolefins such as ethylene, propylene *etc.*, are heterogeneous. Even the classical Ziegler system, $TiCl_4 + (C_2H_5)_3Al$, which is composed of two soluble components, becomes heterogeneous immediately upon mixing the two components, due to reduction of Ti(IV) to unsoluble Ti(III), and partly Ti(II). Other important heterogeneous catalysts for the polymerizations under discussion are transition metal oxides, in particular chromium oxide (Phillips catalyst). Although much of the general knowledge gained with soluble catalysts probably applies also to the heterogeneous systems, the latter are not made up of well defined individual transition metal complexes, and hence do not form part of the subject matter of this book.

However, certain soluble systems based on bis(cyclopentadienyl)titanium dichloride and bis(cyclopentadienyl)ethyltitanium chloride, in combination with an alkylaluminum, are known to be catalysts for ethylene polymerization [62–64]. From a technical point of view, these are poor catalysts, because they decompose thermally to paramagnetic, soluble Ti(III) species which, in this case, are inactive. From a scientific point of view however, these systems have been of great utility in the elucidation of many mechanistic details of the polymerizations in particular, and of homogeneous catalysis with transition metal complexes in general [64]. Much of this principal knowledge has been treated in Chapter 7, and will not be repeated here.

8.4.4. Reactions of Polar Monomers

It should be born in mind that complex formation with an alkylaluminum halide is not the only way by far to activate a M−H or M−R bond for the insertion of one or more monomer molecules. It has proved to be merely the most effective and convenient way for simple, unsubstituted olefins.

Substituted, polar monomers such as acrylonitrile, methyl methacrylate, *etc.*, on the other hand, may react also with other transition metal hydride or alkyl complexes. Thus, acrylonitrile has been dimerized by Misano *et al.* [67] to 1,4-dicyanobutene-1, with $(CH_2 = CHCN)_3RuCl_2$ as catalyst $(150\,°C, 20$ atm. H_2 pressure):

$$2\,CH_2 = CHCN \xrightarrow{Cat.} NCCH = CHCH_2CH_2CN$$

The polymerization of acrylonitrile, methacrylonitrile and methyl methacrylate at room temperature has been achieved by A. Yamamoto *et al.* [68a] using complexes such as $(C_2H_5)_2Fe(dipyridyl)_2$, $CoH(N_2)[P(C_6H_5)_3]_3$, or $Ru(H)H[P(C_6H_5)_3]_4$. Copolymerization of acrylonitrile with methyl methacrylate [68b] or methyl acrylate [68c] has proved that the polymer is not generated by a free radical or ionic mechanism, because the composition of the copolymer is quite different. One of the neutral ligands is assumed to dissociate thus providing a site for monomer coordination. The molecular weights obtained, *ca.* 10^5, are quite high. No β hydrogen transfer takes place, but spontaneous termination renders the metal complex inactive. This last step is still ill-defined.

On the other hand, polymerization of a polar monomer (vinylidene chloride) with a "modified Ziegler system" $(TiCl_4/(C_2H_5)_2Al(OC_2H_5)/pyridine)$ at $25\,°C$ has been reported to take a free radical course [69].

8.5. Metathesis (Disproportionation)

8.5.1. General Aspects

One of the most remarkable catalytic reactions is olefin metathesis*) in which the net transformation comprises complete scission of two carbon-carbon double bonds with simultaneous formation of two new double bonds or, in other words, a redistribution of alkylidene moieties:

$$
\begin{array}{c}
RHC = CHR' \\
+ \\
RHC = CHR'
\end{array}
\underset{}{\overset{Cat.}{\rightleftarrows}}
\begin{array}{c}
RCH \\
\| \\
RCH
\end{array}
+
\begin{array}{c}
R'CH \\
\| \\
R'CH
\end{array}
\tag{8.77}
$$

The reaction was first reported in 1964 by Banks and Bailey [71], who used supported molybdenum and tungsten catalysts at elevated temperature $(100–300\,°C)$. Application to propylene gives the expected olefins, butene-2 and ethylene:

$$2\,CH_2 = CHCH_3 \rightleftarrows CH_3CH = CHCH_3 + H_2C = CH_2 \tag{8.78}$$

With higher olefins, product distribution indicates that double bond migration is also important. The name "olefin disproportionation" was first chosen by the inventors for this process, and is still used by some authors, although it was stated later [72b] that metathesis is a less ambiguous and more appropriate term.

---

* From the Greek "μετατίθεγαι": to transpose, to change. In chemistry, the term is used to denote the interchange of atoms or groups of atoms between two molecules, the structure of the molecules being not otherwise altered [70]. Some authors prefer the anglicized Latin translation of metathesis, *i.e.* dismutation.

Shortly afterwards, Calderon *et al.* [72] discovered that soluble, modified Ziegler catalysts based upon tungsten salts (*e.g.* $WCl_6/C_2H_5AlCl_2/C_2H_5OH$) would bring about metathesis of acyclic olefins at room temperature, in benzene solution, at remarkable rates. These and related systems have been investigated widely in the meantime and, as so often with transition metal catalysis, have provided more insight into the mechanistic aspects of metathesis than the heterogeneous systems.

In the early nineteen sixties, Natta, Dall'Asta *et al.* [72, 108] discovered the "ring-opening polymerization" of cyclic olefins, yielding polyalkenamers with the same type of Ziegler catalysts:

$$p\ (CH_2)_n \overset{\displaystyle\overset{CH}{\|}}{\underset{\displaystyle CH}{}} \longrightarrow [-CH{=}CH(CH_2)_n]_p$$

Calderon *et al.* [72b] recognized that this important reaction proceeds according to the same principles as does metathesis, and it was assumed that in this case chain growth *via* macrocycles of increasing size was involved. Formally this may be depicted as follows (*cf.*, however, Section 8.5.6):

$$(CH_2)_n \overset{CH}{\underset{CH}{\|}} + \overset{HC}{\underset{HC}{\|}} (CH_2)_n \rightleftharpoons (CH_2)_n \overset{HC=CH}{\underset{HC=CH}{}} (CH_2)_n$$

$$(CH_2)_n \overset{HC=CH}{\underset{HC=CH}{}} (CH_2)_n \rightleftharpoons (CH_2)_n \overset{HC=CH}{\underset{HC\quad CH}{\underset{\|\quad\|}{}}} (CH_2)_n \tag{8.79}$$

Cyclic oligomers containing up to 120 monomer units actually have been found in the low molecular weight fraction of a polyoctenamer [74]. Proof of macrocyclic structure for the high molecular weight polyalkenamers is lacking due to obvious analytical difficulties. Moreover, even traces of an acyclic olefin, present as an impurity in the monomer or deriving from the decomposition of the organometallic catalyst, would transform macrocycles into linear polymers (*cf.* Section 8.5.5). Hence, high molecular weight polyalkenamers are generally considered to be open chain macromolecules.

8.5.2. The Catalysts

The original heterogeneous catalysts reported by Banks and Bailey [71] consisted either of molybdenum oxide supported on alumina or were prepared by impregnating preactivated alumina with solutions of molybdenum (or tungsten) hexacarbonyl in cyclohexane with subsequent removal of solvent. The hexacarbonyl catalysts have only minimal activity until subjected to mild thermal activation *in vacuo*. From X-ray photoelectron spectra of such catalysts it was concluded [75] that activation is accompanied by loss of CO ligands, that the active molybdenum species are noncarbonyl containing entities with an oxidation number greater than zero, but less than six, and that these active species are attached to electron withdrawing sites on the alumina.

A number of heterogeneous catalysts are now known, based not only on Mo and W, but also on other metals such as Re, V, Te, and using a variety of high surface area supports such as SiO_2, or MgO. But MO_3 and WO_3 formulations still appear to be the most effective ones [107]. Several modifications also have been reported; thus the incorporation of minor amounts of alkali or alkaline earth metal ions reduces double bond migration, and treatment of the catalyst with HCl increases the activity in certain cases [107]. Enhanced activity has also been reported for a molybdenum catalyst which was treated at 500 °C in a CO stream [76]. Magnetic measurements allowed correlation of the activity with the amount of reduced [Mo(V)] species in the catalyst. The addition of TiO_2 brought about a further activity improvement, so that the heterogeneous metathesis of linear olefins could be carried out readily at ambient temperature [1]. Differences in the Mo(V) EPR signal indicated that the environment of the molybdenum centers changes on addition of TiO_2. This observation as well as the strong synergistic effect of TiO_2 led to the tentative formulation of the active centers as bimetallic surface complexes of Mo(V) and Ti(IV), presumably with oxygen bridges:

The most effective soluble catalysts are based on tungsten halides, but Mo, Re and Ta also yield quite active systems. Another group of metals, comprising Ti, Zr, V, Nb and some lanthanides, have been reported to catalyze metathesis in suitable combinations with an organometallic, but they generally yield only poorly active systems. Several group VIII metal complexes, generally in the absence of an alkylaluminum or other organometallic compound, have been reported as catalysts, but also of low activity [108].

As already mentioned, most soluble catalysts are of the Ziegler type, i.e. they involve interaction of the transition metal halide with an alkylaluminum halide. Many of the early catalyst combinations caused considerable Friedel-Crafts type side reactions due to the strong Lewis acid character of, e.g., WCl_6 and $C_2H_5AlCl_2$; these side reactions include alkylation of aromatic solvents by the olefinic substrate, branching and, with cyclic olefins, crosslinking and other transformations leading to partial loss of double bonds. In particular the so-called alkyl-free catalysts ($WCl_6/AlBr_6$) for the ring opening polymerization [77] were shown by Höcker et al. [78] to operate predominantly according to a cationic mechanism with olefinic double bond opening, rather than proceeding via the metathesis mechanism. Only after the addition of a strong alkylation agent, e.g. C_4H_9Li, are polyalkenamers obtained.

These earlier catalysts suffered also from partial insolubility in the reaction medium, instability of the active species, and hence reproducibility was unsatisfactory. Considerable improvement has been achieved by several groups of workers. Dall'Asta et al. [108] introduced three-component catalyst systems, adding oxygen-containing compounds such as alcohols, phenols, water or hydroperoxides to W/Al systems; an important factor is the reaction of the additive with the transition metal compound prior to the addition of the organometallic component. Presumably, the additive plays a multiple role. In some cases, complex formation with the donor ligand will provide solubility or stability, in others ligand exchange (e.g.

Cl *versus* OC_2H_5) may take place. In any case the additive depresses the Friedel-Crafts activity, at least partially. Zuech *et al.* [79] used less acidic W(II) and Mo(II) complexes such as $L_2(NO)_2MCl_2$ (M = W, Mo; L = triphenylphosphine or pyridine), in combination with alkylaluminum halides, thus making further additives unnecessary, but the metathesis reaction rate is markedly slowed. Pampus *et al.* [80] disclosed the system $WCl_6/(C_2H_5)_2O/$ $(C_2H_5)_4Sn$ which appears to be one of the most stable so far investigated. Moreover interference from side reactions of the Friedel-Crafts type is minimized.

In all cases it is assumed that the active catalyst contains one or several reduced metal species. The reduction is thought to be brought about by the organometallic component. The alkylaluminums are considered generally to serve a dual function by analogy with other Ziegler catalysts (*cf.* Section 7.5.2). Apart from their action in alkylation and reduction of the transition metal center, they are thought to form part of the active complex [72b, 81]. In Pampus' system, the two functions appear to be separated into alkylation [$(C_2H_5)_4Sn$, unlike alkylaluminums, has no empty orbital available for easy complex formation], and complexation (ether).

8.5.3. Thermodynamic Considerations

The metathesis of acyclic olefins [Eq. (8.77)] is thermoneutral *i.e.* the reaction enthalpy is negligible because analogous bonds are broken and remade. Therefore this type of metathesis leads to an entropy-determined equilibrium. In the absence of steric hindrance and double bond shifts, a statistical distribution of all possible combinations of alkylidene fragments should be expected. Actually, in propene metathesis [Eq. (8.78)], an equilibrium characterized by a 2:1:1 molar ratio of propylene:ethylene:butene-2 is achieved readily with heterogeneous catalysts [71]. For higher olefins, more products are found generally because of double bond migration. With certain soluble systems, however, double bond shift is negligible, and statistical primary product composition is obtained also with higher olefins [72b, 82b]. If steric hindrance becomes important, the equilibrium may be shifted strongly in favour of the less hindered olefin.

In ring opening polymerization the situation is somewhat different [108]. Although the number and type of bonds do not change during polymerization, these reactions are favoured by an enthalpy term due to release of ring strain. This term is relatively large (>20 kJ/mol) in four- and five-membered rings, and smaller (12–20 kJ/mol) in higher-membered cycles, with the exception of cyclohexene which is essentially strainfree. The entropy terms involved in ring opening polymerization are somewhat complicated. To a first approximation one would expect reactions as depicted in Eq. (8.79) to be antientropic (diminution in the number of molecules), but for larger rings (>C_7) there is a predominant, favorable contribution from torsional and vibrational entropy [73c]. Thus metathesis of cyclopentene is strongly enthalpy favored, although slightly antientropic; cyclohexene cannot be polymerized by this mechanism (enthalpy zero, entropy unfavorable), whereas the metathetic polymerization of cyclic olefins >C_7 is enthalpy as well as entropy favored.

8.5.4. Metathesis Stereospecificity

Thermodynamically controlled *cis-trans* equilibra of all component olefins are obtained in the course of acyclic olefin metathesis, irrespective of whether pure *cis*, pure *trans*, or a mixture of both isomers is used as starting material. The *trans* content in equilibrated mixtures is *ca.* 80–85% at 0–20 °C [82a, 49]. Pure *trans* olefins react, of course, more slowly, because of their lower coordinative ability. It remains an open question as to whether or not the metathesis step itself provides for equilibrium concentrations of *cis* and *trans* isomers [72b], or if the reaction is stereospecific, and equilibration is the result of an independent, simultaneous isomerization [82a].

In ring opening polymerization the situation is again different. Several cycloolefins, and particularly cyclopentene, may be polymerized with a high degree of stereospecificity. Thus, $MoCl_5/(C_2H_5)_3Al$ converts cyclopentene (without solvent) into a polypentenamer containing more than 99% *cis* double bonds. This is the most stereospecific ring opening polymerization yet reported [73d]. The catalyst $WCl_6/CHOH(CH_2Cl)_2/(i-C_4H_9)_3Al$, on the other hand gives 80–85% of *trans* double bonds [83]. Control of the *cis/trans* ratio may often be achieved by varying the molar ratio of catalyst components, or the temperature [83, 108]. Thus, the system $WCl_6/(C_2H_5)_4Sn/(C_2H_5)_2O$ [80] produces polypentenamer with 80% *trans* double bonds at room temperature, if the catalyst components are brought together at room temperature. This catalyst is inactive at -30 °C. If, however, the catalyst is prepared at -30 °C and the polymerization is carried out at this temperature, a product with 92% *cis* double bonds is obtained. The conclusion is that two different active species are effective in the two cases, which have been formulated tentatively as $W(IV)[(C_2H_5)_2O]Cl_4$ and $W(V)$ $(C_2H_5)Cl_4$ respectively. Presumably the W(V)alkyl complex is stable only at $T \leq -30$ °C, and decomposes homolytically at room temperature. The *cis* polymer is isomerized readily to the thermodynamically favored *trans* isomer with the *trans* producing catalyst, but not *vice versa*. The formation of *cis* polypentenamer at -30 °C must arise from steric effects, the ligands of the respective catalyst forming a matrix which permits incorporation of the monomer into the polymer only in the *cis* producing manner. Presumably the *trans* polymer cannot become coordinated to the *cis* active species.

8.5.5. Practical Applications

The reaction indicated in Eq. (8.78) has been used industrially since 1966 to convert propylene into polymerization grade ethylene and high purity butenes, with heterogeneous catalysts (Triolefin Process [84]). Metathesis of short chain olefins, with reiteration of the cycle: double bond shift to the outer positions of the hydrocarbon chain, separation of α olefins and metathesis, provides a method for increasing molecular weight, and may become attractive for commercial production of long chain olefins [85]. Application of metathesis to olefins with masked functional groups may provide an interesting route to bifunctional compounds; *e.g.* metathesis of methyl esters of certain unsaturated fatty acids has been achieved, giving alkenes and dicarboxylic acid dimethyl esters, under mild conditions, with soluble catalysts [86].

The polymerization of cyclic olefins *via* metathesis is limited by the fact that only a few starting materials are available at a commercially attractive price. The most favourable situation occurs with cyclopentene. This monomer not only can be supplied at a price close to that of, *e.g.*, butadiene, but also its polymer, in the all *trans* form, is the most promising elastomer (for synthetic rubbers) among all polyalkenamers prepared by ring opening polymerization [108].

The molecular weight of polyalkenamers may be regulated, *i.e.* reduced, by adding small amounts of acyclic low molecular weight olefins:

$$
\begin{array}{ccc}
H_2C = CH_2 & & H_2C \quad CH_2 \\
+ & \longrightarrow & \| \; + \; \| \\
RHC = CHR & & RHC \quad CHR
\end{array}
$$

This reduction in molecular weight may, however, become a handicap, if the monomer is not free of linear olefinic impurities which would lead to irreproducible average molecular weights. The complete degradation of a polyalkenamer by equimolar amounts of acyclic low molecular weight olefins is of increasing interest for analytical and synthetic purposes. The degradation products reveal the monomer units originally present in the unsaturated polymer; *e.g.* the metathesis of polypentenamer with butene-2 yields 2,7-nonadiene [73c]:

$$
\text{\small wwwCH}_2\text{CH} \doteq \text{CHCH}_2\text{CH}_2\text{CH}_2\text{CH} \doteq \text{CHCH}_2\text{CH}_2\text{ww}
$$

$$
\text{CH}_3\text{CH} \doteq \text{CHCH}_3 \qquad \text{CH}_3\text{CH} \doteq \text{CHCH}_3
$$

On the other hand, such reactions provide a method for the synthesis of unusual unsaturated hydrocarbons which may not easily be accessible by other routes.

8.5.6. Reaction Mechanism

Mechanisms involving a cyclobutene intermediate [87], a tetramethylenemetal complex [88], a five-membered ring including the metal [89], and a carbene complex of the transition metal [90], have been suggested. Although the actual mechanism may not yet be established definitively, a large body of evidence accumulated during the past few years favors the metal-carbene mechanism, first proposed by Chauvin *et al.* in 1970 [90]. (For the electronic structure of transition metal carbenoid ligands see Section 7.3.)

Kroll and Doyle [91] reported the use of carbene complexes of molybdenum and tungsten as catalyst components for the metathesis of acyclic olefins. Treatment of these complexes (which are not catalytically active by themselves) with an alkylaluminum halide leads to a species which is an active catalyst. Dolgoplosk *et al.* [92] conducted a similar experiment, initiating the ring-opening polymerization of cyclopentene with a carbene complex formed *in situ* by decomposition of phenyldiazomethane in the presence of WCl_6.

A related observation, although not in the context of olefin metathesis, was made by E. O. Fischer and coworkers [93], who found α-methoxystyrene as the main product from the reaction of a carbenechromium complex and a vinyl ether, *e.g.*:

$$
(CO)_5Cr=C(OCH_3)C_6H_5 + CH_2=CHOC_2H_5 \longrightarrow CH_2=C(OCH_3)C_6H_5 + \text{other products}
$$

Evidently, the carbene complex breaks the double bond of the vinyl ether. Casey and Burkhardt [94] extended this discovery to the unactivated alkene *trans*-butene-2, by using a more reactive carbenemetal species*):

---

* Carbene-transition metal complexes are particularly stable if the carbene ligand contains a heteroatom α to the carbenoid carbon; for a Review see [95].

$$(CO)_5W=C(C_6H_5)_2 + CH_3CH=CHCH_3 \xrightarrow[4h]{50\,°C} CH_3CH=C(C_6H_5)_2 + \text{other products}$$

$$(54\%)$$

Applying this same reaction to an olefin which would give a potentially stable carbene ligand, these authors were able to demonstrate that in the alkene scission one fragment is incorporated into the new alkene, while the other fragment forms a new metalcarbene complex:

$$(CO)_5W=C(C_6H_5)_2 + CH_2=C(OCH_3)C_6H_5 \xrightarrow[6h]{32\,°C}$$

$$(CO)_5W=C(OCH_3)C_6H_5 + CH_2=C(C_6H_5)_2 + \text{other products}$$

$$(24\%) \qquad\qquad (26\%)$$

These alkene scissions were explained in terms of a mechanism involving a four membered ring including the metal, as intermediate*):

$$
\begin{array}{c}
(CO)_5W=CR_2 \\
+ \\
R'_2C=CH_2
\end{array}
\rightleftarrows
\begin{array}{c}
(CO)_5W \text{—} CR_2 \\
| \qquad\quad | \\
(R')^2C \text{—} CH_2
\end{array}
\rightleftarrows
\begin{array}{c}
(CO)_5W \qquad CR_2 \\
\| \; + \; \| \\
R'_2C \qquad CH_2
\end{array}
\qquad (8.80)
$$

Lappert and coworkers [96] were able to isolate a metalcarbene intermediate in an actual metathesis starting with a simple Rh(I) complex, $Rh[C_6H_5)_3P]_3Cl$, although only with very electronrich olefins, providing for particularly stable carbene-metal bonds:

$R = C_6H_5;\; R' = p\text{-}CH_3C_6H_4$

Katz and McGinnis [97] contributed additional evidence to a carbene mechanism for olefin metathesis. They showed that during the metathesis of cyclooctene in the simultaneous presence of *trans*-butene-2 and *trans*-octene-4, with a soluble Mo/Al system, the "cross-product" [C_{14}, Eq. (8.81)] is relatively frequent in the very early stages of the reaction ($C_{14}/C_{12} = 1.3$, $C_{14}/C_{16} = 3.3$, extrapolated to zero time):

* Lappert *et al.* [96] regard this intermediate as the product of an oxidative addition, increasing the formal oxidation number of the metal by two (the carbenoid ligand is generally considered as neutral, *cf.* Section 7.3).

$$\text{(cyclooctene)} + CH_3CH = CHCH_3 + C_3H_7CH = CHC_3H_7 \longrightarrow$$
$$\hspace{3cm} C_4 \hspace{3cm} C_8$$

$$\hspace{9cm} (8.81)$$

This observation is evidently best explained by a mechanism analogous to that depicted in Eq. (8.80); none of the earlier suggested routes would account for this finding. The carbene mechanism would also allow ready interpretation of ring opening polymerization of cyclic olefins [Eq. (8.82)], and the observed formation of large rings [74], under certain conditions [Eq. (8.83)], as well as the molecular weight regulating action of acyclic olefins [Eq. (8.84)].

$$\hspace{9cm} (8.82)$$

$$\hspace{9cm} (8.83)$$

$$\hspace{9cm} (8.84)$$

The possible origin of the first carbenemetal species in the usual metathesis catalyst systems (Section 8.5.2), which do not involve carbene ligands from the very beginning requires comment. Some indications are available from reactions other than metathesis. Thus, Pu and Yamamoto [98] suggested the intermediate formation of hydrido(carbene)metal complexes from methylmetal complexes in order to account for the formation of polydeuterated methane in the reaction of such methylmetal species with D_2.

$$CH_3 - M \;\rightleftarrows\; CH_2 = M - H$$

Cooper and Green [99] found evidence for a similar intermediate in the transformation of a methyltungsten complex. With ethylmetal complexes, β hydrogen transfer is probably the preferred reaction, but there might well be a small but kinetically significant amount of α hydrogen transfer to the metal ($k_2 \ll k_1$), providing for a certain number of active carbene-metal centers:

$$CH_3CH_2M \xrightarrow{\quad k_1 \quad} CH_2 = CH_2 + MH$$
$$\hspace{2cm} \xrightarrow{\quad k_2 \quad} CH_3CH = MH$$

In fact, it has been reported [82b] that methylaluminum halides (in combination with molybdenum compounds) give more active catalysts than do ethylaluminum halides, which might be due to the more ready formation of carbenoid species from the former.

In the particular case of the $Re(CO)_5Cl/C_2H_5AlCl_2$ system, the formation of a propylidene carbene ligand *via* reaction of an ethyl group with a CO ligand, and subsequent elimination of oxygen, was suggested by Farona and Greenlee [100], although the detailed mechanism remained obscure. A C_3 carbene species as starting complex was invoked to explain the product distribution in the very early stages of the metathesis with linear olefins.

Yet a different suggestion concerning the initial carbenoid species involves the direct scission of carbon-carbon double bonds by the metal [1]. Evidence for such carbene ligand formation has been found with heterogeneous catalyst (molybdenum oxide on alumina), at $T \geqslant 300\,°C$. When a mixture of ethylene and hydrogen was passed over this catalyst, considerable amounts of methane were formed. Separate experiments showed that ethane is not appreciably cracked under the same conditions, $< 500\,°C$. Hence methane evolution may be considered as strong evidence for the formation of intermediate methylene-carbene metal complexes. Once the first set of such complexes is formed by direct scission of olefinic double bonds, the catalytic cycle, Eq. (8.80), becomes feasible.

Considering all the evidence it would appear highly probable that olefin metathesis proceeds *via* a carbene mechanism, with four-membered metal containing rings as intermediates.

Appendix 8.1. *Determination of K_1'' (from Fig. 8.4)*

The equilibrium constant*) is given by [*cf.* Eq. (8.22)]:

$$K_1'' = \frac{[Rh_2Cl_2L_4][L]^2}{[RhClL_3]^2} \tag{8.85}$$

The concentrations of the three species involved can be obtained, for each of the curves 1–6 in Fig. 8.4, in the following way:
The absorbance at a given wave length is (1 cm light path):

$$A = \varepsilon_1[RhClL_3] + \varepsilon_2[Rh_2Cl_2L_4] \tag{8.86}$$

The extinction coefficients have the values $\varepsilon_1 = 1.42 \times 10^3$ [29a] and $\varepsilon_2 = 0.8 \times 10^3$ [30], at 410 nm. Since $RhClL_3$ and the dimer are the only rhodium species present at measurable concentration, the following equation is valid:

$$[RhClL_3] + 2\,[Rh_2Cl_2L_4] = [Rh]_{total} \tag{8.87}$$

The concentrations of the two rhodium species can be expressed, for each curve, in terms of $[Rh]_{total}$, A, ε_1 und ε_2, with the aid of Eqs. (8.86) and (8.87).

---

* Fig. 8.4 was first erroneously interpreted as a consequence of equilibrium Eq. (8.20) [29a], and later reevaluated along the lines indicated here [29b, 30].

The equilibrium concentration of free ligand [L] is obtained from a mass balance of the phosphine in the system:

$$3\,[RhClL_3]_0 + [L]_{added} = [L]_{eq} + 3\,[RhClL_3]_{eq} + 4\,[Rh_2Cl_2L_4]_{eq}$$

Evaluation of the curves 1–6 in Fig. 8.4 gives an average value of K_1'' ca. 10^{-4}.

References

[1] G. Henrici-Olivé and S. Olivé, unpublished results. [2] T. Otsu, A Shimizu and M. Imoto, J. Polym. Sci., Part A-1, *4*, 1579 (1966). [3] T. A. Manuel, J. Org. Chem., *27*, 3941 (1962). [4] J. E. Lyons, J. Org. Chem., *36*, 2497 (1971). [5] R. Cramer and R. V. Lindsey, J. Amer. Chem. Soc., *88*, 3534 (1966). [6] F. Asinger, B. Fell and P. Krings, Tetrahedron Lett. *1966*, 633. [7] H. Bönnemann, Angew. Chem. Internat. Edit. *9*, 736 (1970). [8] J. F. Nixon and B. Wilkins, J. Organometal. Chem., *44*, C25 (1972). [9] C. P. Casey and C. R. Cyr, J. Amer. Chem. Soc., *95*, 2248 (1973). [10] G. Henrici-Olivé and S. Olivé, J. Organometal. Chem., *29*, 307 (1971).

[11] J. A. Kerr, Chem. Rev., *66*, 465 (1966). [12] H. Kanai, J. Chem. Soc. Chem. Commun., *1972*, 203. [13] D. Bingham, D. E. Webster and P. B. Wells, J. Chem. Soc., Dalton Trans., *1972*, 1928. [14] P. M. Maitlis: The Organic Chemistry of Palladium, Academic Press, New York, 1971, Vol. 2, p. 137. [15] M. Orchin and W. Rupilius, Catal. Rev., *6*, 85 (1972). [16] W. Strohmeier, Topics in Current Chem., *25*, 71 (1972), and literature therein. [17] W. Strohmeier and R. Endres, Z. Naturforsch. *27b*, 1415 (1972). [18] P. S. Hallman, B. R. Mc Garvey and G. Wilkinson, J. Chem. Soc. A, *1968*, 3143. [19] F. H. Jardine, J. A. Osborn and G. Wilkinson, J. Chem. Soc. A, *1967*, 1574. [20] R. E. Harmon, J. L. Parsons, D. W. Cooke, S. K. Gupta and J. Schoolenberg, J. Org. Chem., *34*, 3684 (1969).

[21] G. Henrici-Olivé and S. Olivé, J. Mol. Catal., *1*, 121 (1976). [22] T. E. Paxson and M. F. Hawthorne, J. Amer. Chem. Soc., *96*, 4674 (1974). [23] K. A. Klinedinst and M. Boudart, J. Catal., *28*, 322 (1973). [24] T. A. Weil, S. Metlin and I. Wender, J. Organometal. Chem., *49*, 227 (1973). [25] E. L. Muetterties and F. J. Hirsekorn, J. Amer. Chem. Soc., *96*, 4063 (1974); F. J. Hirsekorn, M. C. Rakowsky and E. L. Muetterties, J. Amer. Chem. Soc., *97*, 237 (1975). [26] W. Strohmeier, R. Fleischmann and W. Rehder-Stirnweiss, J. Organometal. Chem., *47*, C37 (1973). [27] R. L. Augustine and J. F. Van Peppen, Chem. Commun., *1970*, 495. [28] J. Halpern, Accounts Chem. Res., *3*, 386 (1970). [29] a) H. Arai and J. Halpern, Chem. Commun., *1971*, 1571; b) J. Halpern and C. S. Wong, J. Chem. Soc. Chem. Commun., *1973*, 629; c) J. Halpern, in: Y. Ishii and M. Tsutsui, Eds., *Organotransition-Metal Chemistry*, Plenum Publ. Corp., New York, 1975; d) J. Halpern, personal communication. [30] C. A. Tolman, P. Z. Meakin, D. L. Lindner and J. P. Jesson, J. Amer. Chem. Soc., *96*, 2762 (1974).

[31] J. A. Osborn, F. H. Jardine, J. F. Young and G. Wilkinson, J. Chem. Soc. A, *1966*, 1711. [32] R. W. Mitchell, J. D. Ruddick and G. Wilkinson, J. Chem. Soc. A, *1971*, 3224. [33] C. O'Connor and G. Wilkinson, Tetrahedron Lett., *18*, 1375 (1969). [34] C. O'Connor and G. Wilkinson, J. Chem. Soc. A, *1968*, 2665. [35] N. Tamiya and S. L. Miller, J. Biol. Chem., *238*, 2194 (1963). [36] D. Kleiner and R. H. Burris, Biochim. Biophys. Acta, *212*, 417 (1970). [37] J. E. Lyons, L. E. Rennick and J. L. Burmeister, Ind. Eng. Chem. Prod. Res. Develop., *9*, 2 (1970). [38] L. Horner, H. Büthe and H. Siegel, a) Tetrahedron Lett., *1968*, 4023; b) Angew. Chem. Internat. Edit. *7*, 942 (1968). [39] a) W. S. Knowles and M. J. Sabacky, Chem. Commun., *1968*, 1445; b) W. S. Knowles, M. J. Sabacky and B. D. Vineyard, J. Chem. Soc. Chem. Commun., *1972*, 10; c) W. S. Knowles, M. J. Sabacky, B. D. Vineyard and D. J. Weinkauff, J. Amer. Chem. Soc., *97*, 2567 (1975). [40] T. P. Dang and H. B. Kagan, Chem. Commun., *1971*, 481; W. Dumont, J. C. Poulin, T. P. Dang and H. B. Kagan, J. Amer. Chem. Soc., *95*, 8295 (1973).

[41] a) J. Smidt, W. Hafner, R. Jira, J. Sedlmeier, R. Sieber, R. Rüttinger and H. Kojer, Angew. Chem., *71*, 176 (1959); b) J. Smith, W. Hafner, R. Jira, R. Sieber, J. Sedlmeier and A. Sabel, Angew. Chem. Internat. Edit., *1*, 80 (1962); c) R. Jira, J. Sedlmeier and J. Smidt, Ann. Chem., *693*, 99 (1966). [42] W. H. Clement and C. M. Selwitz, J. Org. Chem., *29*, 241 (1964). [43] I. I. Moiseev, O. G.

Levanda and M. N. Vargaftik, J. Amer. Chem. Soc., 96, 1003 (1974), and earlier references therein. [44] P. M. Henry, J. Amer. Chem. Soc., 86, 3246 (1964); 88, 1595 (1966); J. Org. Chem., 38, 2415 (1973). [45] a) G. Bracher, P. S. Pregosin and L. M. Venanzi, Angew. Chem., Internat. Edit., 14, 563 (1975); b) H. C. Clark, C. Jablonski, J. Halpern, A. Mantovani and T. A. Weil, Inorg. Chem., 13, 2213 (1974). [46] a) K. Jonas and G. Wilke, Angew. Chem. Internat. Edit., 8, 519 (1969); b) G. Henrici-Olivé and S. Olivé, J. Polym. Sci., Polym. Chem. Edit., 11, 1953 (1973). [47] G. Hata, Chem. Ind. (London), 1965, 223. [48] a) G. Wilke et al., Angew. Chem. Internat. Edit. 5, 151 (1966); b) B. Bogdanović, B. Henc, H. G. Karmann, H. G. Nüssel, D. Walter and G. Wilke, Ind. Eng. Chem., 62(12), 34 (1970); c) B. Bogdanović, Angew. Chem. Internat. Edit., 12, 954 (1973); d) B. Bogdanović, personal communication. [49] D. M. Golden, K. W. Egger and S. W. Benson, J. Amer. Chem. Soc., 86, 5416 (1964). [50] a) J. Ewers, Angew. Chem. Internat. Edit., 5, 584 (1966); b) Jap. Pat. 7,224,523 (1972); c) US Pat. 3,651,111 (1972).

[51] P. Longi, G. Mazzanti, A. Roggero and A. M. Lachi, Makromol. Chem., 61, 63 (1963). [52] H. Bestian and K. Clauss, Angew. Chem. Internat. Edit., 2, 704 (1963). [53] P. Taylor and M. Orchin, J. Amer. Chem. Soc., 93, 6504 (1971). [54] R. F. Heck, J. Amer. Chem. Soc., 91, 6707 (1969). [55] N. H. Phung and G. Lefebvre, Compt. Rend., 265, 519 (1967). [56] J. Chatt, R. S. Coffey, A. Gough and D. T. Thompson, J. Chem. Soc. (A), 1968, 190. [57] a) M. G. Barlow, M. J. Bryant, R. N. Haszeldine and A. G. Mackie, J. Organometal. Chem., 21, 215 (1970); b) US Pat. 3,709,955 (1973); c) G. Henrici-Olivé and S. Olivé, Angew. Chem. Internat. Edit., 14, 104 (1975); d) G. Henrici-Olivé and S. Olivé, Topics in Current Chem., 67, 1 (1967). [58] G. Natta, P. Corradini and I. W. Bassi, J. Amer. Chem. Soc., 80, 755 (1958); G. Natta and G. Mazzanti, Tetrahedron, 8, 86 (1960). [59] G. Henrici-Olivé and S. Olivé, Chem.-Ing.-Tech., 43, 906 (1971). [60] E. H. Immergut, G. Kollman and A. Malatesta, J. Polym. Sci., 51, S-57 (1961).

[61] G. Henrici-Olivé and S. Olivé, J. Polym. Sci., Part B, 8, 205 (1970). [62] J. C. W. Chien, J. Amer. Chem. Soc., 81, 86 (1959). [63] W. P. Long and D. S. Breslow, J. Amer. Chem. Soc., 82, 1953 (1960). [64] G. Henrici-Olivé and S. Olivé, Angew. Chem. Internat. Edit., 6, 790 (1967); Advan. Polym. Sci., 6, 421 (1969). [65] A. T. Cocks and K. W. Egger, J. Chem. Soc. Faraday Trans 1972, 423 (Vol. 68). [66] K. W. Egger and A. T. Cocks, in: S. Patai, Ed., The Chemistry of the Carbon-Halogen Bond, J. Wiley & Co., New York, 1973. [67] A. Misono, Y. Uchida, M. Hidai and H. Kanai, Chem. Commun., 1967, 357. [68] a) A. Yamamoto, Ann. N.Y. Acad. Sci., 239, 60 (1974), and earlier references therein; b) A. Yamamoto, T. Shimizu and S. Ikeda, Makromol. Chem., 136, 297 (1970); c) S. Komiya, A. Yamamoto and S. Ikeda, Bull. Chem. Soc. Jap., 48, 101 (1975). [69] J. Ulbricht and M. Arnold, Plaste und Kautschuk, 18, 166 (1971). [70] The Shorter Oxford Dictionary, 3rd Edit., Clarendon Press, Oxford, 1969.

[71] R. L. Banks and G. C. Bailey, Ind. Eng. Chem. Prod. Res. Develop., 3, 170 (1964). [72] a) N. Calderon, H. Y. Chen and K. W. Scott, Tetrahedron Lett., 1967, 3327; b) N. Calderon, E. A. Ofstead, J. P. Ward, W. A. Judy and K. W. Scott, J. Amer. Chem. Soc., 90, 4133 (1968). [73] a) G. Natta, G. Dall'Asta and G. Mazzanti, Angew. Chem. Internat. Edit., 3, 723 (1964); b) G. Dall'Asta, Makromol. Chem., 154, 1 (1972); c) G. Dall'Asta, Pure Appl. Chem., XXIVth Internat. Congress, Vol. 1, 133 (1974); d) G. Dall'Asta and G. Motroni, Angew. Makromol. Chem., 16/17, 51 (1971). [74] E. Wasserman, D. A. Ben-Efraim and R. Wolovsky, J. Amer. Chem. Soc., 90, 3286 (1968). [75] a) D. A. Whan, M. Barber and P. Swift, J. Chem. Soc. Chem. Commun., 1972, 199; b) J. Smith, W. Mowat, D. A. Whan and E. A. Ebsworth, J. Chem. Soc. Dalton Trans., 1974, 1742. [76] G. Henrici-Olivé and S. Olivé, Angew. Chem. Internat. Edit., 12, 153 (1973). [77] P. R. Marshall and B. J. Ridgewell, Europ. Polym. J., 5, 29 (1969). [78] H. Höcker and F. R. Jones, Makromol. Chem., 161, 251 (1972). [79] E. A. Zuech, W. B. Hughes, D. H. Kubicek and E. T. Kittleman, J. Amer. Chem. Soc., 92, 528 (1970). [80] a) G. Pampus, G. Lehnert and D. Maertens, 164th National Meeting Amer. Chem. Soc., New York, Fall 1972; Reprints Polym. Div., p. 880; b) G. Pampus and G. Lehnert, Colloque sur la Polymerisation, Groupe Française des Polymères, Lyon, France, June 1974.

[81] R. Musch, Dissertation, Mainz, Germany, 1974. [82] a) W. B. Hughes, Chem. Commun., 1969, 431; b) J. Amer. Chem. Soc., 92, 532 (1970). [83] P. Günther, F. Haas, G. Marwede, K. Nützel, W. Oberkirch, G. Pampus. N. Schön and J. Witte, Angew. Makromol. Chem., 14, 87 (1970). [84] Chem. Week, July 23, 1966, p. 70. [85] US Pat. 3,491,163 (1970) to Phillips Petroleum. [86] P. B. van Dam, M. C. Mittelmeijer and C. Boelhouwer, J. Chem. Soc. Chem. Commun. 1972, 1221. [87] C. P. Bradshaw, E. J. Howman and L. Turner, J. Catal., 7, 269 (1967). [88] G. S. Lewandos and R. Pettit, J. Amer. Chem. Soc., 93, 7087 (1971). [89] R. H. Grubbs and T. K. Brunck, J. Amer. Chem. Soc., 94,

2538 (1972). [90] J. L. Herisson and Y. Chauvin, Makromol. Chem., *141*, 161 (1970); J.-P. Soufflet, D. Commereuc and Y. Chauvin, Compt. Rend., *276*, 169 (1973).
[91] W. R. Kroll and G. Doyle, Chem. Commun. *1971*, 839. [92] B. A. Dolgoplosk, T. G. Golenko, K. L. Makovetskii, I. A. Oreshkin and E. I. Tinyakova, Dokl. Akad. Nauk SSSR, *216*, 807 (1974).
[93] E. O. Fischer and K. H. Dötz, Chem. Ber., *105*, 3966 (1972). E. O. Fischer and B. Dorrer, Chem. Ber., *107*, 1156 (1974). [94] C. P. Casey and T. J. Burkhardt, J. Amer. Chem. Soc., *96*, 7808 (1974); see also C. P. Casey, H. E. Tuinstra and M. C. Saeman, J. Amer. Chem. Soc., *98*, 608 (1976). [95] D. J. Cardin, B. Cetinkaya and M. F. Lappert, Chem. Rev., *72*, 545 (1972). [96] D. J. Cardin, M. J. Doyle and M. F. Lappert, J. Chem. Soc. Chem. Commun. *1972*, 927. [97] T. J. Katz and J. McGinnis, J. Amer. Chem. Soc., *97*, 1592 (1975); see also T. J. Katz *et al.*, J. Amer. Chem. Soc., *98*, 605, 606 (1976). [98] L. S. Pu and A. Yamamoto, J. Chem. Soc. Chem. Commun. *1974*, 9. [99] N. J. Cooper and M. L. H. Green, J. Chem. Soc. Chem. Commun. *1974*, 761. [100] M. F. Farona and W. S. Greenlec, J. Chem. Soc. Chem. Commun. *1975*, 759.

Suggested Additional Reading

[101] R. Cramer, *Transition Metal Catalysis Exemplified by some Rhodium – Promoted Reactions of Olefins*, Accounts. Chem. Res., *1*, 186 (1969). [102] B. R. James, *Homogeneous Hydrogenation*, John Wiley & Sons, New York, 1973. [103] P. M. Maitlis, *The Organic Chemistry of Palladium*, Academic Press, New York, 1971; Vol. II, Ch. II B. [104] G. Lefebvre and Y. Chauvin, *Dimerization and Co-dimerization of Olefinic Compounds*, in: R. Ugo, Ed., *Aspects of Homogeneous Catalysis*, Carlo Manfredi, Editore, Milano 1970; Vol. 1. [105] G. Henrici-Olivé and S. Olivé, *Oligomerization of Ethylene with Soluble Transition Metal Catalysts*, Advan. in Polym. Sci., *15*, 1 (1974). [106] G. Henrici-Olivé and S. Olivé, *Polymerisation – Katalyse, Kinetik, Mechanismen*, Verlag Chemie, Weinheim 1969; Chapter 3. [107] R. L. Banks, *Catalytic Olefin Disproportionation*, Topics in Current Chemistry, *25*, 39 (1972). [108] G. Dall'Asta, *Preparation and Properties of Polyalkenamers*, Rubber Chem. and Technol. *47*, 511 (1974).

9. Reactions of Conjugated Diolefins

9.1. General Aspects

Conjugated diolefins such as butadiene or isoprene differ in several respects from the olefins with isolated double bonds, discussed in Chapter 8. The partial double bond character of the bond in between the two double bonds (*cf.* Section 6.2.3) restricts free rotation about this bond, and gives rise to distinct *cis* and *trans* conformations. In butadiene the *trans* conformation is energetically favored by 9.6 kJ/mol; for isoprene the *cis* conformation prevails. The conjugated dienes may coordinate to a transition metal center *via* only one double bond (as a monodentate ligand), or *via* both (as a bidentate ligand). In the former case, they may or may not react like a monoolefin. In the latter different reaction mechanisms are generally to be expected, which are characterized by the tendency of coordinated diolefins to be converted into a π allyl ligand. There are two different modes for this [1–3]. Thus, insertion of a coordinated diolefin into a M—R bond (where R may be, for example, an alkyl group or hydrogen) gives an allyl ligand which in most cases stabilizes in the π allyl form:

$$
\begin{array}{ccc}
& \text{CH}_2 & \\
\text{R} & \overset{\diagup}{\text{CH}} & \\
\text{L}_n\text{M} & & \\
& \text{CH} & \\
& \text{CH}_2 &
\end{array}
\longrightarrow
\text{L}_n\text{M} \longrightarrow
\begin{array}{c}
\text{RCH}_2\ \text{H} \\
\text{C} \\
\text{CH} \\
\text{C} \\
\text{H}\quad \text{H}
\end{array}
\tag{9.1}
$$

Reactions of this type do not change the formal oxidation number of the metal center. If a terminal vinyl group is present in the conjugated diolefin, the addition of R is invariably to C^1; inner double bonds react more slowly. In substituted dienes such as isoprene, R adds preferentially to the non-substituted double bond. Thus, the more frequent insertion modes are, respectively (formulated for R = H) [3]:

$$
\begin{array}{ccc}
& \text{CH}_2 & \\
\text{H} & \overset{\diagup}{\text{CH}} & \\
\text{M} & & \\
& \text{CH} & \\
& \text{CH} & \\
& \text{CH}_3 &
\end{array}
\longrightarrow
\text{M} \longrightarrow
\begin{array}{c}
\text{CH}_3 \\
\text{CH} \\
\text{CH} \\
\text{CH} \\
\text{CH}_3
\end{array}
\tag{9.2}
$$

$$
\begin{array}{ccc}
& \text{CH}_2 & \\
\text{H} & \overset{\diagup}{\text{CH}} & \\
\text{M} & & \\
& \text{C–CH}_3 & \\
& \text{CH}_2 &
\end{array}
\longrightarrow
\text{M} \longrightarrow
\begin{array}{c}
\text{CH}_3 \\
\text{CH} \\
\text{C–CH}_3 \\
\text{CH}_2
\end{array}
\tag{9.3}
$$

Two steric arrangements of the π allyl ligand are possible after the insertion of a conjugated diolefin into a metal-hydrogen bond:

$$
\begin{array}{cc}
& \text{CH}_3\ \text{H} \\
& \text{C} \\
\text{M} \longrightarrow & \text{CH} \\
& \text{C} \\
& \text{H}\quad \text{H}
\end{array}
\qquad
\begin{array}{cc}
& \text{H}\quad\text{CH}_3 \\
& \text{C} \\
\text{M} \longrightarrow & \text{CH} \\
& \text{C} \\
& \text{H}\quad \text{H}
\end{array}
$$

$$\textit{anti}\qquad\qquad\textit{syn}$$

Utilizing a hydridonickel complex, Tolman [3] has shown that the *anti* form is the initial product, as one would expect for insertion of a *cisoid* (bidentate) diene. With time, however, and especially at elevated temperature, equilibration takes place, since the sterically less hindered *syn* form is thermodynamically favored. This may become important for the stereochemistry of the product of further reaction of such a π allylic ligand. (For structure and bonding in π allyl complexes, *syn-anti* equilibria, *etc.*, *cf.* Section 7.2.2.)

The second mode of π allyl ligand formation from 1,3-diolefins arises if a catalyst center permits simultaneous coordination of two butadiene molecules. A dimeric ligand with two π allyl ends may be formed by a simple electronic rearrangement [1, 2]:

$$L_n M + 2\ CH_2{=}CHCH{=}CH_2 \longrightarrow L_n \underset{H_2C \diagdown \underset{CH}{|} \diagup CH-CH_2}{\overset{\overset{CH}{H_2C \diagup \overset{|}{} \diagdown CH-CH_2}}{\underline{\hspace{1cm}} M \underline{\hspace{1cm}}}} \tag{9.4}$$

In this case the formal oxidation number of the metal is increased by two. Such complexes have been isolated as intermediates in certain dimerizations and oligomerizations. Both types of intermediates [Eqs. (9.1) and (9.4)] will repeatedly be encountered in the following Sections.

9.2. Selective Hydrogenation to Monoenes

Several of the hydrogenation catalysts described in Section 8.2 have been found effective also for the reduction of conjugated diolefins. Of particular interest here are catalysts which cause selective hydrogenation of only one of the two conjugated double bonds. Thus, $RuHCl[P(C_6H_5)_3]_3$ gives very selectively pentene-2 (95%) from pentadiene. In this case, selectivity is attributed to the negligible activity of the catalyst towards internal olefins (*cf.* Section 8.2.1). In other cases, selectivity is brought about by a thermodynamic effect: after the first hydrogenation, the monoene is displaced from the catalyst center by the better co-ordinating (bidentate) diolefin [45].

Hydrogenation with pentacyanocobaltate, $[Co(CN)_5]^{3-}$, provides a particularly interesting example from a mechanistic point of view. This catalyst is able to reduce a great variety of organic substrates [4], but carbon-carbon double bonds are only reduced when forming part of a conjugated system, and conjugated dienes are selectively reduced to monoenes. In the case of butadiene it was found [5] that in the presence of an excess of cyanide ion ($CN^-/Co = 10$) the reaction could be directed to butene-1 (90%), whereas with a ratio $CN^-/Co = 5$, most of the product (85%) is *trans*-butene-2. The mechanism of this reaction has been studied by Kwiatek and his group [6], as well as by Burnett *et al.* [7]. Work with pentacyanocabaltate is carried out generally in aqueous media at room temperature and atmospheric pressure. With molecular hydrogen, a hydrido compound is formed reversibly, in an oxidative addition reaction (*cf.* Section 7.4.3):

$$2\ [Co(CN)_5]^{3-} + H_2 \ \rightleftarrows\ 2\ [HCo(CN)_5]^{3-} \tag{9.5}$$

The hydrogen can be transferred to a suitable compound containing conjugated double bonds, such as butadiene, isoprene, sorbic acid, etc. It was found that the hydrido compound [Eq. (9.5)] inserts a butadiene; the intermediate was isolated:

$$[HCo(CN)_5]^{3-} + C_4H_6 \; \rightleftarrows \; [Co(CN)_5(C_4H_7)]^{3-} \tag{9.6}$$

This complex does not react with H_2, but with another hydridocobalt it forms the butene stoichiometrically and irreversibly:

$$[Co(CN)_5(C_4H_7)]^{3-} + [HCo(CN)_5]^{3-} \; \rightarrow \; 2\,[Co(CN)_5]^{3-} + C_4H_8 \tag{9.7}$$

In a catalytic cycle (with an excess of butadiene and H_2 present), Eq. (9.7) is assumed to be followed by Eq. (9.5), in order to restore the active species.

The cyanide ion dependent selectivity towards butene-1 or butene-2 was explained in terms of a $\sigma\pi$ equilibrium of the allyl ligand formed in reaction (9.6). At low cyanide ion concentrations, dissociation of a cyanide ligand would provide more coordination sites, thus favoring the π allyl form:

The σ allyl complex is assumed to give butene-1 in a bimolecular reaction with $[HCo(CN)_5]^{3-}$, in the course of which the H ligand of the latter is transferred to the α carbon of the former. The π complex, on the other hand, gives *trans*-butene-2 which follows directly from the thermodynamically more stable *syn* isomer of the π allyl complex.

9.3. Dimerization and Oligomerization to Open-Chain Products

The dimerization and oligomerization of butadiene and other conjugated diolefins to yield open chain products (as opposed to cyclic dimers and oligomers, *cf.* Section 9.5) has been achieved with a number of group VIII transition metal compounds, mostly in combination with reducing agents (*e.g.* alkylaluminums or $NaBH_4$), and sometimes in the presence of a neutral ligand (*e.g.* phosphine or phosphite) [8]. At least two different mechanisms may be discerned. They are based on the two types of π allyl intermediate shown in Eqs. (9.1) and (9.4), and lead to different products. Using butadiene as an example, we shall discuss these mechanisms in the following two sections.

9.3.1. Mechanism Involving a Metal Hydride as Carrier of the Kinetic Chain

Cobalt containing catalyst systems, which dimerize butadiene selectively (80–90%) to yield 3-methyl-1,3,6-heptatriene, with some 1,3,6-octatriene, were reported in the nineteen sixties. These systems comprise not only $CoCl_2/NaBH_4$ [9] and $Co_2(CO)_8/(C_2H_5)_3Al$ [10], but also $(\eta^3\text{-}C_3H_5)_3Co$ without any other component [1b, 11]. Working with the first of these systems in the presence of butadiene, Natta and coworkers [9b] were able to isolate an intermediate which turned out to be a key compound from a mechanistic point of view. A complex of formula $C_{12}H_{19}Co$ was isolated after treating $CoCl_2$ with butadiene and $NaBH_4$ in alcohol at $-30\,°C$. The geometry of this complex, which has been determined by X-ray crystallography, is shown in Fig. 9.1.

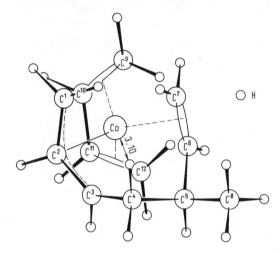

Fig. 9.1. Molecular geometry of the $C_{12}H_{19}Co$ complex (according to Natta *et al.* [9b]).

The cobalt center is surrounded by a 3-methylheptadienyl ligand (C^1-C^8) and by a butadiene molecule (C^9-C^{12}). The former ligand is π bonded to the metal *via* a vinyl double bond C^6-C^7, and by an η^3-allyl group (π allyl) C^1-C^3; the butadiene is coordinated in its *cis* configuration as a bidentate ligand. The original chloride anions associated with cobalt have disappeared. The methyl group (C^8) in the methylheptadienyl ligand strongly supports the intermediate formation of a metalhydride. Presumably reduction of Co(II) to Co(I), hydride formation and coordination of a butadiene molecule occurs first [Eq. (9.8), complex I]. This is then followed by insertion of butadiene into the Co−H bond, resulting in the π allyl complex II. Under the influence of a second, coordinated butadiene molecule, the π allyl group transforms to a σ allyl ligand (complex III). Amazingly, the σ bond to the metal is formed at the substituted end of the allyl group; we shall return to this point below. Insertion of butadiene and coordination of a third molecule leads to complex IV, which is identical with the complex shown in Fig. 9.1.

$$
\begin{array}{ccc}
\text{(I)} & \text{(II)} & \text{(III)} \\
& \xrightarrow{C_4H_6} & \\
\end{array}
\tag{9.8}
$$

$$
\xrightarrow{C_4H_6} \quad \text{(IV)} \qquad \longrightarrow \quad CH_2{=}CHCH{=}CHCH(CH_3)CH{=}CH_2
$$

$$
+\ \text{I}
$$

Complex IV, which is relatively stable at low temperature, catalyzes the dimerization of butadiene, without any other additive, at 60 °C [9a]. This species is therefore reasonably assumed to be an intermediate in the catalytic cycle. Fig. 9.1 indicates the unique position of one hydrogen atom β to the π allyl group, only 3.1 Å from the cobalt center. Transfer of this hydrogen to the metal with simultaneous release of methylheptatriene, regenerates complex I. We may consider Eq. (9.8) as representing the actual catalytic cycle. Note that a cobalt hydride (complex I) continues the kinetic chain after release of the dimer product.

As mentioned above, the dimer product consists of *ca.* 10–20% of the linear isomer 1,3,6-octatriene. Its formation requires the πσ rearrangement [II→III, Eq. (9.8)] with σ bond formation at the nonsubstituted end of the allyl group:

$$II \xrightarrow{C_4H_6} \quad
\begin{array}{c}
CH_2CH=CHCH_3 \\
| \\
Co \\
H_2C \diagdown_{CH-CH}\diagup CH_2
\end{array}
\quad (III\ a)$$

$$\xrightarrow{C_4H_6} \quad
\begin{array}{c}
H_2C \diagup^{CH-CH}\diagdown CH_2 \\
Co \\
H_2C \diagdown_{\underset{H}{C}}\diagup CHCH_2CH_2CH=CHCH_3
\end{array}
\qquad (9.9)$$

$$(IV\ a)$$

$$\longrightarrow \quad CH_2=CHCH=CHCH_2CH=CHCH_3 + I$$

One question remains, namely why does the branched dimer occur much more frequently than the linear dimer, although the former requires the sterically more hindered σ complex III and the latter the more favorable σ complex IIIa. A simple suggestion has been made by van Leeuwen: the less stable σ complex IIIa might be more prone to further reaction (*i.e.* insertion) [12]. If this were so, one might anticipate the relative frequency of branched and linear molecules to depend on the metal itself, and on the other ligands in the complex, which have an influence on the stability of the metal-carbon bond (*cf.* Section 7.3). This is actually the case (*vide infra* and Section 9.6).

A different suggestion concerning the predominant formation of branched dimer was made by Hughes and Powell [13]. These authors investigated the reaction products of substituted π-allylpalladium chloride with substituted butadienes, and found it impossible to reconcile the constitution and stereochemistry of these products with the classical insertion mechanism. They proposed instead a concerted electrocyclic process, during the course of which the carbon-carbon bond formation occurred outside the coordination sphere of the metal. Translated to the present problem the following reaction sequence would result:

$$II \longrightarrow \underset{(IIIb)}{\overset{\overset{\displaystyle C-C=C-C}{\underset{\displaystyle Co\cdots}{|}}}{}} \longrightarrow \quad Co \begin{matrix} C-C \\ \diagdown \\ C-C \end{matrix} \begin{matrix} C-C \\ C \end{matrix}$$

$$\longrightarrow \quad Co \begin{matrix} C=C \\ | \\ C \\ C=C \end{matrix} \begin{matrix} C-C \\ | \\ C \end{matrix} \longrightarrow \quad Co \begin{matrix} C \\ \| \\ C \\ C-C-C-C \end{matrix} \overset{C_4H_6}{\longrightarrow} IV \tag{9.10}$$

The proposed intermediate IIIb differs from IIIa in that the butadiene molecule is co-ordinated in its *trans* conformation, only *via* one double bond. It is, however, assumed to adopt a *cis* orientation in order to enter into the electrocyclic process depicted in Eq. (9.10). With this mechanism, the $\pi\sigma$ rearrangement resulting in the sterically less hindered metal-carbon σ bond would give the branched dimer, 3-methylheptatriene. Although this electro-cyclic mechanism is not generally accepted it might be an alternative to insertions of conjugated dienes. Whether the two mechanisms run parallel [e.g. reaction (9.10) giving the branched and reaction (9.9) giving the linear dimer], or whether the linear dimer is formed also in an analogous electrocyclic reaction starting from the sterically more hindered σ allyl-cobalt complex, must evidently remain an open question at the moment.

In conclusion, it should be noted that the various cobalt systems mentioned initially are assumed to operate according to the same mechanism, *i.e.* going through a cobalt hydride stage; the original ligands are replaced by butadiene and reaction products thereof. With $(\eta^3$-allyl$)_3$Co, intermediate formation of complex IV [Eq. (9.8)] has even been proved [11]. Several iron based systems have also been used for the dimerization of butadiene, giving the same products, but in different relative amounts. Thus, $FeCl_3$ in the presence of trialkyl-aluminum and triphenylphosphine gives a catalyst which dimerizes butadiene at room temperature, resulting in 70% 1,3, 6-octatriene, and only 30% of the branched dimer. Other iron systems tend to give higher oligomers and polymers [8].

9.3.2. Mechanism Involving a Zerovalent Metal Center as Carrier of the Kinetic Chain

The dimerization of butadiene with palladium catalysts proceeds by a different mechanism [2, 14, 15]. The products are characterized by the absence of methyl groups, excluding hence a hydride route as exemplified in the previous section. Thus, the catalyst bis(triphenyl-phosphine)(maleic anhydride)palladium(0) converts butadiene, in benzene or acetone solution, into 1,3,7-octatriene (7 h, 85%):

$$2\,H_2C = CHCH = CH_2 \quad \xrightarrow[115\,°C]{Pd} \quad H_2C = CHCH = CHCH_2CH_2CH = CH_2$$

In alcohol solution, the reaction proceeds more rapidly, even at lower temperatures, but the alcohol is incorporated into the product. 1-Alkoxy-2,7-octadiene and 1,3,7-octatriene are formed, the relative ratio depending on the alcohol used, and on the reaction conditions [15]. In methanol, and with $Pd(0)[(C_6H_5)_3P]_4$ as catalyst, essentially exclusive formation of methoxyoctadiene is obtained, with the following isomer distribution [2]:

$$2\,H_2C = CHCH = CH_2 + ROH \left\{ \begin{array}{l} ROCH_2CH = CHCH_2CH_2CH_2CH = CH_2 \quad (90\%) \\[2ex] H_2C = CHCH(OR)CH_2CH_2CH_2CH = CH_2 \quad (10\%) \end{array} \right.$$

With phenol, the corresponding phenoxyoctadienes, with the same isomer distribution, are formed; in this case a catalyst system consisting of $(\eta^3\text{-}C_3H_5PdCl)_2$, or $PdCl_2$ plus triphenyl-phosphine and sodium phenoxide was used [2].

The alkoxy- and phenoxyoctadienes can be decomposed to octatrienes plus alcohol with the same catalysts, under reduced pressure. The presence of phosphine is required to maintain the palladium in solution. Highest yield and turnover in the synthesis of 1,3,7-octatriene has been achieved using hexafluoro-2-phenylpropanol-2, because of the easy decomposition of the alkoxyoctadiene in this case. After distilling off the octatriene product, fresh butadiene may be added and the cycle repeated [2]. Essentially no trimers and higher oligomers are formed in all these reactions. The trimer can, however, be obtained, if 1,3,7-octatriene is used (in excess) as one of the components in the synthesis.

The mechanism suggested [2, 14b] for these interesting reactions has been inspired by Wilke's fine work concerning the cyclooligomerization of butadiene with nickel catalysts [1] (cf. Section 9.5). Wilke has shown that the active species is simply a Ni(0) atom ("naked nickel") maintained in solution by neutral ligands which are easily displaced by butadiene molecules. A key intermediate in Wilke's reactions is complex I below, for the formation of which two butadiene molecules have combined in the coordination sphere of the Ni(0) center, giving a bis π allyl ligand:

(I)   (II)   (III)

L = tertiary phosphine

The corresponding Pd(0) complex (II) has been suggested as active species in the systems under discussion, although attempts to isolate such a species failed [2, 14b]. In the particular case of the $PdCl_2/NaOC_6H_5$ catalyst, a binuclear palladium complex III was isolated, and was suggested alternatively as active species. The sodium phenoxide was assumed to transform $PdCl_2$ to a more active compound (e.g. to the bridged bimetallic complex in III). For the present discussion we prefer to use the monometallic complex II as the key catalytic chain carrier, in particular because the necessity for phosphine in order to maintain palladium in solution has been stressed repeatedly [2, 14].

The active species II once formed [e.g. Eq. (9.11)], the octatriene synthesis merely requires an intramolecular hydrogen shift from C^4 to C^6, possibly with a π allyl σ allyl intermediate, as complex IV in Eq. (9.12). Replacement of the triene by fresh butadiene restores species II.

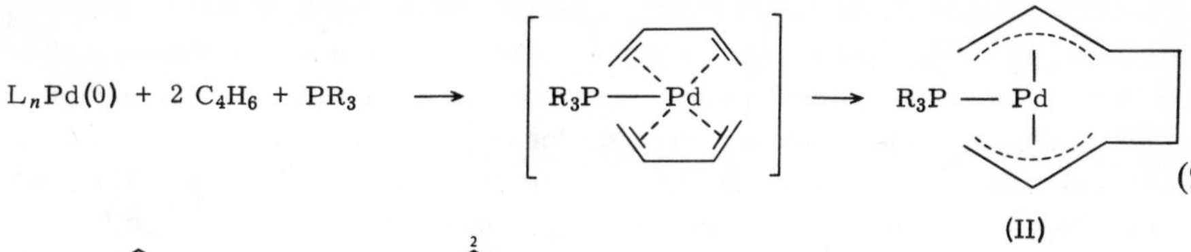

$L_nPd(0)$ + 2 C_4H_6 + PR_3
R_3P — Pd
R_3P — Pd
(II)

(9.11)

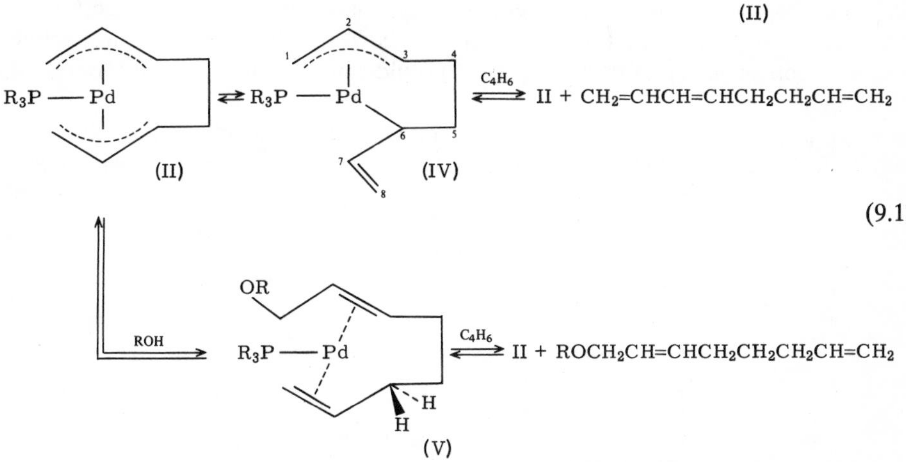

R_3P — Pd
(II)
R_3P — Pd
(IV)
C_4H_6
II + CH_2=CHCH=CHCH$_2$CH$_2$CH=CH$_2$
ROH
OR
R_3P — Pd
H
H
(V)
C_4H_6
II + ROCH$_2$CH=CHCH$_2$CH$_2$CH$_2$CH=CH$_2$

(9.12)

In the presence of alcohol or phenol, nucleophilic attack of RO^- at the 1-position of one of the π allyl groups leads to intermediate V. Work with CH_3OD has indicated that the proton is added to C^6. Displacement of the 1-alkoxy- or 1-phenoxy-octadiene by fresh butadiene restores complex II. The small amount of 3-alkoxyoctadiene (10%) stems from attack of RO^- in the 3-position. The observed decomposition of the octadienylethers on the same catalyst requires all steps in Eq. (9.12) to be reversible.

9.3.3. The Case of Styrene

Styrene is not strictly a 1,3-diene but it has a double bond in conjugation with the aromatic ring system. Actually, it can be dimerized selectively to *trans*-1,3-diphenylbutene-1, with π-allylnickel iodide as catalyst [16]. A hydride mechanism similar to Eq. (9.8), including π allyl stabilization of the intermediates has been suggested [17]:

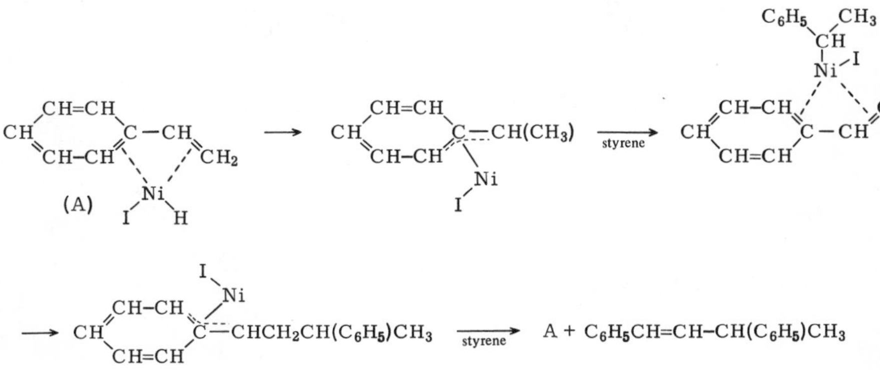

(A)
styrene
styrene
A + C_6H_5CH=CH–CH(C_6H_5)CH$_3$

This proposal was based on the fact that styrene, like butadiene (*cf.* Section 9.6), is able to maintain the nickel catalyst in solution, whereas other olefins and vinylic compounds such as ethylene, propylene or methyl methacrylate, cause rapid precipitation of Ni^0 sponge. It is assumed that the first styrene molecule, after insertion into the $Ni-H$ bond, loses its aromaticity, and thus makes a π allyl group available for the stabilization of the nickel. The same occurs after insertion of the second styrene molecule, but at this stage β-hydrogen transfer takes place, giving the product and restoring the active site. [Evidently, an electrocyclic mechanism corresponding to that in Eq. (9.10) could equally well have been formulated, with the same result.]

There is a precedent for the suggested π allyl intermediate in the following molybdenum complex, for which this kind of allylic structure has been demonstrated by X-ray analysis [18]:

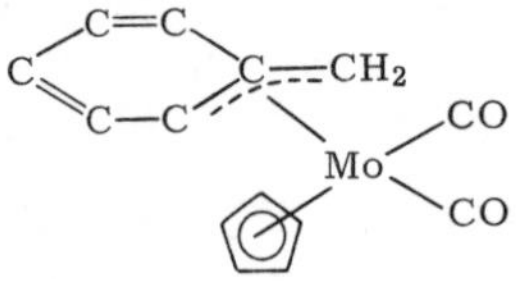

9.4. Codimerization with Monoenes

Butadiene and ethylene combine to form 1,4-hexadiene in the presence of a variety of iron, cobalt or nickel Ziegler type catalysts. Generally the addition of a tertiary phosphine or phosphite is necessary, to maintain the low valent transition metal species in solution [19]. But also $RhCl_3 \cdot 3 H_2O$ or $[RhCl(C_2H_4)_2]_2$, in the presence of HCl [20] or organic chlorides with labile chlorine [21], can be used as catalysts. No ethylene or butadiene dimers are formed, which is amazing since some of the catalysts (in particular those based on rhodium) would dimerize ethylene to butenes at comparable rates, in the absence of butadiene.

In all cases where mechanistic studies have been made, a metal hydride is assumed to be the chain carrier, and there is good evidence for it (*vide infra*). A general mechanism may be summarized as follows (using the butadiene/ethylene couple as an example, and omitting all other ligands):

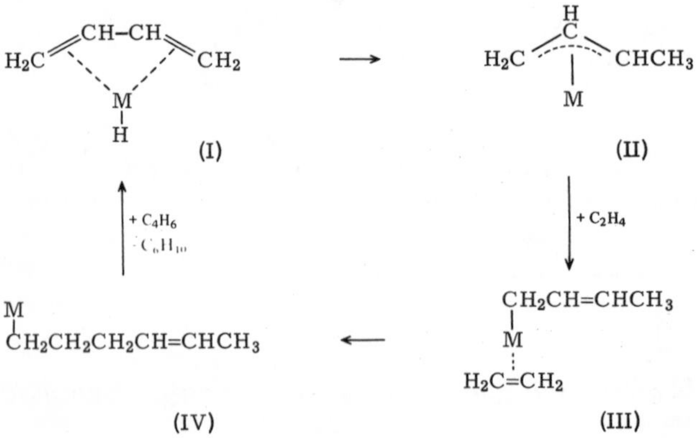

(9.13)

One may ask why the synthesis takes just this course. The reason is probably a combination of thermodynamic and steric effects. The π allyl complex II in Eq. (9.13) may safely be assumed to be more stable than the corresponding ethyl complex. On the other hand, the coordination of a second butadiene molecule to complex II is, presumably, hindered by the other ligands. The last step in the cycle (IV→I) is understood easily as a strong tendency to β hydrogen transfer, which takes place as soon as a β hydrogen becomes available.

Note that the almost exclusive formation of 1,4-hexadiene requires the πσ rearrangement of the allyl complex (II→III) to occur in the expected way, i.e. forming the metal-carbon σ-bond at the nonsubstituted end of the π allyl group. This contrasts with the situation observed in butadiene dimerization [Eq. (9.8)], and appears to lend some credence to the electrocyclic mechanism suggested for the particular case of the butadiene-butadiene reaction [Eq. (9.10)]. However, steric reasons (hindrance of the metal-carbon σ-bond formation at the substituted end of the π allyl group, by the ligands) would provide an alternative explanation.

We shall discuss now a few codimerization systems in some more detail, with the aim of corroborating the general reaction scheme, Eq. (9.13). Tolman [22] used a cationic metal hydride catalyst, $[HNiL_4]^+$ [with $L = P(OC_2H_5)_3$], which was obtained by protonation of NiL_4 with H_2SO_4 in methanol. At room temperature this complex reacts with butadiene to give the cationic π-crotylnickel species $[\pi\text{-}C_4H_7NiL_3]^+$. Its formation is independent of the butadiene concentration, and inhibited by added phosphite ligand, indicating ligand dissociation from $[HNiL_4]^+$ before reaction with butadiene. The π-crotylnickel species, which may be isolated using $^-PF_6$ as anion, was shown by proton NMR spectroscopy to react with ethylene to give a complex containing both a π crotyl group and a coordinated ethylene. At 100 °C, and in presence of both monomers, the codimerization process becomes catalytic; the *trans* isomer of 1,4-hexadiene is the major product. It is concluded reasonably that the insertion of ethylene into the allyl-metal bond is the rate determining step.

The system $CoCl_2/(C_2H_5)_3Al/DPE$ [DPE = 1,2-bis(diphenylphosphino)ethane, $(C_6H_5)_2P\text{-}CH_2CH_2P(C_6H_5)_2$] shows high activity and selectivity for the production of the *cis* isomer of 1,4-hexadiene from butadiene and ethylene [23, 24]. The reaction is preferably carried out in 1,2-dichloroethane at 80–110 °C where conversions of $\geq 95\%$ and selectivities of $\geq 98\%$ can be achieved. At higher temperature, isomerization to the thermodynamically favored *trans, trans*-2,4-hexadiene was observed. A mechanistic study (Henrici-Olivé and Olivé [24]) has shown that the active species is a Co(I) hydride. Cobalt(II) chloride dissolves in 1,2-dichloroethane only in the presence of DPE (DPE/Co = 2). Addition of an excess of $(C_2H_5)_3Al$ results in the formation of a dark brown complex, showing broad absorption peaks at 14 100 and 17 900 cm^{-1}; very low extinction coefficients ($\varepsilon < 100$) indicate the presence of an octahedral complex (*cf.* Section 5.2.1). With an excess of tetrahydrofuran, the brown complex decomposes, and the red $[(DPE)_2Co(I)H]$ [25] is formed. Further addition of $(C_2H_5)_3Al$ restores the brown complex. A paramagnetic Co(II) hydride has been observed as the precursor of the active species [24 b]. Presumably the latter is formed by the following reaction sequence: ligand exchange Cl/C_2H_5, β-hydrogen transfer with formation of the observed Co(II) hydride, further ligand exchange Cl/C_2H_5, and reduction Co(II)→ Co(I) by homolytic splitting of the Co(II)−C bond.

The red Co(I)hydride alone is not a catalyst for the codimerization of butadiene and ethylene; complex formation with the alkylaluminum is required for activity. Hence:

$$(DPE)_2Co(I)hydride \quad \underset{THF}{\overset{(C_2H_5)_3Al}{\rightleftharpoons}} \quad catalyst$$

Based on these observations, an octahedral structure has been proposed for the brown complex:

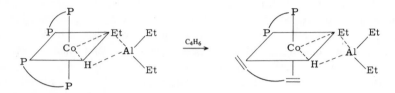

The bridged structure was formulated by analogy with other Ziegler systems. In the presence of both substrates, one of the DPE ligands is probably displaced by butadiene. The hydride hydrogen then adds to the butadiene molecule giving the π allylic crotyl ligand, and a catalytic cycle as depicted in Eq. (9.13) can start. Presumably the alkylaluminum does not remain within the complex during the catalytic cycle, since no ligand able to form a bridge is left.

The mechanism of codimerization with the system $[RhCl(C_2H_4)_2]_2/HCl$ was investigated by Cramer [20]. (The dimeric rhodium complex can be formed *in situ* by reducing $RhCl_3 \cdot 3 H_2O$ with ethylene.) The active species is formed by oxidative addition of HCl to the dimer, with simultaneous splitting of the latter, and increase of the oxidation number of the rhodium from I to III. One of the ethylene molecules is inserted into the Rh—H bond [Eq. (9.14)]. The σ ethylrhodium complex next coordinates a butadiene molecule, and a reversible hydrogen shift from the ethyl group to the diene occurs forming a π-ethylene-π-crotyl complex [Eq. (9.15)]. This complex then undergoes the addition step: the crotyl group adds to the ethylene. A 4-hexenylrhodium complex is produced.

$$[RhCl(C_2H_4)_2]_2 + 2 HCl \rightarrow 2 RhCl_2(C_2H_5)(C_2H_4) \tag{9.14}$$

$$RhCl_2(C_2H_5)(C_2H_4) + C_4H_6 \rightleftharpoons H_2C \overset{\overset{H}{C}}{\underset{\underset{H_2C=CH_2}{Rh}}{\cdots\cdots}} CH-CH_3 + C_2H_4 \tag{9.15}$$

At this point we have arrived at a species corresponding to complex IV in Eq. (9.13), and the usual cycle can follow. Note that for this catalyst the sequence of the first codimerization is reversed (first insertion of an ethylene into a metal-hydrogen bond, then insertion of butadiene), because it starts with a preformed ethylene complex.

The stereospecificity of the reaction in certain cases (Tolman's nickel catalyst yielding predominantly *trans*-1,4-hexadiene; the cobalt catalyst giving selectively the *cis*-isomer) is related evidently to the *syn-anti* equilibrium in the π allyl complex [complex II, Eq. (9.13)]. As mentioned in Section 9.1, the first product of the insertion of butadiene into a metal-hydrogen bond is in the *anti* form. If further reaction occurs before equilibration, the *cis*-

hexadiene is expected, whereas the thermodynamically favored *syn* form of the π allyl group leads to the *trans* isomer.

We cited butadiene ethylene codimerization as an example, but the codimerization has been extended also to substituted 1,3-diolefins and monoenes. A particularly interesting example is the combination of 1,3-cyclooctadiene with ethylene, using a π-allylnickel chloride plus $(C_2H_5)_3Al_2Cl_3$ catalyst, in conjunction with an optically active phosphine, which leads to optically active 3-vinylcyclooctene [26]. An optical purity of 70% was obtained with (−)-dimenthylisopropylphosphine at 0 °C.

9.5. Cyclodimerization and -Oligomerization

Nickel(0) catalysts, extensively studied by Wilke, Heimbach and their coworkers [1, 27, 28], have proved to be the most versatile and useful cyclization catalysts, although a number of other catalysts also have been reported in the literature [8].

We shall limit ourselves to the extremely active catalysts containing zero-valent nickel. They are prepared most easily by reducing nickel acetylacetonate in benzene solution, with an organoaluminum compound (*e.g.* triisobutylaluminum), in the presence of a suitable ligand [27b]. Cyclic oligoolefins (*e.g.* 1,5-cyclooctadiene or 1,5,9-cyclododecatriene) are particularly effective stabilizing ligands, but butadiene itself can also be used. In the absence of such a ligand, the reduction of nickel acetylacetonate leads to precipitation of metallic nickel. The reduction Ni(II)→Ni(0) with an alkylaluminum proceeds in the usual way, through exchange of ligands between the nickel and the aluminum centers, homolytic splitting of the nickel-carbon bonds, and/or β-hydrogen transfer [29].

The Ni(0) complex of all-*trans*-1,5,9-cyclododecatriene (CDT) can be isolated crystalline (red needles); its structure has been determined by X-ray analysis [30]:

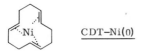 CDT−Ni(0)

The complex has a 16 electron configuration; it is very reactive, extremely air-sensitive, but relatively stable in water. It can be stabilized by addition of a phosphine ligand. CDT-Ni(0) catalyzes the cyclizations to be discussed in the following sections. Alternatively, Ni(II) can be reduced *in situ*.

The cyclizations of butadiene to yield C_8 and C_{12} rings, as well as the cocyclization with ethylene to yield a C_{10} ring, are quite remarkable, if one considers the difficulty usually found in the formation of medium nonaromatic rings from α, ω bifunctional linear molecules, which requires extreme dilution [1a]. The Ni(0) catalyzed cyclooligomerization, on the other hand, can be carried out at high butadiene concentrations, or even in the absence of a solvent. Evidently the catalyst forces the required number of monomer molecules into a position favorable for ring closure.

9.5.1. Cyclodimerization of Butadiene

If the reduction of nickel acetylacetonate, [Ni(acac)$_2$] is carried out in the presence of an excess of butadiene and of a phosphine or phosphite ligand (P/Ni = 1), the Ni(0) species is maintained in solution by the butadiene and the phosphorus containing ligand. At the same time butadiene is converted catalytically into the cyclic dimers *cis,cis*-1,5-cyclooctadiene (COD), vinylcyclohexene (VCH), and *cis*-1,2-divinylbutane (DVB):

COD   VCH   DVB

The fraction of each dimer formed depends on the phosphorus containing ligand used (see Table 9.1), and on the conversion. At ≤85% conversion, considerable amounts of DVB are found (up to 40% of the total product). At higher conversion the amount of this dimer is negligible; evidently, it is converted into one of the other isomers, as soon as the concentration of monomer becomes small.

Table 9.1. Cyclodimerization of butadiene with (C$_8$H$_{12}$)NiL. Dependence of the reaction product composition on L. T = 80 °C, normal pressure, 3 h [27 b].

L	COD (%)	VCH (%)	>C$_8$ (%)
(C$_6$H$_{11}$)$_3$P[a)	41	40	19
(C$_6$H$_5$)$_3$P	64	27	9
(C$_6$H$_5$O)$_3$P	81	7.4	11.6
(*o*−CH$_3$C$_6$H$_4$O)$_3$P	92	5.7	2.3
(*o*−C$_6$H$_5$C$_6$H$_4$O)$_3$P	96	3.1	0.9

[a)] Tricyclohexylphosphine.

The isolation, at low temperature, of intermediate I, bearing an open chain C$_8$ ligand [27 b] led to the following formulation for the dimerization:

The influence of the ligand L may then be rationalized in terms of the equilibria $I \rightleftharpoons II \rightleftharpoons III$. Strong electron acceptors such as triphenylphosphite stabilize the bis-π-allylnickel complex I which preferentially produces cyclooctadiene. Tertiary phosphines, on the other hand (alkyl-phosphine > arylphosphine) appear to transform one of the π allyl ends to a σ allyl group, which then tends to give vinylcyclohexene. Actually, intermediate III, with L = tricyclo-hexylphosphine, has been detected in solution and identified by its NMR signal [30].

9.5.2. Cyclotrimerization

In the absence of a phosphine or phosphite ligand, the Ni(0) catalyst transforms butadiene into the cyclic trimer CDT. The most abundant isomer contains all double bonds in the *trans* configuration (IV), which is easily obtained with 80–90% selectivity at $T \leqslant 100\,°C$ [27a], accompanied by the *trans, trans, cis*-(V) and *trans, cis, cis*-(VI) isomers. (The *cis,cis,cis*-isomer is not formed.)

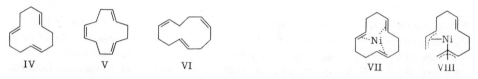

IV   V   VI   VII   VIII

Again, identification of a reaction intermediate has shed some light on the mechanism. The catalyst complex CDT-Ni(0) (VII) was dissolved in butadiene at $-40\,°C$. The excess of butadiene was then distilled off, also at $-40\,°C$. The residue contained all CDT originating from VII, plus a crystalline nickel complex with the same composition as VII, but with a different melting point ($+1\,°C$ as compared to $+102\,°C$ for VII). Evidently butadiene has replaced CDT from the nickel center, but at this low temperature no catalytic cycle has been established; the reaction stopped at intermediate VIII, the structure of which was inferred from NMR data.

This proposed structure is similar to that of a ruthenium complex (Fig. 9.2), which is formed from $RuCl_4$ and butadiene, and which has been studied by X-ray structural analysis [31]:

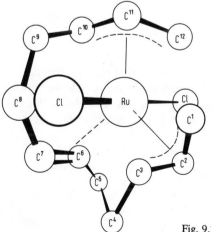

Fig. 9.2. Structure of the butadiene trimer $RuCl_2$ complex [31].

Based on the structure of intermediate VIII, the following mechanism may be formulated for the catalytic cyclotrimerization of butadiene [27a]:

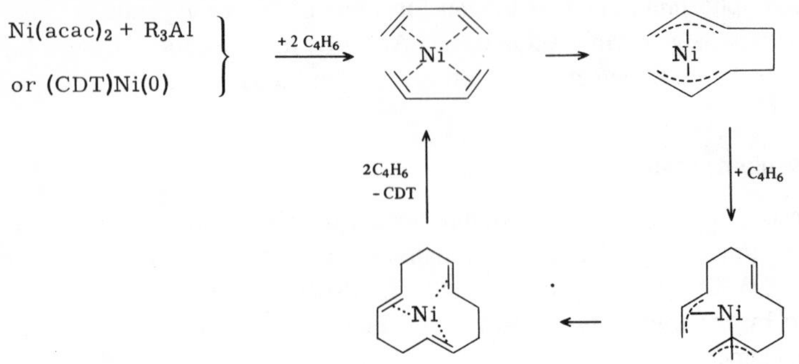

Although a nickel species with an α, ω-bisallyl, dimeric ligand has been isolated only in the presence of a phosphite ligand (complex I, Section 9.5.1), a stepwise process is assumed in the course of which the dimeric ligand is formed, before the third monomer molecule enters the active complex. Note that the catalytic cycle is closed when CDT is displaced from the Ni(0) center by butadiene molecules. Presumably the chelating effect of two conjugated dienes is more favorable than interaction with the three isolated double bonds of CDT. The real active site is a bare nickel atom ("naked nickel" [1a]).

9.5.3. Extensions of the Cyclization

The formation of the cyclic trimer CDT is suppressed if ethylene as well as butadiene are brought in contact with one of the Ni(0) catalysts mentioned in the preceding section. Instead a cocyclization of two monomers takes place. At $T \leqslant 40\,°C$, predominantly *cis, trans*-1,5-cyclodecadiene is formed, accompanied by some linear decatriene [27c]:

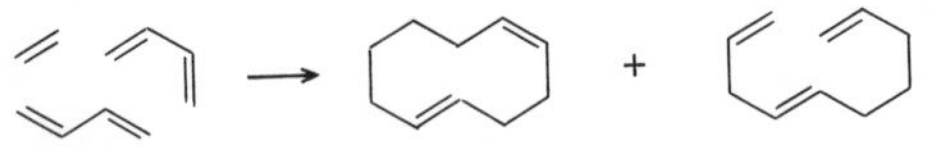

A further extension comprises the use of substituted butadienes (isoprene, piperylene, 2,3-dimethylbutadiene, *etc.*) in cyclodimerizations, cyclotrimerizations and cocyclooligomerizations. A great number of substituted cyclic compounds have become available in this way, although generally a mixture of isomers is formed [28]. A few of these reactions are exemplified below:

a) Cyclodimerization of *trans*-piperylene

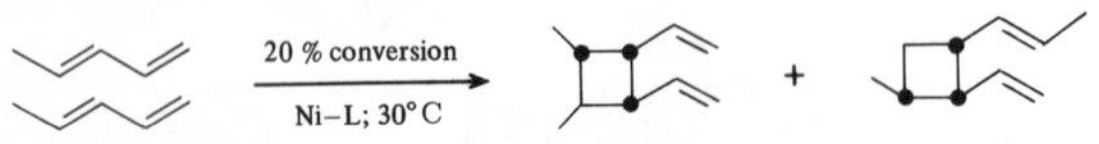

b) cocyclotrimerization of butadiene, isoprene and ethylene

c) cocyclotrimerization of butadiene and isoprene

9.6. Polymerization

By far the most important technical use of conjugated dienes is their transformation to elastomers. Although, in the synthesis of 1,4-*cis*-polyisoprene [32], a product was found which is essentially identical with natural rubber, the production of polybutadiene and styrene-butadiene rubbers is far advanced as regards both, quantity and elaboration [33].
We shall restrict our discussion to polybutadiene. Structural and geometric isomerism leads to four different polybutadiene structures (see Table 9.2). In all four cases, $\geq 95\%$ steric

Table 9.2. The four polybutadiene isomers and associated soluble catalysts.

Denotation	Structural Formula	Melting point and properties	Catalyst System	Ref.
1,4-*cis*		2 °C Elastomer	$CoCl_2 \cdot 2$ pyridine/$(C_2H_5)_2AlCl$ $Co_2(CO)_8/AlBr_3/Al(C_2H_5)_3$ $TiI_4/Al(i-C_4H_9)_3$ $[(\eta^3\text{-}C_3H_5)Ni(O_2CCF_3)]_2$	[34] [34] [34] [47]
1,4-*trans*		140 °C Thermoplast	$VCl_4/Al(C_2H_5)_3$ $[(\eta^3\text{-}C_3H_5)NiI]_2$ $[(\eta^3\text{-}C_3H_5)Ni(O_2CCF_3)]_2/P(OR)_3$ $RhCl_3/C_2H_5OH$ (or H_2O)	[34] [47] [47] [34]
1,2-*isotactic*		126 °C Thermoplast	$Cr(acac)_3/R_3Al$ [a]	[34]
1,2-*syndiotactic*		156 °C Thermoplast	$V(acac)_3/R_3Al$	[34]

[a] acac = acetylacetonate anion.

purity has been achieved with soluble transition metal systems, some of the Ziegler type, whereas others are monometallic catalysts.

It turns out that the course of the polymerization reaction is extremely sensitive to all factors influencing the overall geometry of the active metal site, such as nature of the metal itself, its formal oxidation state, and the number and type of ligands bonded to it [34, 47]. Although it is not yet possible to understand the interdependence between structure of the active site and stereoregularity of the polymer in complete detail, a number of relationships has been elucidated.

9.6.1. Structural Isomerism

Present knowledge indicates that the structural isomerism (1,2- or 1,4-polymer) is determined essentially by the transition metal itself. 1,2-Addition, where only one of the double bonds is used for polymerization, prevails with Ziegler systems based on group V or VI transition metals, in particular V, Cr [34], and Mo [35]. With chromium based catalysts it has been observed that the tacticity varies with the ratio of alkylaluminum to chromium compound [36]. At Al/Cr > 10, pure isotactic polymer is obtained, whereas Al/Cr = 2 results in pure syndiotactic polymer; at intermediate values of Al/Cr, both types of structure are formed simultaneously. Evidently at least two active sites with different geometry exist, presumably with varying valency state for the metal and with different ligands.

1,2-Addition may, in principle, be understood as the nondiene specific reaction of only one double bond, comparable to Ziegler type polymerization of α olefins. The availability of only one coordination site would appear to be a decisive prerequisite. It has, however, been suggested also [13] that 1,2-addition might well be diene specific, proceeding in a concerted, electrocyclic way, with C−C bond formation outside the coordination sphere of the metal, as discussed in Section 9.3, in the context of Eq. (9.10). A solution to this question requires further work.

1,4-Addition, on the other hand, is evidently specific for conjugated dienes. Ziegler type catalysts based on titanium or group VIII metals, as well as monometallic group VIII catalysts have been used. Although the industrial production (in particular that of the important 1,4-*cis*-polybutadiene) is confined essentially to the more active (and frequently heterogeneous) bimetallic systems, mechanistic research has concentrated on definite monometallic complexes of known structure.

Natta and coworkers reported the polymerization of butadiene with π-allylnickel bromide as catalyst in 1964, and revealed that the growing chain end might be stabilized as a π allyl ligand to the metal [37]. Recent proton NMR studies in the polymerizing system [38, 39] have supported strongly a mechanism involving π allyl intermediates, in 1,4-polymerization. Presumably a π allyl ligand transforms to a σ allyl group under the influence of coordinated butadiene; after insertion of the monomer molecule into the metal-carbon σ bond, the growing chain end may stabilize again as a π allyl ligand. Disregarding the other ligands for the moment, such a mechanism is depicted in Eq. (9.16):

$$
\underset{\substack{CH \\ CH_2P_n}}{\overset{CH_2}{HC{\Large<} \!\!-\!\!- M}} \underset{C_4H_6}{\rightleftharpoons} P_nCH_2CH=CHCH_2M \longrightarrow \underset{\substack{CH \\ CH_2P_{n+1}}}{\overset{CH_2}{HC{\Large<} \!\!-\!\!- M}} \qquad (9.16)
$$

Initiation of 1,4-polymerization by π allyl (or π crotyl) metal complexes, then, is straight-forward. In Ziegler type catalysts, the alkylation of the metal is presumed, as usual, to be the first step. Insertion of a monomer molecule into the metal-carbon bond, or into a metal-hydrogen bond formed after β-hydrogen abstraction from the alkyl group, would then give the first π allyl complex:

$$R{-}M + C_4H_6 \longrightarrow R{-}M \langle\!\!\langle\,\rangle\!\!\rangle \longrightarrow HC\Big\langle\!\!\begin{array}{c} CH_2 \\ \text{---} \\ CH \\ | \\ CH_2R \end{array} M$$

R = alkyl or H

In the $RhCl_3/C_2H_5OH$ (or H_2O) system (cf. Table 9.2), the catalytic reaction might be preceded by oxidative addition of an alcohol (or water) molecule, providing the required hydride ligand. Actually, the polymerization of butadiene with $RhCl_3$ in tritiated water gives tritium labeled polymer [34].

We may thus conclude that Eq. (9.16) describes roughly a general mechanism for the propagation step in 1,4-polymerization, independently of the particular initiation step.

9.6.2. Geometric Isomerism

The amazing influence of the ligands on the geometric isomerism (1,4-cis or 1,4-trans) which, in the particular case of the π allylnickel catalysts, may be estimated from Table 9.2, has been discussed thoroughly by Teyssié [47]. The dimeric catalyst complexes are slightly dissociated to monomers, in solution, see Eq. (9.17) where X represents an anionic ligand such as Cl^- or CF_3COO^- and the organic ligands may be π allyl (R = H or π crotyl (R = CH_3) groups.

$$HC\Big\langle\!\!\begin{array}{c} CHR \\ \text{---} \\ CH_2 \end{array} Ni\Big\langle\!\!\begin{array}{c} X \\ \\ X \end{array}\!\!\Big\rangle Ni\begin{array}{c} H_2C \\ \text{---} \\ HCR \end{array}\!\!\Big\rangle CH \rightleftharpoons 2\; HC\Big\langle\!\!\begin{array}{c} CHR \\ \text{---} \\ CH_2 \end{array} NiX \qquad (9.17)$$

The mononuclear complex appears to be the catalytically active one. This has been inferred from kinetic studies showing that the polymerization rate is proportional to the square root of the catalyst concentration [17, 40, 41, 47]. (See, however, Section 9.6.3.)

The most straightforward explanation of trans or cis specificity is based simply on monoden-tate or bidentate coordination of the diolefin, respectively. This hypothesis was originally forwarded by Arlman [42] for heterogeneous Ziegler catalysts based on different crystalline modifications of $TiCl_3$: α-$TiCl_3$, which offers only one free coordination site per active center, promotes the formation of 1,4-trans-polybutadiene, whereas β-$TiCl_3$, which is assumed to have active centers with one, and others with two vacancies, gives a mixture of all-trans and all-cis polymer molecules. Fig. 9.3 may help to explain how monodentate coordination can actually give rise to 1,4-insertion into a metal-carbon bond. The butadiene molecule is coordinated through one of its double bonds, in its energetically favored s-trans-conformation. The plane of the butadiene is supposed to be parallel to that defined by the metal and the other ligands. C^α represents the first carbon atom of a growing polymer chain (which is σ-allylic under the influence of the coordinated monomer). The essential argu-

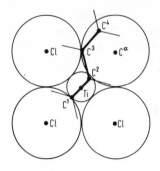

Fig. 9.3. Reactive position of butadiene at an α-TiCl$_3$ active center offering only one coordination site (Arlman [42]).

ment is that the distance $C^\alpha - C^4$ is shorter than $C^\alpha - C^2$. This situation is assumed to lead to preferential reaction of C^α with C^4. It follows that 1,4-*trans* polybutadiene is formed. (Evidently other metal complexes, with other geometries, may favour 1,2-addition.) This general picture once accepted, the ligand influence observed with the π allylnickel catalysts may now be discussed. Varying the anionic ligand, the geometrical isomerism ranges from 98% *cis* ($X = CF_3COO^-$) to 99% *trans* ($X = I^-$) [47]. (These data refer to polymerization at 30–40 °C, in aliphatic hydrocarbon solvents.) On the other hand, the pure *cis*-catalyst bis(π-allylnickel trifluoroacetate) is transformed to a pure *trans*-catalyst by the mere addition of a strongly coordinating ligand, such as a tertiary phosphite (*cf.* Table 9.2). The mononuclear form of the catalyst [Eq. (9.17)] offers sufficient space for bidentate coordination of butadiene. With $X = CF_3COO^-$ such coordination is enhanced by the marked electron-acceptor properties of this anionic ligand. But, if a strongly coordinating ligand such as a tertiary phosphite is added to the system, this ligand occupies one coordination site on the mononuclear catalyst species, thus preventing bidentate coordination of butadiene, and giving rise to 1,4-*trans*-polybutadiene. The iodide ligand, on the other hand, is a poor electron acceptor; hence the $Ni-X$ bond is considerably more covalent with $X =$ iodide. Higher electron density at the metal center leads to rejection of the second double bond, again with *trans*-insertion resulting.

This very appealing, simple interpretation of the experimental facts is, however, not yet quite satisfactory as far as 1,4-*cis* insertion is concerned. If we assume that the growing chain, after each insertion step, stabilizes as a π allyl complex, as indicated in Eq. (9.16), the *cis* configuration, still present in the σ bonded chain, would be lost, and the thermodynamically favored *syn* form of the π allyl complex would be formed*). This latter, however, would most probably generate the 1,4-*trans* configuration after the insertion of the next monomer. To circumvent this contradiction, Teyssié [47] has suggested that the σ allyl form of the growing chain has a certain life time, during which many monomer units are inserted before the complex rearranges to the π allyl form, which was called the "dormant" state of the active complex. This mechanism would, of course, apply also to *trans* polymerization.

A somewhat different explanation for *cis* polymerization was given by Furukawa [44] who suggested that intramolecular coordination of the double bond of the penultimate monomer unit (called the back-biting coordination) forces the last monomer unit into the *cis* form, for steric reasons.

---

* Indeed NMR measurements have indicated that various π allylmetal catalysts, giving *cis* or *trans* polymer, are always present in the most probable *syn* form [47].

9.6.3. Equibinary Polydienes

Teyssié *et al.* discovered a new type of specific control in the polymerization of butadiene with π allylnickel catalysts, in particular with the very active $[(\eta^3\text{-}C_3H_5)Ni(O_2CCF_3)]_2$ [43, 47]. While work with this catalyst in aliphatic hydrocarbons yields pure 1,4-*cis* polymer, and the addition of strong ligands leads to pure 1,4-*trans* (preceding Section), the addition of certain ligands of medium coordinating power, such as trifluoracetic acid, an aromatic or olefinic hydrocarbon, or a chlorinated hydrocarbon, leads to exactly equal amounts of *cis* and *trans* polymer. In most cases the *cis* and *trans* units are distributed randomly along the polymer chain, but with chlorinated hydrocarbons a marked tendency to alternating *cis* trans placement has been observed. The name "equibinary" polydienes was given to these polymers. The equibinary composition is attained asymptotically, as a function of the amount of ligand added, different ligands being more or less effective. The result is shown in Fig. 9.4.

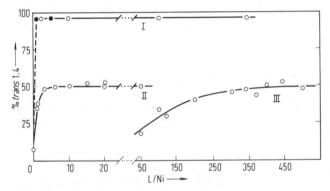

Fig. 9.4. Isomeric composition of 1,4-polybutadiene (Teyssié *et al.* [47]). Catalyst: $[(\eta^3\text{-}C_3H_5)\text{-}Ni(O_2CCF_3)]_2$; *n*-heptane solvent; added ligands: $(C_6H_5O)_3P$, C_2H_5OH (I); CF_3CO_2H (II); C_6H_6 (III).

As one may expect on the basis of Fig. 9.4, polymerization in solvent benzene gives also equibinary polybutadiene. There is, however, a mechanistically significant restriction: at very low catalyst concentrations the *cis* isomer content increases. Fractionation of a polymer made under these conditions reveals the presence of two types of polymer molecules, equibinary and pure *cis*, evidently due to two different active sites. Under the conditions of pure equibinary polymerization, the overall rate is no longer proportional to the square root, but to the first power of the catalyst concentration. From these observations it was concluded convincingly that equibinary polymerization takes place at the binuclear form of the catalyst. Why the weakly coordinating ligands (or solvents) prevent the binuclear catalyst from dissociating, and which is the actual control mechanism, is not yet clear. More insight into this specific catalysis would perhaps help us to understand certain aspects of steric control in transition metal catalyzed reactions; the more so, since other examples of equibinary structures have been found. Thus, equibinary (1,4-1,2)-polybutadiene, (*cis*-1,4-3,4)-polyisoprene, and (1,2-3,4)-polyisoprene have been reported; also in catalytic cyclopropanation of olefins, as well as in the isomerization of allylamido derivatives, 1:1 mixtures of *cis* and *trans* isomers have been observed under certain limiting conditions [47].

9.6.4. Kinetics and Molecular Weights

The 1,4-*trans* polymerization of butadiene using π-allylnickel iodide in different solvents has been investigated by the dilatometric method [17, 40, 41]. The rate of polymerization is first order with respect to the monomer concentration up to relatively high conversion (see Fig. 9.5), indicating that the concentration of active catalyst centers remains constant.

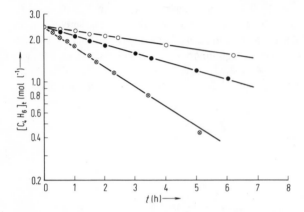

Fig. 9.5. Polymerization of butadiene with π-allylnickel iodide. $[C_4H_6]_0 = 2.4$ mol l^{-1}; $T = 50\,°C$; ○: $[Ni] = 11.3 \times 10^{-3}$ mol l^{-1}, benzene; ●: $[Ni] = 17.0 \times 10^{-3}$ mol l^{-1}, tetrahydrofuran; ⊕: $[Ni] = 10.8 \times 10^{-3}$ mol l^{-1}, 1,2-dichloroethane. [17]

As already mentioned in Section 9.6.2, the rate increases with the square root of the catalyst concentration. This finding is interpreted in the following way: Since the number of active sites is constant (no chain termination), the square root relationship over the whole polymerization time can hold only if the active mononuclear complex, after the insertion of one monomer unit (or of several units, *cf.* Section 9.6.2) recombines with another active site to regenerate the binuclear complex. In other words, equilibrium Eq. (9.17) is valid also for the polymerizing system, with R representing the growing chain.

The mechanism so far envisaged should give living polymers. In fact, polybutadiene obtained with π allylnickel iodide shows the characteristics of this type of polymer, i.e. linear increase in molecular weight with conversion, (at least up to some 50% conversion, see Fig. 9.6). Assuming that all Ni centers are able to insert butadiene, the number average molecular weight should be given by:

$$\text{Molecular weight} = \frac{[C_4H_6]_0 - [C_4H_6]_t}{[Ni]_{total}}$$

The dotted lines in Fig. 9.6 are calculated using this equation. The deviations from the straight lines, in a range of conversion where the rate of polymerization still follows first order kinetics with respect to monomer concentration, indicates that the molecular weight is lowered by a chain transfer process (or processes), without perturbation of the overall rate. Two chain transfer processes having this effect are conceivable: transfer to monomer [which

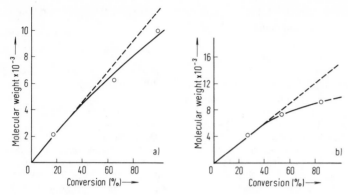

Fig. 9.6. Molecular weight (vapor pressure osmometry) as a function of conversion. $[C_4H_6]_0 = 2.4$ mol l^{-1}; $T = 50\,°C$; (a) $[Ni] = 11.3 \times 10^{-3}$ mol l^{-1}, benzene; (b) $[Ni] = 8.5 \times 10^{-3}$ mol l^{-1}, 1,2-dichloroethane. [17]

is analogous to that observed in linear dimerization, see *e.g.* Eq. (9.8)], or intramolecular chain transfer by copolymerization of an isolated double bond of the chain:

$$RCH_2CH=CHCH_2 \quad \begin{array}{c} \\ Ni \\ | \\ CH_2CH=CHCH_2 \end{array} \longrightarrow \quad \begin{array}{c} RCH_2CH-CHCH_2 \\ |\quad | \\ Ni\quad CH_2CH=CHCH_2 \end{array}$$

(9.18)

$$\longrightarrow \quad NiH \quad + \quad \begin{array}{c} RCH_2CH=CCH_2 \\ | \\ CH_2CH=CHCH_2 \end{array}$$

Note that after the incorporation of an isolated double bond, no allylic stabilization is possible, thus β-hydrogen transfer is assumed to lead to a hydridonickel species which then continues the kinetic chain. Actually, the same effect permits regulation of the polybutadiene molecular weight by addition of ethylene. (Intermolecular chain transfer by insertion of an isolated double bond of another polymer molecule would not lower the number average molecular weight because the number of polymer molecules would remain at one per nickel site.)

Both suggested transfers involve the insertion of butadiene into a nickel-hydrogen bond in order to continue the kinetic chain, and hence lead to polymer molecules with terminal methyl groups. Such end groups have actually been observed [47]. Thus, some transfer to the monomer cannot be excluded, but the progressive deviation of the molecular weight from the linear relationship given above, at higher conversion (more polymer, less monomer) points more to the transfer depicted in Eq. (9.18).

Similar results have been obtained also with the $[(\eta^3\text{-}C_3H_5)Ni(O_2CCF_3)]_2$ catalyst, under conditions where equibinary polybutadiene was the product [43b]. Additionally it was shown by NMR experiments that all π-allyl-Ni centers are active for chain initiation (disappearance of the clear doublets of the *syn* and *anti* protons of the $\eta^3\text{-}C_3H_5$ group), and that chain initiation, *i.e.* the insertion of the first butadiene into the allyl-metal bond, and the further growth steps, proceed at comparable rates. The "living" character of the polymer was confirmed by seeding experiments at low monomer/catalyst ratio: after consumption of a first lot of monomer, introduction of additional monomer causes further polymerization, with the expected increase of the molecular weight.

References

[1] a) G. Wilke, Angew. Chem. Internat. Edit., 2, 105 (1963); b) G. Wilke et al., Angew. Chem. Internat. Edit., 5, 151 (1966).   [2] E. J. Smutny, Paper presented at the New York Academy of Science Meeting on "Catalytic Hydrogenation and Analogous Pressure Reactions", New York, Sept. 1972.   [3] C. A. Tolman, J. Amer. Chem. Soc., 92, 6785 (1970).   [4] See, e.g.: B. R. James, Homogeneous Hydrogenation, John Wiley & Sons, New York, 1973.   [5] M. S. Spencer and D. A. Dowden, U. S. Pat. 3,0009; German Pat. 1,114,183; through Chem. Abstr., 56, 8558 (1962).   [6] J. Kwiatek and J. K. Seyler, J. Organometal. Chem., 3, 421, 433 (1965).   [7] M. G. Burnett, P. J. Connolly and C. Kemball, J. Chem. Soc. A, 1967, 800; ibid., 1968, 991.   [8] R. Baker, Chemical Rev., 73, 487 (1973).   [9] a) G. Natta, U. Giannini, P. Pino and A. Cassata, Chem. Ind. (Milan), 47, 524 (1965); b) G. Allegra, F. L. Giudice, G. Natta, U. Giannini, G. Fagherazzi and P. Pino, Chem. Commun., 1967, 1263.   [10] S. Otsuka, T. Taketomi and T. Kikuchi, J. Amer. Chem. Soc., 85, 3709 (1963); S. Otsuka and T. Taketomi, Eur. Polym. J., 2, 289 (1966).
[11] H. Bönnemann, Ch. Grard, W. Kopp and G. Wilke, XXIIIrd Internat. Congr. of Pure and Applied Chemistry, Boston 1971, Vol. 6; H. Bönnemann, Angew. Chem. Internat. Edit., 12, 965 (1973).
[12] P. W. N. M. van Leeuwen, private communication.   [13] R. P. Hughes and J. Powell, J. Amer. Chem. Soc., 94, 7723 (1972).   [14] E. J. Smutny, a) U. S. Pat. 3,267,169 (Aug. 16, 1966); b) J. Amer. Chem. Soc., 89, 6794 (1967).   [15] a) S. Takahashi, T. Shibano and N. Hagihara, Tetrahedron Lett., 1967, 2451; b) S. Takahashi, H. Yamazaki and N. Hagihara, Bull. Chem. Soc., Jpn., 41, 254 (1968).
[16] L. I. Redkina, K. L. Marcovezkii, E. I. Tinyakova and B. A. Dolgoplosk, Dokl. Acad. Nauk SSSR, 186, 397 (1969).   [17] G. Henrici-Olivé, S. Olivé and E. Schmidt, J. Organometal. Chem., 39, 201 (1972).   [18] F. A. Cotton and M. D. LaPrade, J. Amer. Chem. Soc., 90, 5418 (1968).   [19] G. Lefebvre and Y. Chauvin, in: R. Ugo, Edit., Aspects of Homogeneous Catalysis, Carlo Manfredi Editore, Milano, 1970, vol. 1, p. 184.   [20] R. Cramer, J. Amer. Chem. Soc., 89, 1633 (1967).
[21] A. C. L. Su and J. W. Collette, J. Organometal. Chem., 46, 369 (1972).   [22] C. A. Tolman, J. Amer. Chem. Soc., 92, 6777 (1970).   [23] a) A. Miyake, G. Hata, M. Iwamoto and S. Yuguchi, 7th World Petroleum Congr., P. D. No. 22 (3), Mexico 1967; b) M. Iwamoto and S. Yuguchi, Bull. Chem. Soc. Jpn., 41, 150 (1968).   [24] G. Henrici-Olivé and S. Olivé, a) J. Organometal. Chem., 35, 381 (1972); b) Chem. Commun., 1969, 1482.   [25] F. Zingales, F. Canziani and A. Chiesa, Inorg. Chem., 2, 1303 (1963).   [26] B. Bogdanović, B. Henc, B. Meister, H. Pauling and G. Wilke, Angew. Chem. Internat. Edit., 11, 1023 (1972).   [27] G. Wilke et al., Ann. Chem. a) 727, 143 (1969); b) 727, 161 (1969); c) 727, 183 (1969).   [28] a) P. Heimbach, in: R. Ugo, Edit., Aspects of Homogeneous Catalysis, Vol. 2, D. Reidel Publ. Comp., Dordrecht, Holland, 1974; b) P. Heimbach, Angew. Chem. Internat. Edit., 12, 975 (1973).   [29] K. Fischer, K. Jonas, P. Misbach, R. Stabba and G. Wilke, Angew. Chem. Internat. Edit., 12, 943 (1973).   [30] P. W. Jolly, I. Tkatchenko and G. Wilke, Angew. Chem. Internat. Edit., 10, 329 (1971).
[31] J. E. Lydon, J. K. Nicholson, B. L. Shaw and M. R. Truter, Proc. Chem. Soc. (London) 1964, 421.
[32] S. E. Horne, J. P. Kiehl, I. I. Shipman, V. L. Folt and C. F. Gibbs, Ind. Eng. Chem. 48, 784 (1956).
[33] Chem. & Engng. News, Sept. 16, 8 (1974); June 3, 27 (1974).   [34] W. Marconi, The Polymerization of Dienes by Ziegler-Natta Catalysts, in: A. D. Ketley, Edit., "The Stereochemistry of Macromolecules", Marcel Dekker Inc., New York, 1967, vol. 1.   [35] e.g. Ger. Offen. 2,157,004; May 31, 1972; Ger. (East) 89, 706; May 5, 1972.   [36] G. Natta, L. Porri, G. Zanini and A. Palvarini, Chim. Ind. (Milan), 41, 526 (1959).   [37] L. Porri, G. Natta and M. C. Gallazi, Chim. Ind. (Milan), 46, 428 (1964).
[38] R. Warin, Ph. Teyssié, P. Bourdaudurg and F. Dawans, J. Polym. Sci., Polymer Lett. Edit., 11, 177 (1973).   [39] V. I. Klepikova, G. P. Kondratenkov, V. A. Kormer, M. I. Lobach and L. A. Churlyaeva, J. Polymer Sci., Polym. Lett. Edit., 11, 193 (1973).   [40] B. D. Babitskii, B. A. Dolgoplosk, V. A. Kormer, M. I. Lobach, E. I. Tinyakova and V. A. Yakovlev, Izv. Akad. Nauk SSSR, 1965, 1478.
[41] J. F. Harrod and L. R. Wallace, Macromol. 2, 449 (1969); 5, 682 (1972).   [42] E. J. Arlman, J. Catal. 5, 178 (1966).   [43] a) Ph. Teyssié, F. Dawans and J. P. Durand, J. Polym. Sci., C 22, 221 (1968); b) J. M. Thomassin, E. Walckiers, R. Warin and Ph. Teyssié, J. Polym. Sci., 13, 1147 (1975).   [44] J. Furukawa, Pure Appl. Chem., 42, 495 (1975).

Suggested Additional Reading

[45] A. Andreetta, F. Conti and G. F. Ferrari, *Selective Homogeneous Hydrogenation of Dienes and Polyenes to Monoenes*; in R. Ugo, Edit., "Aspects of Homogeneous Catalysis", Carlo Manfredi Editore, Milano, 1970, vol. 1.   [46] P. Heimbach, P. W. Jolly and G. Wilke, *π-Allylnickel Intermediates in Organic Synthesis,* Chapt. IV; Advan. Organometal. Chem., *8,* 29 (1970).   [47] Ph. Teyssié, M. Julémont, J. M. Thomassin, E. Walckiers and R. Warin, *The Specific Polymerization of Diolefins by η^3-Allylic Coordination Complexes*; in J. C. W. Chien, Edit., *"Coordination Polymerization, A Memorial to K. Ziegler"*, Academic Press, New York 1975.

10. Reactions of Carbon Monoxide

10.1. General Aspects

Among the most selective methods for the catalytic introduction of oxygen into organic substrates are those involving carbon monoxide. In most cases, transition metal carbonyl complexes such as $HCo(CO)_4$, $HRh(CO)$ $(PR_3)_3$ and $Ni(CO)_4$ have been identified, or are assumed, to be catalysts.

In general, these reactions involve insertion of carbon monoxide into a metal-carbon bond (*cf.* Eq. (7.2), Section 7.4.1), with subsequent decomposition of the acylmetal complex by hydrogenolysis, hydrolysis, alcoholysis or related reactions with a nucleophile HY containing an active hydrogen. From a mechanistic view-point one may distinguish reactions where the metal-carbon bond is formed by insertion of an olefin (or acetylene) into a metal-hydride bond, from those where it stems from the oxidative addition of an alkyl halide (RX) to a coordinatively unsaturated metal species. These two routes are summarized below in Schemes I and II respectively. Scheme II includes substrates such as alcohols, ethers, esters, *etc.* Halide ions usually are required as cocatalysts; presumably RX is formed *in situ* from the substrate RZ and HX.

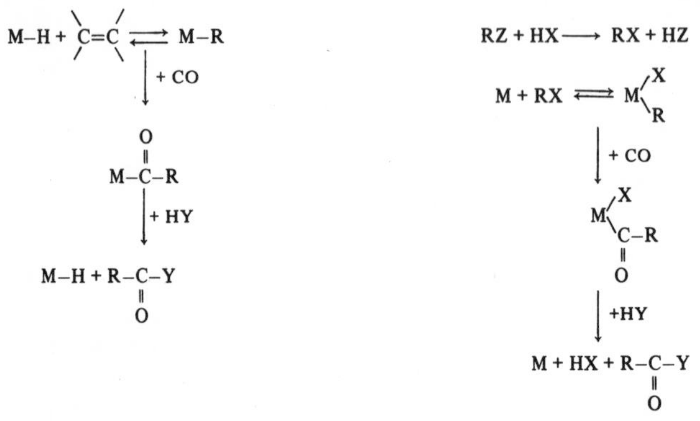

Scheme I   Scheme II

The reaction products range from aldehydes, alcohols, acids, esters, to anhydrides, acid chlorides, amides, *etc.*, depending upon the substrate and the HY component [56].

The basic discoveries in this technically important field of homogeneous catalysis were made in 1938 in Germany by O. Roelen (Ruhrchemie) [1], and by W. Reppe (I.G. Farben) [2]. Roelen discovered the cobalt catalyzed reaction of olefins with carbon monoxide and hydrogen, under pressure, at elevated temperature which leads to aldehydes:

$$C=C \ + CO + H_2 \ \longrightarrow \ HC - CCHO \tag{10.1}$$

The process was named hydroformylation [3], indicating that it comprises the formal addition of the components of formaldehyde $(H-CHO)$ to an olefinic double bond. The older technical name "oxo-reaction" [1] is used also.

The aldehydes can be reduced further to alcohols, with the same catalysts, but at elevated temperature, which constitutes one of the most important technical routes to alcohols, particularly in the C_3 to C_{13} range required for the production of plasticizers, detergents, synthetic lubricants, and as solvents. The detailed mechanism of hydroformylation was not unraveled until 1960, some twentytwo years after its discovery, when Heck and Breslow [4] suggested a reaction scheme, the most important features of which were anticipated in Scheme I above ($HY = H_2$).

The "Reppe reactions" include a wide variety of carbonylations developed in the years 1938 to 1945; due to World War II, a comprehensive account of this work was not published until 1953 [2]. Olefins, acetylenes, alcohols, ethers and esters, *etc.* have been used as substrates, whilst water, alcohols, ammonia, amines, mercaptanes and carboxylic acids act as the nucleophilic component HY [56]. Some examples are given below. The reaction of olefins (and acetylenes) with CO and H_2O [Eqs. (10.2) and (10.3)] is sometimes called "hydrocarboxylation". Under suitable conditions, alcohols are obtained instead [Eq. (10.4), "hydrohydroxymethylation", see Section 10.3.1].

$$RCH = CH_2 + CO + H_2O \longrightarrow RCH_2CH_2CO_2H \tag{10.2}$$

$$HC \equiv CH + CO + H_2O \longrightarrow H_2C = CHCO_2H \tag{10.3}$$

$$RCH = CH_2 + 3CO + 2H_2O \longrightarrow RCH_2CH_2CH_2OH + 2CO_2 \tag{10.4}$$

$$RCH = CH_2 + CO + R'OH \longrightarrow RCH_2CH_2CO_2R' \tag{10.5}$$

$$RCH = CH_2 + CO + H_2NR' \longrightarrow RCH_2CH_2CONHR' \tag{10.6}$$

$$RCH = CH_2 + CO + R'CO_2H \longrightarrow RCH_2CH_2CO_2COR' \tag{10.7}$$

$$RCH_2OH + CO \xrightarrow{I^-} RCH_2CO_2H \tag{10.8}$$

As with hydroformylation, the mechanistic aspects of these homogeneous carbonylations were elucidated only many years after their discovery, and actually are not yet secure in all details. However, it appears clear to-day that, depending upon substrates, catalysts and reaction conditions, most carbonylations proceed according to either Scheme I or II [5–7, 56].

In the following sections we shall first discuss the hydroformylation (Section 10.2). It may have become clear from the above introduction that this reaction is formally closely related to the other carbonylations, nevertheless it is discussed separately for historical reasons and because of its outstanding technical importance. In Section 10.3, we shall treat the formation of alcohols from olefins, carbon monoxide and water as an example of a carbonylation according to Scheme I, and the synthesis of acetic acid from methanol as a process following Scheme II.

10.2. Hydroformylation of Olefins

10.2.1. Reaction Conditions and Catalysts

Most present industrial hydroformylations are conducted at 100–180 °C, with pressures of 100–300 atm, and a CO/H_2 ratio of 1.0 to 1.3, in the presence of cobalt catalysts. Many solvents are suitable, in particular aliphatic, cycloaliphatic or aromatic hydrocarbons. Frequently the substrates and products themselves provide the reaction medium [56]. It was first suggested [1], and later proved [8] that the actual catalyst is a hydridocarbonylcobalt complex, and that under the conditions of hydroformylation almost any form of cobalt, from Raney cobalt to solutions of cobalt salts, is able to form the active species. Cobalt ions are reduced by hydrogen to zerovalent cobalt, whereupon the reaction with carbon monoxide leads to octacarbonyldicobalt, $Co_2(CO)_8$, in which the cobalt is formally in a zero oxidation state:

$$2\,Co + 8\,CO \longrightarrow Co_2(CO)_8 \tag{10.9}$$

This reaction which is exothermic (with 460 kJ/mol octacarbonyldicobalt formed) proceeds equally well with solid or metallic cobalt under the usual hydroformylation conditions. In the presence of H_2, $Co_2(CO)_8$ is then converted into the true catalyst, hydridotetracarbonylcobalt:

$$Co_2(CO)_8 + H_2 \rightleftharpoons 2\,HCo(CO)_4 \tag{10.10}$$

The compound $HCo(CO)_4$ is a gas at room temperature, is very toxic and great caution must be exercised in working with it. On slow cooling it condenses to a light yellow solid with melting point *ca.* −26 °C. The hydride is reasonably soluble in hydrocarbons, slightly soluble in water, and has the assumed structure shown in Fig. 10.1. This conclusion is based on infrared and n.m.r. spectra and by analogy with the chemically related stable complex $HCo(PF_3)_4$ [58].

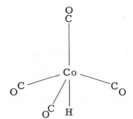

Fig. 10.1. The structure of $HCo(CO)_4$ [58].

A certain minimum partial pressure of CO is required to maintain $HCo(CO)_4$ stable in solution. At 120 °C this is *ca.* 10 atm, at 200 °C about 100 atm [56]. Below the partial pressure indicated, metallic cobalt precipitates.

An interesting catalyst modification consists in replacing a CO ligand by a tertiary phosphine, arsine or phosphite molecule [9]. When $Co_2(CO)_8$ is treated with PR_3 (R = $n\text{-}C_4H_9$) at elevated temperatures under CO and H_2 pressure, a red dimer $Co_2(CO)_6(PR_3)_2$ can be isolated [9]. Its structure has been determined by a three dimensional X-ray study [10]. In contrast to the parent complex, $Co_2(CO)_8$, which in the crystalline state*) contains two carbonyl bridges [12], the phosphine containing complex has a metal-metal bond, see Fig. 10.2.

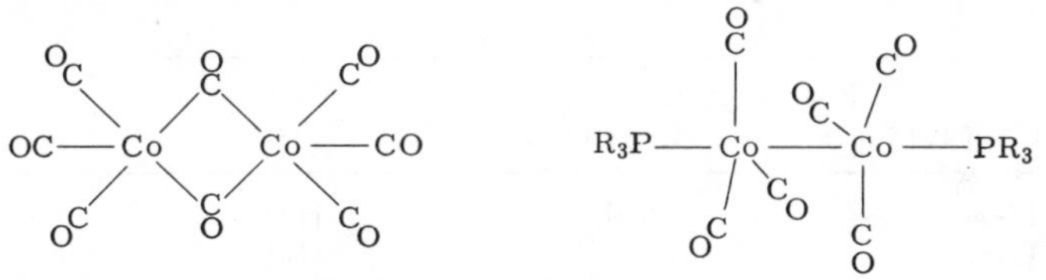

Fig. 10.2. The structures of $Co_2(CO)_8$ [12] and $Co_2(CO)_6[P(n\text{-}C_4H_9)_3]_2$ [10].

The same red phosphine containing compound is also the principal complex isolated from the products recovered after hydroformylation, if $Co_2(CO)_8/PR_3$, in a 1:1 molar ratio, is used as catalyst. Nevertheless, a hydridocobalt complex is assumed to be the active species, by analogy with the parent catalyst [*cf.* Eqs. (10.9) and (10.10)]:

$$Co_2(CO)_8 + 2\,PR_3 \;\rightleftharpoons\; Co_2(CO)_6(PR_3)_2 + 2\,CO \qquad (10.11)$$

$$Co_2(CO)_6(PR_3)_2 + H_2 \;\rightleftharpoons\; 2\,HCo(CO)_3(PR_3) \qquad (10.12)$$

Several other related equilibria have been detected at low cobalt concentrations [22b], and it was suggested that an important intermediate might be $Co_2(CO)_7(PR_3)$ which then reacts with H_2 to yield $HCo(CO)_4$ and $HCo(CO)_3(PR_3)$.

The replacement of a CO by a phosphine stabilizes the catalyst complex. This is primarily because a phosphine (or arsine or phosphite) has less π acceptor capability than CO. Hence, there is more electron density at the metal center for which the remaining CO ligands will compete (*cf.* Fig. 7.16). Consequently, electron back-donation $Co \rightarrow CO$ is enhanced in $HCo(CO)_3(PR_3)$ as compared with $HCo(CO)_4$. Taking into account that the π bond is the most important feature in the bonding between a transition metal and CO (*cf.* Section 6.3.1), such ligand influence would be expected to strengthen the metal-carbonyl bonding. Although this effect cannot be measured directly by bond length determination on the active species, X-ray data from other complexes fully confirm this expectation; an example is given in Table 10.1. The lower the π acceptor ability of the ligand L, $R_3P < (RO)_3P < CO$, the stronger are the bonds between the central metal and CO. As usual, the effect is more marked in the *trans* than in the *cis* position.

The higher stability of the catalyst species $HCo(CO)_3(PR_3)$ as compared with $HCo(CO)_4$ is consequential. With the modified catalyst, CO partial pressures as low as 10–20 atm are sufficient to prevent catalyst decomposition, in the temperature range 150–200 °C. However, because of their greater thermal stability, the phosphine modified catalysts are less reactive than $HCo(CO)_4$. (A CO ligand dissociates in the course of the catalytic cycle, see Section 10.2.2.) Thus, the reaction rate at 145 °C with the nonmodified catalyst is *ca.* five

---

* In solution, a temperature dependent equilibrium between the bridged form and a metal-metal bonded form of octacarbonyldicobalt was detected by infrared sprectroscopy [11].

Table 10.1. Bond lengths in the complex

$$
\begin{array}{c}
\text{CO} \\
\text{OC} \underset{\displaystyle \underset{\text{L}}{|}}{\overset{\displaystyle \overset{|}{}}{\text{—}}} \text{CO} \\
\text{OC} \text{———} \text{CO} \\
\text{Cr}
\end{array}
$$

L	Bond length (Å)		Ref.
	Cr−CO (*trans*)	Cr−CO (*cis*)	
CO	1.909	1.909	[13]
$P(OC_6H_4)_3$	1.861	1.896	[14]
$P(C_6H_5)_3$	1.845	1.880	[14]

times higher than that obtained at 180 °C with the modified one, at equal concentrations. This means that a larger reactor volume is required for the same throughput. However, this inconvenience is balanced by other properties. Thus, the higher stability facilitates separation of the catalyst from the products which can be done simply by distilling off the products and recycling the catalyst to the reactor, whereas with the unmodified cobalt catalyst the metal carbonyl has to be decomposed by chemical methods [56].

The phosphine modified cobalt catalysts are more effective in hydrogenating the primary reaction products of hydroformylation, namely the aldehydes, to alcohols. At 180 °C, the alcohols are obtained directly, in one step. Unfortunately, the higher hydrogenation capacity also affects the olefin feedstock. For propylene, 10–15 % of the olefin is hydrogenated to propane and hence lost for the hydroformylation process. With unmodified cobalt catalyst only 2–3 % of propylene is lost [15]. A further important difference concerns the product distribution, in particular the ratio of linear to branched compounds. The modified catalysts show higher selectivity towards straight chain products which are generally of more value. This last point will be reviewed further in Section 10.2.4.

Besides the cobalt carbonyls mentioned, iron and nickel carbonyls, a number of noble metal complexes, also of chromium, manganese and rhenium, have been used with varying success as catalysts for the hydroformylation of olefins [56, 57, 59]. Rhodium is by far the most active metal, surpassing even cobalt in this respect. All other metals are inferior. Carbonyl-rhodium catalysts have been reported to be several orders of magnitude more reactive than $Co_2(CO)_8$ under comparable conditions [16], and the same trend is observed with phosphine modified carbonylrhodium complexes [57].

Several patents refer to carbonylrhodium catalysts, which can be applied as such (*e.g.* $Rh_6(CO)_{16}$), or made *in situ* from a variety of starting materials [59]. Most recent work, however, concentrates on the use of phosphine modified rhodium catalysts which have several outstanding properties. The phosphine ligands can be applied either by using rhodium complexes already containing them, *e.g.* $RhH(CO)[P(C_6H_5)_3]_3$ [17] or $Rh(CO)$-$(Cl)[P(C_6H_5)_3]_2$ [18], or by adding them, usually in great excess, to a rhodium carbonyl system. A remarkable number of such combinations has already been reported [59]. The hydroformylation of olefins with the modified rhodium catalysts is carried out typically at 80–150 °C, or even at room temperature, with a 1:1 CO/H_2 mixture at 5–50 atm, *i.e.* under relatively mild conditions. The modified rhodium catalysts are characterized by a low hydrogenation capability under hydroformylation conditions (no olefin loss, and over 99 % selectivity towards aldehydes [57]). Moreover the double bond migration is strongly inhibited. On hydroformylation of terminal linear olefins, a very favorable linear to branched ratio can be obtained. With inner olefins, on the other hand, the carbonylation takes place predominantly at the site of the original double bond (see Section 10.2.4). Another important feature is the unique thermal stability of some of the rhodium complexes which greatly facilitates catalyst recycling [18]. These evident advantages will certainly lead to an increasing importance of modified rhodium catalysts in industrial application, despite the high price of the noble metal [19].

10.2.2. The Mechanism

A generally agreed reaction scheme for the hydroformylation of olefins is given in Eqs. (10.13)–(10.18), formulated here for a carbonylcobalt complex (L = CO or PR_3). The most important features of this scheme have been proposed by Heck and Breslow [4]. The individual steps will be discussed subsequently, together with experimental corroboration contributed by several authors.

$$HCo(CO)_3L \rightleftharpoons HCo(CO)_2L + CO \tag{10.13}$$

$$HCo(CO)_3L + RCH = CH_2 \rightleftharpoons RCH_2CH_2Co(CO)_3L \tag{10.14a}$$

$$HCo(CO)_2L + RCH = CH_2 \rightleftharpoons RCH_2CH_2Co(CO)_2L \tag{10.14b}$$

$$RCH_2CH_2Co(CO)_2L + CO \rightleftharpoons RCH_2CH_2Co(CO)_3L \tag{10.14c}$$

$$RCH_2CH_2Co(CO)_3L + CO \rightleftharpoons RCH_2CH_2COCo(CO)_3L \tag{10.15}$$

$$RCH_2CH_2COCo(CO)_3L \rightleftharpoons RCH_2CH_2COCo(CO)_2L + CO \tag{10.16}$$

$$RCH_2CH_2COCo(CO)_2L + H_2 \longrightarrow RCH_2CH_2CHO + HCo(CO)_2L \tag{10.17}$$

$$RCH_2CH_2COCo(CO)_3L \\ + HCo(CO)_3L \longrightarrow RCH_2CH_2CHO + Co_2(CO)_6L_2 \tag{10.18}$$

The importance of the hydridocarbonylcobalt species $HCo(CO)_3L$ in the catalytic cycle is well supported by the fact that this compound (with L = CO) reacts with olefins to produce aldehydes even in the absence of hydrogen and carbon monoxide [58]. In such a reaction $HCo(CO)_4$ furnishes both hydrogen and carbon monoxide; see Eq. (10.19). (Note that in this particular case the branched aldehyde predominates, cf. Section 10.2.3.)

$$RCH = CH_2 + 2 HCo(CO)_4 \longrightarrow RCH(CHO)CH_3 + Co_2(CO)_7 \tag{10.19}$$

Confirmation of the presence of the hydridocarbonyl complex, even under the high temperature and pressure conditions required for hydroformylation on a technical scale, as well as support for several other mechanistic details of the reaction, has been achieved by the direct observation of the reaction in a high pressure infrared spectrophotometric solution cell [20–22]. With $Co(CO)_8$ or the $Co(CO)_8/PR_3$ system as catalyst precursors, it was shown that, under the usual pressure and temperature and in the absence of an olefin, the equilibria (10.10) and (10.12) are displaced well to the right [21a, b]. Spectra obtained actually during hydroformylation, *i.e.* in the presence of an olefin, revealed varying amounts of $HCo(CO)_3L$, depending upon the olefin and L; we return to this point later on.

The dissociation of one carbon monoxide ligand from the relatively stable 18 electron compound $HCo(CO)_3L$ [Eq. (10.13)] is assumed to give a more reactive 16 electron species, $HCo(CO)_2L$. Although there is no direct infrared spectroscopic proof for the formation of the tetracoordinated cobalt species, kinetic evidence supports a dissociative mechanism for the thermolysis of $HCo(CO)_4$ in solution; the first step follows Eq. (10.13) [23]. On the other hand, no experimental evidence is available to decide whether or not the 18 electron complex is at all active, or whether olefin coordination and insertion necessarily require the 16 electron species. In any case, the concentration of the tetracoordinated cobalt species is assumed to be very low.

In the context of Eq. (10.13) it is noteworthy that the overall rate of hydroformylation using unmodified cobalt carbonyl catalyst is inversely proportional to the CO partial pressure. Kinetic measurements by Natta *et al.* [24] led to the following rate equation, which may be used with fair approximation even on an industrial scale [56]:

$$\frac{d\,[\text{aldehyde}]}{dt} \approx k\,[\text{olefin}]\,[\text{Co}]\,p_{\text{H}_2}p_{\text{CO}}^{-1} \tag{10.20}$$

(Note that this relationship implies that the rate is independent of the overall pressure when $H_2/CO = 1$.) Thus, although a certain CO pressure is required to maintain the stability of the catalyst in solution (*cf.* preceding Section), higher CO pressure is detrimental to the rate. This is in agreement with Equilibrium (10.13): the higher the carbon monoxide pressure, the lower the concentration of the more active tetracoordinated species. However, other steps also depend upon the CO pressure, and could be responsible for the CO rate dependence (*vide infra*).

Eqs. (10.14 a) and (10.14 b) describe the insertion of the olefin into the CoH bond of the penta- and tetracoordinated species respectively. These equations are in fact simplifications because, according to all previous knowledge, coordination of the olefin to the metal center is required to precede the insertion step. Whether the insertion can actually take place at the pentacoordinated 18 electron species remains open to discussion (*cf.* Section 10.2.4). After insertion into the CoH bond according to Eq. (10.14 b), the more stable pentacoordinated structure is assumed to be adopted immediately [Eq. (10.14 c)].

No spectral evidence has been found to support the presence of the alkylcobalt species, but in a hydroformylation system involving an iridium complex instead of cobalt, an alkylmetal species may be identified clearly according to a recent report [21 c]. Presumably, the alkyl species in the more active cobalt (and rhodium) based systems are extremely shortlived.

The next step in the reaction is the insertion of one of the CO ligands into the metal-alkyl bond, resulting in an acylcobalt complex [Eq. (10.15)]. A precedent for this insertion in connection with a manganese complex has been reported in Section 7.4.1. Markó *et al.* [25] have shown that the benzylcobalt complex $C_6H_5CH_2Co(CO)_3[P(C_6H_5)_3]$, which is a relatively stable crystalline compound, also reacts in solution with ^{13}CO (20 °C, 1 atm) according to:

$$C_6H_5CH_2Co(CO)_3L + {}^{13}CO \longrightarrow C_6H_5CH_2COCo(CO)_2(^{13}CO)L$$

Thus it appears fairly well documented that ligand CO is used for the formation of the acyl-metal group.

Acylcobalt species have been observed in high pressure infrared spectroscopy under hydro-formylation conditions [21a, b, 22]. For the very rapid hydroformylation of terminal olefins with $HCo(CO)_4$ they represent the dominating cobalt species, showing that the insertion of CO is followed by a relatively slow step. If internal olefins are hydroformylated with $HCo(CO)_4$, or if phosphine modified cobalt catalysts are used with terminal or internal olefins, no acyl complexes are observed, but the hydridocobalt species is predominant. In these latter cases the hydroformylation rate is relatively low, and the infrared results indicate that this is so because the coordination and/or insertion of the olefin [Eq. (10.14a) or (10.14b)] is the slow, rate determining step. Evidently with $HCo(CO)_4$ and terminal olefins this step is much more rapid.

The acylcobalt complex has to react with hydrogen in order to give the aldehyde. This step is assumed to proceed *via* oxidative addition of hydrogen. Evidently the pentacoordinated acylcobalt species is unable to tolerate this oxidative addition which requires two free co-ordination sites. Hence, the reaction is assumed to be preceded by equilibrium, Eq. (10.16) which, then, provides a further reason for the observed detrimental effect of elevated CO partial pressure on the reaction rate.

Eq. (10.17), representing the (irreversible) hydrogenolysis of the acylcobalt complex, is again a shorthand notation since this step includes the oxidative addition of H_2 and the re-ductive elimination of the aldehyde, with the formation of $HCo(CO)_2L$ which will reform $HCo(CO)_3L$ by means of equilibrium (10.13).

Finally, aldehyde formation may also occur, under certain conditions, *via* reaction of the acylcobalt complex with a hydridocobalt species, Eq. (10.18). In the stoichiometric reaction, Eq. (10.19), this is the only way to aldehyde. It has been suggested, that this reaction might even be a major pathway to aldehyde in the catalytic reaction [22a]. From infrared data it was concluded that the establishment of the equilibrium, Eq. (10.10), requires several hours at $T = 100\,°C$, $p_{CO} = 11$ atm, $p_{H_2} = 49$ atm. Most of the cobalt (84%) is then present as $HCo(CO)_4$. Addition of octene-1 causes a sudden decrease of the peaks characteristic for $HCo(CO)_4$, whereas those of $Co_2(CO)_8$ increase. The concentration change gradually reverts as conversion proceeds. These observations were explained assuming partial (or exclusive) aldehyde formation according to Eq. (10.18), with a relatively slow subsequent delivery of $HCo(CO)_4$ from the equilibrium, Eq. (10.10), at least under the applied experimental conditions.

Similar reaction schemes have been suggested for rhodium based catalysts. Markó and his coworkers [26] investigated the hydroformylation of terminal olefins with $Rh_4(CO)_{12}$, and found the following rate law operative:

$$\frac{d\,[\text{aldehyde}]}{dt} = k\,[\text{olefin}]^0\,[\text{Rh}]\,p_{H_2}p_{CO}^{-1}$$

i.e. the (initial) overall rate is independent upon the olefin concentration. Based on a reac-tion scheme analogous to that given in Eqs. (10.13)–(10.18), with $L = CO$, and assuming $RhH(CO)_3$ to be the catalyst formed from $Rh_4(CO)_{12}$ under hydroformylation conditions, the authors interpreted their findings by suggesting that most of the rhodium is transformed after a short induction period to the relatively stable acyl derivative $RCORh(CO)_4$ and that the rate determining step is the reaction of $RCORh(CO)_3$ with H_2 [cf. Eqs. (10.17) and (10.16)].

Hydroformylation of olefins with the well defined complex $HRh(CO)L_3$ where $L = P(C_6H_5)_3$ has been investigated most thoroughly by Wilkinson and his coworkers [17]. (In the absence of CO this complex is a powerful hydrogenation catalyst, cf. Chapter 8.2.2.) Hydroformylation proceeds even at room temperature and 1 atm total pressure (CO: $H_2 = 1$) with remarkable speed, and most of the studies have been made under these conditions. The action of CO on a solution of $HRh(CO)L_3$ involves rapid conversion into $HRh(CO)_2L_2$ (which is proposed as the active catalyst) and $[Rh(CO)_2L_2]_2$. Under hydroformylation conditions (CO and H_2 present) the equilibrium

$$[Rh(CO)_2L_2]_2 + H_2 \; \rightleftharpoons \; 2\,HRh(CO)_2L_2 \tag{10.21}$$

is assumed to lie mainly on the side of the hydrido complex. Although numerous other equilibria appear to be involved [17, 59], the major pathways as suggested by Wilkinson et al. may be summarized as follows:

$$(10.22)$$

At catalyst concentrations $> ca.\ 6 \times 10^{-3}$ mol l^{-1}, steps a–d are assumed to represent the most important route (called the "associative pathway"). The tetracoordinated acylrhodium complex is crucial for the oxidative addition of H_2 (included in step d). The main reason for the higher activity of rhodium catalysts as compared to cobalt is probably associated with the greater ease of this oxidative addition in the case of rhodium [59]. Equilibrium (c) accounts for the observed inhibition by increasing CO partial pressure. At lower rhodium concentrations, dissociation of a further phosphine ligand gives the more active species $HRh(CO)_2L$ [cf. Eq. (8.36)]. The more rapid "dissociative pathway" f–h, b–d, is assumed to predominate under these conditions.

This was inferred from kinetic measurements (in particular from an unusual dependence of rate upon the rhodium concentration), and from the inhibiting effect of an excess of phosphine [17]. It will be recognized easily that the most important individual steps of this cycle, as well as their sequence (coordination, followed by insertion of the olefin into a MH bond, alkyl to acyl rearrangement, oxidative addition of hydrogen, reductive elimination of aldehyde) are essentially the same as those given in Eqs. (10.13)–(10.18).

The role of dicarbonyl complexes as key intermediates is supported by the observation that, if $HRh(CO)_2L_2$ and $HRh(CO)L_2$ are present together in solution, only the former reacts with ethylene at 25 °C, as shown by the NMR spectrum [27].

Acylrhodium intermediates have been observed spectroscopically only if the reaction of $HRh(CO)L_3$ with CO and a terminal olefin is carried out in the absence of H_2. With ethylene at one atm pressure ($C_2H_4 : CO = 1$), complete conversion into a propionylrhodium species occurred [28]. Under technical hydroformylation conditions (high pressure, high temperature) none of the suggested intermediates such as acyl- or alkylrhodium species was detected in the infrared spectrum of the reaction solution, when using the high pressure cell [20]. Instead, the spectrum of an as yet unidentified carbonylrhodium species has been observed and it was suggested that this might be in a rapid equilibrium with one of the complexes involved in Wilkinson's scheme, Eq. (10.22).

Another frequently used rhodium catalyst precursor is $Rh(CO)(Cl)L_2$ [$L = P(C_6H_5)_3$] [18, 19, 25]. Under comparable conditions the rate of hydroformylation is considerably lower than with $HRh(CO)L_3$ [20], and an induction period is observed which has been ascribed to formation of a hydridorhodium species. Addition of an organic base such as triethylamine eliminates the induction period, probably because the hydride species is formed according to Eq. (10.23) [29]:

$$Rh(CO)(Cl)L_2 + H_2 + CO + R_3N \longrightarrow HRh(CO)_2L_2 + R_3NHCl \qquad (10.23)$$

Morris and Tinker [20] have shown that under high pressure and in the absence of amine, the only rhodium species present in amounts detectable by infrared spectroscopy is the starting complex $Rh(CO)(Cl)L_2$. Under these conditions the rate at 100 °C is *ca.* four times lower than with $HRh(CO)L_3$ as catalyst precursor at 80 °C. Evidently the active species is present only in very low concentration. On addition of triethylamine, the rate increases considerably, and at the same time the infrared spectrum of the unidentified rhodium species mentioned in connection with the $HRh(CO)L_3$ system is observed. We may thus safely conclude that the hydroformylation mechanism is the same whether $HRh(CO)L_3$ or $Rh(CO)(Cl)L_2$ is used and it seems likely that the numerous other rhodium complexes reported as more or less active in hydroformylation [59] differ only in the mode by which the common active species is formed.

A very instructive breakdown of the hydroformylation cycle in its individual steps, and the infrared spectroscopic observation of all important intermediates, was achieved recently by Whyman [21c], who used an iridium catalyst. When a heptane solution of $HIr(CO)_3L$ [$L = P(i-C_3H_7)_3$] was treated with ethylene (*ca.* 14 atm, 50 °C) complete transformation into $C_2H_5Ir(CO)_3L$ was observed after 30 minutes. On venting the excess of ethylene and replacing it with CO at 14 atm, conversion into $C_2H_5COIr(CO)_3L$ was complete after a further 30 minutes. On replacing the excess of CO by H_2, the initial complex $HIr(CO)_3L$ reappeared and propionaldehyde was formed quantitatively. This sequence of reactions can be repeated almost indefinitely, the intensity of the aldehyde absorption increasing at the end of each cycle.

Before concluding this section, some consecutive reactions of the aldehydes formed and, in particular, their transformation into alcohols requires discussion. Further hydrogenation takes place with the same hydroformylation catalysts at elevated temperatures. In principle, it is possible to obtain alcohols as the main product in a one step process, by performing the hydroformylation at 180–230 °C with phosphine modified cobalt or rhodium catalysts. But generally it is more practicable, if the reaction is carried out in two steps by first carrying out hydroformylation under the usual conditions, then raising the temperature only after the olefin has been transformed into aldehyde. The main advantage is that olefin loss by hydrogenation to alkane is minimized. Alcohol yields up to 90% can be realized [59]. The reaction is assumed to proceed *via* coordination of the aldehyde to the metal center (M), insertion into the $M-H$ bond and, finally, hydrogenolysis [56, 59]:

$$R-CHO + HM(CO)_n L_m \rightleftharpoons \begin{matrix} H \\ R-C=O \\ \cdot HM(CO)_n L_m \end{matrix}$$

$$\longrightarrow RCH_2OM(CO)_n L_m \xrightarrow{H_2} RCH_2OH + HM(CO)_n L_m$$

Another (mostly undesired) consecutive reaction of the aldehyde product, namely aldol condensation, is found frequently to a small extent. If in special cases this result is required, it can be promoted by addition of a corresponding catalyst [*e.g.* $Mg(OCH_3)_2$]. Thus, 2-ethylhexanol, an important plasticizer intermediate, is formed from propylene by hydroformylation, aldol condensation, and subsequent hydrogenation.

Hydroformylation using primary or secondary alcohols as solvents at moderate temperatures leads to acetals. This process may become important if very sensitive aldehyde products have to be masked by acetal formation [56].

10.2.3. Product Isomer Distribution and its Origins

So far we have neglected the regiospecificity of hydroformylation with regard to an olefinic double bond, and that additional isomers may be expected if double bond migration precedes hydroformylation. The product distribution obtained depends not only on the catalyst and the substrate olefin, but also on the reaction conditions, in particular on the CO partial pressure and on the temperature. Although the mechanism of isomeric distribution control is far from being completely understood, it now appears clear that steric factors play a more important role than do electronic effects.

Let us consider first the hydroformylation of linear terminal olefins. Apart from certain extreme cases mentioned below, these olefins give generally more linear than branched aldehydes [56–58]. At first sight one may be inclined to interpret this observation simply as a favored olefin insertion according to electronic[*] expectations with subsequent CO insertion.

$$\begin{matrix} \overset{\delta+}{M}-\overset{\delta-}{H} \\ \underset{H_2C=CHR}{\overset{\delta-}{} \quad \overset{\delta+}{}} \end{matrix} \longrightarrow MCH_2CH_2R$$

---

* Note that even with $HCo(CO)_4$, a strong acid in aqeuous solution, the hydrogen has predominantly hydride character in nonpolar solvents and in the gas phase [56]; see also Ref. [24] in Chapter 7.

(In Section 7.4.1 we defined the electronically expected mode of insertion as the "Markownikoff mode".) However, certain experimental facts cast some doubt on this simple picture. Thus, in the stoichiometric reaction of propylene with $HCo(CO)_4$ under nitrogen [*cf.* Eq. (10.19)], *ca.* 70% of the aldehyde product is branched, indicating predominant insertion according to the "anti-Markownikoff mode". We are reminded that group VIII transition metals favor this mode of insertion, presumably because strong electron back-donation reverses the polarity of the double bond of the coordinated olefin molecule (*cf.* Table 8.5). Under the high pressure and high temperature conditions of technical hydroformylation, the same catalyst is well able to give the linear isomer predominantly. However, the fraction of linear product then strongly depends upon the CO partial pressure, as shown in Table 10.2 [30a]. (These data were obtained in the presence of ethyl orthoformiate in order to block the carbonyl groups as soon as they were formed, thus avoiding reduction or condensation of the primary reaction product.)

Table 10.2. Hydroformylation of propylene with $Co_2(CO)_8$ [30a]; $T = 108\,°C$, $p_{H_2} = 80$ atm.

p_{CO} (atm)	% Linear isomer	p_{CO} (atm)	% Linear isomer
4	45.4	31	65.0
6.5	49.7	66	70.2
9	51.4	104	72.7
10	53.8	147	73.3
12.5	60.6	224	73.4
21	63.5		

The increase of the linear fraction with increasing CO partial pressure is most easily understood as a consequence of steric hindrance, resulting from incorporation of more CO ligands in the complex at one or more of the individual steps of the catalytic cycle (*e.g.* formation of $RCo(CO)_4$ as opposed to $RCo(CO)_3$ [58]), although an electronic contribution (less electron back-donation to the coordinated olefin because of the additional CO ligand) cannot completely be ruled out. The question of steric hindrance will be discussed in more detail below.

A further observation which casts doubt on the prevalence of electronic effects is that replacement of a CO ligand by a phosphine ligand leads to an increase of the linear fraction, with cobalt as well as with rhodium carbonyl catalysts, *cf.* Table 10.3. The electron density at the metal center, and hence the electron back-donation to the coordinated olefin would be expected to be more marked in complexes containing a phosphine ligand, yet not the branched, but the linear product is favored. Again steric effects resulting from the use of the bulky phosphine ligands appear likely.

Table 10.3. Hydroformylation of octene-1; influence of $P(n\text{-}C_4H_9)_3$ [31]; $p = 200$ atm; $H_2:CO = 1$.

Catalyst	Phosphine/Metal	$T(°C)$	Linear isomer (%)
$HCo(CO)_4$	0	190	65
$HCo(CO)_4$	2	190	83
$HRh(CO)_4$	0	100	52
$HRh(CO)_4$	60	140	65

We come now to the question regarding which individual step of the catalytic cycles [Eqs. (10.13–18) and Eq. (10.22)] might be susceptible to steric influences by ligands. The final product evidently depends on the configuration of the acylmetal species. One suggested control mechanism refers to the alkyl to acyl rearrangement [58]. In the presence of CO, this rearrangement may be assumed to involve a transition state with a coordination number higher than that of the alkyl species, *cf.* Fig. 10.3.

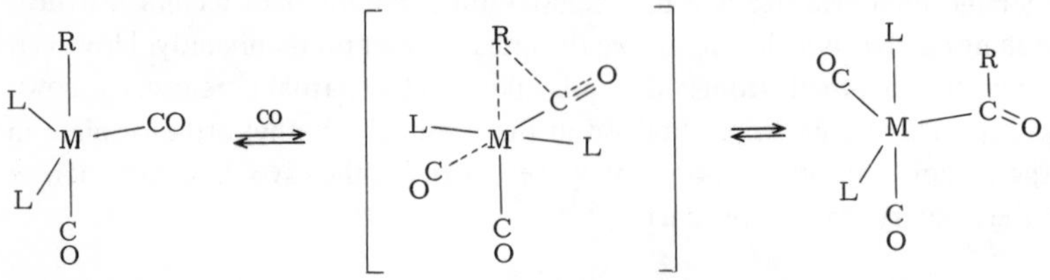

Fig. 10.3. Assumed transition state during alkyl to acyl rearrangement [58]. *Cf.* Eq. (10.15) and step b in Eq. (10.22). M = metal, L = CO or phosphine.

Such a transition state could certainly be sensitive to steric effects, and considerably more favorable for linear alkyl groups than for branched ones. Expressed in terms of reaction kinetics this would mean that the rate constant for the carbonylation reaction of linear alkyl-metal species is higher than that for the branched ones. With regard to the regiospecificity of the stoichiometric reaction, Eq. (10.19), it has to be taken into account that under nitrogen the carbonylation occurs without the incorporation of an additional CO ligand into the coordination sphere of the metal. The steric restrictions are omitted, and the reaction is free to proceed according to electronic requirements. The same might be true with the cobalt carbonyl catalyst at extremely low CO partial pressure (Table 10.2). In correspondence with this view, the unmodified rhodium carbonyl catalyst produces more branched aldehydes from terminal linear olefins than the cobalt species; presumably the rhodium carbonyl complexes have a greater tendency than cobalt complexes to lose carbon monoxide. In fact it has been concluded from kinetic data that the equilibrium

$$HRh(CO)_4 \rightleftharpoons HRh(CO)_3 + CO$$

is not strongly shifted to either side under usual hydroformylation conditions [59].

The amazing preference of cobalt catalysts for the carbonylation of the linear alkyl species becomes evident when one considers the hydroformylation of internal olefins. The cobalt

Table 10.4. Hydroformylation of octenes in a $HCo(CO)_4/P(n-C_4H_9)_3$ system [31]; $P/Co = 2$; $T = 190\,^{\circ}C$; $p = 200$ atm; $H_2:CO = 1$.

n-Octene	Product distribution			
	1-Nonanal	2-Methyl-1-octanal	2-Ethyl-1-heptanal	2-Propyl-1-hexanal
Octene-1	83	11	4	2
Octene-4	76	10	6	7

catalysts have a strong capability of promoting double bond migration under hydroformylation conditions, the remarkable consequence being that internal and terminal olefins may give very similar product distributions, as exemplified in Table 10.4. It should, however, be noted that the overall rate of hydroformylation is generally lower with inner olefins, reflecting their lower coordination ability (cf. Section 7.2.1).

Double bond migration proceeds via the olefin insertion, β-hydrogen abstraction mechanism discussed in Section 8.1.1. Since all steps in the hydroformylation cycle, except the final formation of the aldehyde, are assumed to be reversible, we may now rewrite the most relevant part of the reaction schemes in Section 10.2.2, taking double bond migration into account:

$$
\begin{array}{l}
RCH_2\text{-}CH_2\text{-}CH_2\text{-}M(CO)_mL_n \xrightleftharpoons[(3a)]{CO} RCH_2\text{-}CH_2\text{-}CH_2\text{-}CO\text{-}M(CO)_mL_n \\
\\
HM(CO)_mL_n \\
+ \quad\quad (1a) \quad RCH_2\text{-}CH=CH_2 \quad\quad\quad\quad n\text{-Aldehyde} \\
RCH_2\text{-}CH=CH_2 \\
\quad\quad\quad\quad HM(CO)_mL_n \\
\\
RCH_2\text{-}CH\text{-}M(CO)_mL_n \xrightleftharpoons[(3b)]{CO} RCH_2\text{-}CH\text{-}CO\text{-}M(CO)_mL_n \\
\quad\quad CH_3 \quad\quad\quad\quad\quad\quad\quad\quad CH_3 \\
\\
HM(CO)_mL_n \\
+ \quad\quad (1b) \quad RCH=CH\text{-}CH_3 \quad\quad 2\text{-Methylaldehyde} \\
RCH=CH\text{-}CH_3 \\
\quad\quad\quad\quad HM(CO)_mL_n \\
\\
RCH\text{-}M(CO)_mL_n \xrightleftharpoons[(3c)]{CO} RCH\text{-}M(CO)_mL_n \\
\quad CH_2\text{-}CH_3 \quad\quad\quad\quad\quad CH_2\text{-}CH_3 \\
\\
\quad\quad\quad\quad 2\text{-Ethylaldehyde}
\end{array}
\tag{10.24}
$$

The data of Table 10.4 indicate that extensive double bond migration [Equilibria (2) in Eq. (10.24)] has taken place prior to the product determining CO insertion. When the hydroformylation of octene-1 is interrupted at 50% conversion, isomerized olefin is detected, but the isomer mixture is far from being equilibrated [31]. And yet the aldehyde distributions obtained from internal and terminal olefins are such similar. One has to conclude that most of the double bond migration and subsequent CO insertion takes place in the coordination sphere of the metal without the substrate molecule leaving the complex.

However, the product distribution from internal and terminal olefins is not always as similar as in Table 10.4. Fig. 10.4 depicts the hydroformylation of heptene-1 and of heptene-2 at different temperatures, using unmodified carbonylcobalt catalyst. Evidently double bond migration proceeds more slowly than hydroformylation at lower temperature.

Instructive experiments concerning double bond migration and the predilection for carbonylation at the terminal carbon have been reported by Pino and coworkers [33]. Using 5,5,5-d_3-pentene-2 as substrate and $Co_2(CO)_8$ as catalyst precursor, hydroformylation was car-

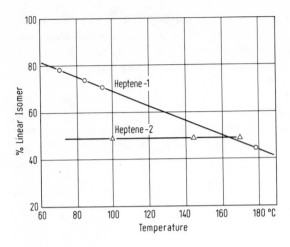

Fig. 10.4. Linear fraction of reaction product from hydroformylation of terminal and internal olefins, as a function of temperature [32]. $H_2/CO = 3$; overall pressure 246 atm; Catalyst: $(Co_2(CO)_8$.

ried out at $100\,°C$ and 200 atm total pressure ($H_2 : CO = 1$). The product was investigated with respect to isomer distribution:

$$CH_3CH=CHCH_2CH_3 \xrightarrow[Co_2(CO)_8]{100\,°C}$$

→ $CH_3CH_2CH_2CH_2CH_2CHO$ 75%   (I)

→ $CH_3CH_2CH_2CH(CHO)CH_3$ 19%   (II)

→ $CH_3CH_2CH(CHO)CH_2CH_3$ 6%   (III)

Furthermore, the linear fraction I was analyzed (by NMR) with regard to the average deuterium content at each carbon atom. This was carried out after transforming the aldehyde end group into a methyl ester group, and the following deuterium distribution was obtained:

$$CH_{1.04}D_{1.96} - CH_{1.91}D_{0.09} - CH_2 - CH_{1.53}D_{0.47} - CH_{1.53}D_{0.47} - COOR$$

From this result it was concluded convincingly that ca. two thirds of the n-hexanal arises from hydroformylation at the terminal carbon atom closest to the original double bond, and one third from reaction at the terminal carbon atom originally bearing the three deuterium atoms. The fact that, within the limits of experimental error, no deuterium is found at carbon atom 4 of the ester, and that carbon 5 had a very low deuterium content, indicates that the double bond migrates along the substrate molecule essentially only once before hydroformylation at the preferred terminal carbon atoms takes place, and that the substrate molecule does not leave the complex in between. If the same experiment is carried out at low CO partial pressures (5.5 atm), a random distribution of deuterium at all positions is found, showing that under these conditions double bond migration is much faster than hydroformylation.

Additional information has been obtained from similar experiments by Piacenti and coworkers [30b], who prepared a linear alkylcobalt carbonyl by a different route, from $NaCo-(CO)_4$ and 1-iodopropane-3-d$_3$. At $80\,°C$, 80 atm CO and 80 atm H_2, the only observed product was n-butanal-4-d$_3$. This result does not vary if the experiment is repeated in the presence of pentene-1. It can be concluded that, at this temperature, the insertion of CO into the n-alkyl-Co bond [forward reaction of Equilibrium (3a), Eq. (10.24)] is much more rapid than β-H abstraction from the alkyl group (2a). This is in perfect agreement with the control mechanism suggested by Orchin (see Fig. 10.3 and context). A further conclusion is that the backward reactions of the CO insertion equilibria (3a–c), although well documented in the absence of CO [58], do not play an important role under hydroformylation conditions.

Double bond migration is generally less frequent with rhodium carbonyls than with cobalt catalysts. With the former, hydroformylation takes place predominantly at the original site of the double bond, unless very low CO partial pressure is applied. The low isomerization capacity is particularly marked with phosphine modified rhodium catalysts.

10.2.4. Some Aspects of Modified Catalysts

Not only phosphines, but also arsines, stibines, phosphites and amines have been used as modifiers, to provide the cobalt, rhodium, and other transition metal carbonyl catalysts with different ligands. But above all, the phosphorus derivatives gave the best results for industrial applications.

The effect of phosphine additives upon cobalt or rhodium carbonyl catalysts is not identical, although there are several common trends. We shall discuss first the phosphine modified cobalt catalysts in some detail, followed by the effects of phosphorus and nitrogen containing ligands on rhodium. Finally, catalysis by some other phosphine modified carbonylmetal complexes will be mentioned briefly.

Tucci [34] investigated the influence of phosphine basicity on activity and selectivity of phosphine modified cobalt carbonyl. Some results on the hydroformylation of hexene-1 are given in Table 10.5. With the exception of the last two examples given in this Table, the product linearity (at $p_{CO} = 1$) does not depend very strongly on the basicity of the phosphines (as expressed by their pK_a values).

Table 10.5. Hydroformylation of hexene-1 with HCo(CO)$_3$L (L = tertiary phosphine). Effect of phosphine basicity [34c, 57]. $T = 160\,°C$; p *ca.* 70 atm. (H$_2$/CO = 1); P/Co = 1.

Nr.	L	pK_a	Product linearity (%)
1	P(i-C$_3$H$_7$)$_3$	9.4	85.0
2	P(C$_2$H$_5$)$_3$	8.7	89.6
3	P(n-C$_3$H$_7$)$_3$	8.6	89.5
4	P(n-C$_4$H$_9$)$_3$	8.4	89.6
5	P(n-C$_8$H$_{18}$)$_3$	8.4	90.2
6	P(C$_2$H$_5$)$_2$(C$_6$H$_5$)	6.3	84.6
7	P(C$_2$H$_5$)(C$_6$H$_5$)$_2$	4.9	71.7
8	P(C$_6$H$_5$)$_3$	2.7	62.4

The overall hydroformylation rate, on the other hand, shows a somewhat more marked dependence on ligand basicity (Figs. 10.5 and 10.6).

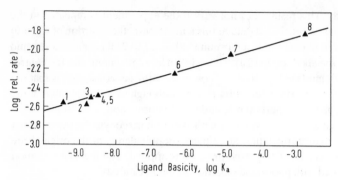

Fig. 10.5. Relationship between relative rate of hydroformylation of hexene-1 and organophosphine basicity [34c]. Numbers refer to Table 10.5; identical reaction conditions.

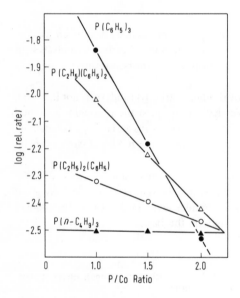

Fig. 10.6. Concentration effects of various phosphines on the rate of hydroformylation of hexene-1 [34c].

All results are best interpreted assuming that in the presence of phosphine L the following equilibrium operates:

$$HCo(CO)_3L + CO \ \rightleftarrows \ HCo(CO)_4 + L \tag{10.25}$$

With more basic phosphines, the equilibrium position lies far to the left, even at P/Co = 1, whereas with less basic ligands some $HCo(CO)_4$ is present. Since the latter is by far the more active catalyst, Figs. 10.5 and 10.6 become easily understandable. In other words, the reported data reflect the influence of ligand basicity on the position of the equilibrium, Eq. (10.25), rather than on the actual activity of the catalytic species $HCo(CO)_3L$. The activity series may even be inverted, the better donor phosphines generating more active catalyst species, as indicated by the crossover of the straight lines in Fig. 10.6.

The optimal selectivity towards linear products appears to be guaranteed as long as the basicity of the phosphine (or its concentration) is high enough to keep the equilibrium, Eq. (10.25), far to the left. No additional effect from the bulkiness of R in PR_3 is indicated by the data of Table 10.5. The relatively low selectivity towards linear product at the foot of the Table is merely a consequence of the presence of the less selective $HCo(CO)_4$, at $L/Co = 1$. The selectivity of 62.4% observed with triphenylphosphine actually approaches that obtained with the unmodified catalyst under comparable conditions [56].

Fig. 10.7 shows that the H_2/CO ratio (at constant overall pressure) has no influence on the product selectivity, and only an insignificant one (at $H_2/CO \geq 1$) on the reaction rate. This is in marked contrast to the findings with the unmodified catalyst (*cf.* Table 10.2 for selectivity, and Eq. (10.20) for the rate). The observed discrepancy between modified and nonmodified systems can be traced back to the fact that the CO ligands are more strongly bonded in the phosphine substituted complexes than in $HCo(CO)_4$ (*cf.* Section 10.2.1). Based on high pressure infrared spectroscopy data, it was concluded in Section 10.2.2 that the hydrogenolysis of the acylmetal complex is the slowest step with the unmodified catalyst, whereas with cobalt/phosphine systems olefin coordination and/or insertion is slowed down by the phosphine ligand, so that it becomes the rate determining step. The fact that in this latter case the rate is nearly independent of p_{CO} (Fig. 10.7) appears incompatible with the necessity of CO dissociation according to Eq. (10.13). Hence, olefin coordination and insertion most probably occur *via* Eq. (10.14a), at least in the modified catalyst system. The inhibitory effect of bulky phosphine ligands on this step is immediately evident. The independence of the reaction rate on p_{H_2} underlines the statement that the H_2 addition to the acylmetal complex is not rate determining. The low rate at $H_2/CO \leq 1$ (low H_2 partial pressure with concomitant high CO partial pressure, see Fig. 10.7) may then be attributed to reduced formation of the active species according to Eq. (10.12).

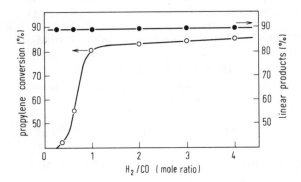

Fig. 10.7. Hydroformylation of propylene with $Co_2(CO)_8/P(n\text{-}C_4H_9)_3$; $p_{CO} = 2$; $T = 160\,°C$; p ca. 70 atm. Effect of CO and H_2 partial pressures on rate and selectivity [34a].

With the unmodified cobaltcarbonyl catalyst, on the other hand, the observed proportionality between rate and p_{H_2} [Eq. (10.20)] may be due either to the rate determining addition of H_2 to the acylmetal complex, or alternatively to a slow formation of the hydrido-metal catalyst, if aldehyde formation proceeds according to Eq. (10.18). The inverted proportionality between rate and p_{CO} is the consequence of Eq. (10.16). The negative influence of low CO partial pressure on selectivity towards linear products (Table 10.2) was ascribed in Section 10.2.3 to release of steric hindrance by CO dissociation at low p_{CO}. The reduced CO dissociation in case of the phosphine modified catalyst readily explains the high selectivity, even at low CO partial pressures.

A further characteristic property of phosphine modified cobalt catalysts is their capacity to hydrogenate the aldehyde product directly to yield alcohols. As one might have expected, the alcohol/aldehyde ratio increases with increasing H_2/CO ratio, as shown in Fig. 10.8.

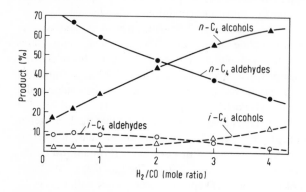

Fig. 10.8. Effect of CO and H_2 partial pressures on aldehyde/alcohol distribution; conditions as in Fig. 10.7 [34b].

With modified rhodium carbonyl catalysts, on the other hand, essentially no conversion of aldehyde to alcohol takes place. Phosphine as well as phosphite ligands are able to direct the hydroformylation of linear terminal olefins towards straight chain aldehydes; 80–90% selectivity can readily be achieved, if the phosphorus containing ligand is applied in excess [57, 59]. With $HRh(CO)L_3$ ($L = P(C_6H_5)_3$), the hydroformylation of hexene-1 was reported to yield even 94% of *n*-heptanal, if the reaction is carried out at 85 °C and at 27 atm in molten triphenylphosphine [17]. Presumably the large excess of phosphine is required to suppress the sterically less hindered f–h route in Eq. (10.22). Interestingly, the hydrogenating ability of the complex $HRh(CO)L_3$ (*cf.* Table 8.2) is completely suppressed in the presence of CO.

A most remarkable property of phosphine modified rhodium catalysts (with a high excess of phosphine) is their very low double bond migration capability, which is in contrast to cobalt catalysts (*cf.* Table 10.4 and Fig. 10.4). Thus, Asinger *et al.* [31] obtained 99% 2-propylhexanal-1 by hydroformylating *trans*-octene-4 with a carbonylrhodium/tributylphosphine system (P/Rh = 60), at 140 °C and 200 atm, $H_2/CO = 1$:

$$C_3H_7CH = CHC_3H_7 \xrightarrow{\text{CO,H}_2} C_3H_7CH_2CH(CHO)C_3H_7 \qquad 99\% \qquad (10.26)$$

Evidently this lack of double bond migration can be of great importance if a particular branched aldehyde is required.

The rhodium amine complex $HRh(CO)L_3'$, where $L' =$ tribenzylamine, catalyzes the hydroformylation of linear terminal olefins under very mild conditions (20–30 °C, 1 atm, $H_2/CO = 1$) with high selectivity towards straight chain aldehydes (80–90%), although the rate of product formation is low [35]. At higher temperatures the rate increases, but the fraction of branched aldehyde, as well as isomerization of the olefin and hydrogenation to alkane become significant. In contrast to the phosphine modified systems, no excess of ligand is required. Addition of tribenzylamine drastically reduces the reaction rate, indicating that the active species is formed by dissociation of at least one amine ligand [*cf.* Eq.

(10.22)]. The fact that the two complexes, $HRh(CO)L'_3$ with the strong σ donor ligand $L' =$ tribenzylamine, and $HRh(CO)L_3$ with the poor σ donor and strong π acceptor ligand $L =$ triphenylphosphine, both permit high selectivity towards straight chain aldehydes, appears to underline once more the predominance of steric effects in selectivity.

Carbonyls of manganese $[Mn_2(CO)_{10}]$ and rhenium $[Re_2(CO)_{10}]$, with an excess of tributyl-phosphine L have also been used as hydroformylation catalysts [36]. The formation of $HMn(CO)_4L$ and $HRe(CO)_4L$, under H_2 and CO pressure, was proved by infrared spectroscopy, indicating a hydroformylation mechanism similar to that with modified cobalt carbonyls. Whereas here the unmodified carbonylmetal complexes are practically inactive, the modified systems are moderately active at $200\,°C$ and $p > 180$ atm ($H_2/CO = 1$). Only terminal olefins are hydroformylated; mixtures of aldehydes and alcohols are obtained; selectivity towards linear hydroformylation products is 80–90%, but side reactions (hydrogenation and isomerization of the starting olefin) are also important.

A remarkable selectivity in favor of linear aldehyde has been achieved also with a platinum complex, $HPt(SnCl_3)(CO)L_2$ where $L = P(C_6H_5)_3$. This catalyst operates at $100\,°C$ and 200 atm ($H_2/CO = 1$) in aromatic solvents, and may be formed *in situ* from *trans*-$PtHClL_2$ with a fivefold excess of $SnCl_2 \cdot 2\,H_2O$, under hydroformylation conditions. The exact structure of the relatively unstable complex is unknown although a four-coordinate ionic salt $[HPt(CO)L_2]^+[SnCl_3^-]$ has been suggested [37]. The catalyst yields $> 95\%$ linear hexanal from pentene-1. The crystalline complex can be recovered unchanged from the reaction mixture and may be reused without loss of activity. The rate of reaction compares favorably with that using cobalt catalysts under similar conditions; phosphine modified rhodium catalysts operate effectively at this temperature, but with less selectivity towards linear product, unless an excess of phosphine is added, which then lowers the rate.

10.2.5. Hydroformylation of Dienes

Unconjugated dienes can be hydroformylated to yield mixtures of mono- and dialdehydes with cobalt [56] as well as with rhodium [59] carbonyl catalysts, with variable yields. The larger the distance between the double bonds, the better the yield of dialdehyde.

The hydroformylation of conjugated dienes, and in particular of butadiene, has been studied extensively in view of the technical importance of the expected 1,6-hexanedial and/or 1,6-hexanediol products. Carbonylcobalt, carbonylrhodium, and phosphine modified carbonyl-cobalt catalysts give only saturated monoaldehydes and alcohols. It has been shown [38a] that the primary step is hydrogenation of one of the double bonds. Presumably a π allyl metal complex is formed which is active for hydrogenation, but not for hydroformylation:

$$CH_2=CHCH=CH_2$$
$$+ \qquad \longrightarrow \quad HC \overset{CHCH_3}{\underset{CH_2}{\diagdown\diagup}} \text{—} M(CO)_mL_n$$
$$HM(CO)_mL_n$$

$$\overset{H_2}{\longrightarrow} \quad CH_3CH=CHCH_3 + HM(CO)_mL_n$$

The resulting monoolefin is then normally hydroformylated. With the phosphine modified cobalt catalyst, mainly the straight chain alcohol, *n*-pentanol, is obtained at $200\,°C$, 200 atm.

Phosphine modified rhodium catalysts, on the other hand, are able to hydroformylate butadiene twice. With a high excess of triphenylphosphine (P/Rh = 40), a 1:1 mixture of dialdehydes and saturated monoaldehydes has been obtained in a rapid reaction, at 135 °C and 200 atm [38]. A relatively high concentration of unsaturated monoaldehyde was detected if the reaction mixture was investigated at incomplete conversion, indicating that for the phosphine modified rhodium catalyst the primary step is hydroformylation to β, γ or γ, δ unsaturated monoaldehydes which then, due to lack of double bond migration with these catalysts [compare Eq. (10.26)], are further hydroformylated to yield mostly branched dialdehydes. (α, β Unsaturated aldehydes would not react [37].)

A relatively high yield of linear 1,6-hexanediol from butadiene was claimed in a recent patent [39], which refers to a three step process. In the first step, butadiene is treated with $Rh(Br)(CO)L_2$ where $L = P(C_6H_5)_3$, at 120 °C, 280 atm, $H_2/CO = 1$, in the presence of an alcohol. The unsaturated monoaldehyde formed is immediately trapped as an acetal. In the second stage, the acetal is hydroformylated at 170 °C, 80–110 atm, with a phosphine modified carbonylcobalt catalyst, which is able to migrate the double bond to the terminal position and to give predominantly terminal aldehyde (cf. Table 10.4). Finally, the dialdehyde acetals are transformed into diols by hydrogenation with Raney nickel at elevated temperatures and pressures. The 1,6-hexanediol fraction constitutes 89% of the diols (42% with respect to the initial butadiene).

10.2.6. Hydroformylation of Substituted Olefins

Unsaturated alcohols, phenols, ethers, aldehydes, carboxylic acids, esters, anhydrides, nitriles, as well as N-substituted amides and imides, have been hydroformylated, in particular with rhodium catalysts, although with variable success [59].

Substituted olefins having strong conjugation between the olefinic and substituent double bonds, such as styrene or α, β unsaturated aldehydes, tend to become hydrogenated rather than hydroformylated. With phosphine modified rhodium catalysts, however, styrene can be transformed into the two isomeric phenylpropanals, with high yield:

These compounds are important intermediates in the technical preparation of scents, cosmetics, pharmaceuticals, etc. [40].

In α, β-unsaturated esters, such as acrylates, methacrylates and crotonates, the conjugation is less pronounced than with α, β unsaturated aldehydes. The esters can generally be hydroformylated with good yields, with cobalt as well as with rhodium catalysts. The olefinic double bond is much more polarized in these compounds than in simple olefins. Interestingly, electronic effects appear to determine the product isomer distribution more forcibly, at least at relatively low temperatures. The α carbon has the higher negative charge, hence the electronically expected mode of insertion leads to the sterically more hindered alkylmetal species:

Actually, the stoichiometric reaction of $HCo(CO)_4$ with acrylates, in the absence of carbon-monoxide and hydrogen, at $0\,°C$, gives $>80\%$ of the α formyl ester [4], indicating that the electronically favored intermediate (I) predominates. In the catalytic reaction, the ratio of α formyl to β formyl product depends strongly upon the temperature as is shown in Fig. 10.9 [41].

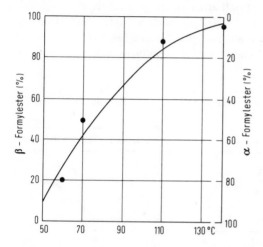

Fig. 10.9. Hydroformylation of acrylic ester with $Co_2(CO)_8$, at 300 atm, $H_2/CO = 1$ [56].

Evidently, the electronic effect on olefin insertion into the $M-H$ bond predominates at lower temperatures, whereas at higher temperatures the sterically less hindered β-product is favored. Even more stringent is the electronic control in the case of hydroformylation of methyl methacrylate with a phosphine modified rhodium catalyst where, at $110\,°C$ and 600 atm, 94% of the sterically hindered α formyl ester is obtained [56]:

$$H_2C=C(CH_3)CO_2R \xrightarrow[\text{Rh/P}]{\text{CO/H}_2} \begin{cases} RO_2CC(CH_3)_2CHO & 94\% \\ RO_2CCH(CH_3)CH_2CHO & 6\% \end{cases}$$

Presumably the high pressure required (high CO as well as high H_2 partial pressure are effective [42]) additionally slows down the isomerization of the electronically favored alkyl-metal species, by blocking coordination sites.

The hydroformylation of α, β unsaturated carboxylates is important in the direct synthesis of lactones. Reaction conditions are chosen such that β formyl esters predominate; during subsequent hydrogenation of the aldehyde functions to yield alcohols, lactones form spontaneously [41]:

$$R-CH=CHCO_2R' \xrightarrow[\text{Catalyst}]{\text{CO/H}_2} RCH(CHO)CH_2CO_2R'$$

$$\xrightarrow{H_2} RCH(CH_2OH)CH_2CO_2R' \xrightarrow{-R'OH} \underset{O}{\overset{R}{\diagdown}}\!\!\bigcirc\!\!=O$$

10.2.7. Asymmetric Hydroformylation

A number of prochiral substrates has been hydroformylated with cobalt or rhodium based catalyst systems involving optically active phosphine ligands. Up to 25% optical purity in the products has been achieved in some cases [43]. Some interesting data of Pino et al., summarized in Table 10.6, were obtained with a system containing $HRh(CO)L_3$ ($L = P(C_6H_5)_3$) as catalyst precursor, and an excess of the optical active ligand $(-)$-2,2-dimethyl-4,5-bis-(diphenylphosphinomethyl)-1,3-dioxolane [$(-)$-DIOP, see Section 8.2.4]. The hydroformylation was carried out at room temperature and pressure, because low temperatures and low pressures greatly favor asymmetric induction. Long reaction times (many days) are, of course, required under these conditions.

Table 10.6. Hydroformylation with the catalyst system $HRh(CO)L_3/(-)$-DIOP (molar ratio 1:4) [43]. Room temp.; 1 atm ($H_2/CO = 1$); aromatic solvents.

Substrate	Isolated optical active compound	optical purity (%)	Chirality
butene-1	2-methylbutanal	18.8	R
octene-1	2-methyloctanal	15.2	R
cis-butene	2-methylbutanal	27.0	S
3-methylpentene-1	4-methylhexanal	4.6	R
styrene	α-phenylpropanal	22.7	R

It is still an open question whether asymmetric induction in these systems is inherently limited to ca. 25% optical purity in the product, or whether part of the catalytically active species did not undergo asymmetric synthesis (e.g. because replacement of L by DIOP was incomplete).

In the reaction of 3-methylpentene-2, which itself contains an asymmetric carbon center, stereoelective hydroformylation is observed. Starting from the racemic olefin, the (R)-antipode is hydroformylated at a higher rate, yielding (R)-4-methylhexanal; the recovered olefin shows some accumulation of the (S)-antipode (3,3% optical purity).

With respect to the individual steps of the catalytic cycle, during which asymmetric induction actually takes place, it is important to note that the 2-methylbutanal formed from butene-1 (ca. 7% of the aldehydes, the remainder being pentanal), and that formed from cis-butene-2 (98% of product) have different chirality (Table 10.6). According to the generally accepted reaction mechanism [cf. Eq. (10.22)], both butenes yield a common intermediate, a sec-butylrhodium complex which, if the asymmetric ligand L* contains an asymmetric center, must exist in two diastereomeric forms, i.e.:

$$
\begin{array}{cc}
\underset{\text{(S)}}{C_2H_5}\diagdown\underset{}{C}\diagup CH_3 & \underset{\text{(R)}}{H}\diagdown\underset{}{C}\diagup CH_3 \\
H\diagup \quad \diagdown Rh\text{-}L^*_{(R)} & C_2H_5\diagup \quad \diagdown Rh\text{-}L^*_{(R)}
\end{array}
$$

If asymmetric induction does not take place in the formation of the diastereomeric alkyl-metal species, but in a later step, then the chiral aldehydes resulting from both butenes should have the same chirality and the same optical purity. That (R) and (S) chirality prevail in the aldehyde obtained from butene-1 and cis-butene respectively implies that the diastereomeric composition of the intermediate alkylmetal complexes depends on the initial C_4 olefin, and that at least for one of them asymmetric induction occurs during the formation of the alkylrhodium complex. Fig. 10.10 illustrates how asymmetric induction may be visualized even with the symmetric *cis*-butene molecule.

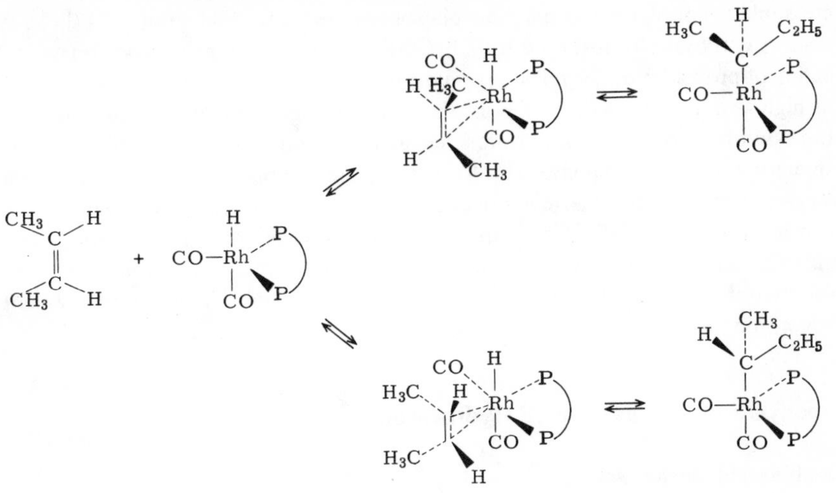

Fig. 10.10. Suggested mode of asymmetric induction during formation of the alkylrhodium complex from *cis*-butene [43]. $P^\frown P = (-)$-DIOP.

The question remains as to which factors control the diastereomeric composition of the alkylmetal species. Free energy differences between the two relevant conformers in the olefin metal complexes would give rise to a higher concentration of one relative to the other (thermodynamic control); on the other hand, insertion of the olefin into the metal-hydrogen bond would proceed at different rates if one or other conformer were to be the starting point, due to steric hindrance (kinetic effect). A somewhat related precedent exists in formation of a chiral product from *cis*-butene in boron chemistry [44]. The (non-catalytic) reaction of optically active diisopinocampheylborane with *cis*-butene gives diisopinocampheyl-2-butylborane; oxidation yields isopinocampheol and optically active 2-butanol, with an optical purity of 87%. Boron does not offer the possibility of double bond formation with the olefin, as is generally assumed to be the case for transition metal centers (*cf.* Fig. 7.1). Hence the very high asymmetric induction appears to be due, at least in this case, primarily to kinetic control.
The complexity of the multi-step cycle in hydroformylation further complicates the understanding of asymmetric induction: it is not yet clear whether induction in the alkylmetal complex formation is decisive for the final configuration of the aldehyde, because CO insertion and/or liberation of the aldehyde may invert the configuration of the chiral center of the product [45].

10.2.8. Heterogeneization of Homogeneous Hydroformylation Catalysts

Several attempts have been made to fix soluble hydroformylation catalysts such as $HRh(CO)L_3$ or $Rh(Cl)(CO)L_2$ where $L = P(C_6H_5)_3$ on insoluble supports, aiming to combine the advantages of homogeneous catalysis, such as high catalyst efficiency and selectivity, with those of heterogeneous catalysts, in particular easy product separation and recovery of the expensive noble metal catalyst. Physical absorption of the catalyst complexes on carbon or alumina carriers [46], as well as chemical bonding to an insoluble polymeric backbone, such as crosslinked polystyrene containing phosphorus as functional groups [47], have proved satisfactory. Physically absorbed $Rh(Cl)(CO)L_2$ can be used for the vapor-phase hydroformylation of propylene under relatively mild conditions. Although the catalyst efficiency is not as high as in homogeneous catalysis, the system is highly stable, has low volatility, long life and essentially the same normal/branched aldehyde ratio is found as at the same reaction temperature in the liquid phase [19]. Insoluble catalysts obtained from incorporation of rhodium complexes into polymeric matrices have been used for continuous liquid phase operation over long periods [47]. Very little rhodium was leached from the support, and relatively high effective concentration of catalytic species was obtained, as compared with the soluble catalyst where solubility limits concentration. Certainly, further improvement will be made within this area.

10.3. Carbonylations of Unsaturated Hydrocarbons

10.3.1. Catalysts and Mechanism

The carbonylations formulated in Eqs. (10.2)–(10.7) are catalyzed by carbonyl complexes of group VIII metals. Carbon monoxide pressures varying from 10 to 400 atm, and temperatures between 100 °C and 300 °C are generally required, depending on substrate and catalyst [56]. $Ni(CO)_4$ was the preferred catalyst for acetylene reactions, which played an important role in the early days of carbonylation when olefins were not yet available at competitive prices. For the carbonylation of olefins Fe, Co, Rh, Pd and Ru catalysts are applied more frequently. Metal carbonyls can also act as catalysts for the carbonylation of amines, giving N-formyl derivatives; cobalt [48] as well as ruthenium [49] based catalysts have been applied.

Although it has been reported that metal carbonyls alone are active, it is known that the addition of acids (HCl, HI) or of halogens (in particular I_2) increases the reaction rate considerably. One may conclude that the actual catalyst is a hydridocarbonylmetal species which might be formed in situ under the reaction conditions, with the cooperation of the hydrogen donor HY (H_2O, ROH, H_2NR, etc; cf. Eqs. (10.2)–(10.7), and Scheme I, Section 10.1). A hydridometal species will, evidently, form more readily by oxidative addition of an acid (e.g. $Pd/P(C_6H_5)_3/HCl$, with subsequent in situ carbonylation); added iodine will react with HY yielding HI, which then could oxidatively add to the metal.

Once hydridocarbonylmetal complexes are accepted as the active catalysts, the mechanism of the carbonylations can be explained easily by a mechanism anologous to that of hydroformylation, with the difference that, in the last step, the acylmetal species reacts with the hydrogen donor HY (e.g. H_2O) to yield product and to reestablish the catalyst:

$$RCH = CH_2 + HM(CO)_m \longrightarrow RCH_2CH_2M(CO)_m \tag{10.27}$$

$$RCH_2CH_2M(CO)_m + CO \longrightarrow RCH_2CH_2COM(CO)_m \tag{10.28}$$

$$RCH_2CH_2COM(CO)_m + H_2O \longrightarrow RCH_2CH_2CO_2H + HM(CO)_m \tag{10.29}$$

The reaction of olefins with CO and H_2O can be directed to different products, changing the catalyst and/or the reaction conditions. Thus from ethylene, one obtains propionic acid in good yield with carbonylnickel at 300 °C. Upon maintaining the concentration of the acid in the reaction medium relatively high, and lowering the temperature to 240 °C, the anhydride prevails. With an iron catalyst in the presence of a base, the main product is propanol (see Section 10.3.2) whereas with octacarbonyldicobalt and at low water concentrations, ethyl propionate is formed [56].

As in hydroformylation, isomeric product mixtures are obtained frequently. Low temperatures and high CO pressures tend to favor the formation of linear products from terminal olefins, although at the expense of the rate. Phosphine modified catalysts have also been used, and the product may vary dramatically with the phosphine ligand. For example, the reaction of α-methylstyrene with CO and alcohol, at 100 °C and 390 atm, using phosphine modified Pd catalysts, gives the following products [50a] (for DIOP, see Section 10.2.7):

The ester obtained with the $PdCl_2$/DIOP system contains an asymmetric carbon center. Utilising the optical active form of the diphosphine, $(-)$-DIOP, the asymmetric synthesis of this ester was achieved with *ca.* 50% optical purity, when $(CH_3)_3COH$ was used instead of C_2H_5OH, and a large amount of benzene was added as diluent (benzene : alcohol : olefin = 400 : 1.5 : 1) [50b].

10.3.2. Hydrohydroxymethylation of Olefins

The synthesis of alcohols according to Eq. (10.4) will now be treated in more detail, by way of example. The reaction proceeds under relatively mild conditions (100 °C, 10–15 atm CO). The catalyst precursor is pentacarbonyliron in the presence of an amine, preferably an N-alkylpyrrolidine; alcohol is used as solvent. Discarding for the moment the formation of CO_2, the alcohol synthesis can be formulated as:

$$RCH = CH_2 \xrightarrow[\text{Fe(CO)}_5/NR_3]{\text{CO/H}_2\text{O}} RCH_2CH_2CH_2OH \tag{10.30}$$

i.e. an alcohol having one carbon more than the starting olefin is produced. Formally, one hydrogen atom has been added to one and a hydroxymethyl group to the other side of the olefinic double bond; from thence comes the name hydrohydroxymethylation*[), by analogy with the hydroformylation*[). Starting from propylene, a mixture of *n*-butanol and *i*-butanol (90% yield) can be obtained easily, even in continuous operation [51].

A deep red iron complex identified as the ionic compound $[R_3NH]^+[HFe_3(CO)_{11}]^-$ can be isolated from the reaction solution of an alcohol synthesis with the $Fe(CO)_5/R_3N$ system as

---

* Strictly speaking both names are incorrect, since hydro means water.

catalyst. It was also shown independently that pentacarbonyliron reacts with amine and water, with the following stoichiometry [51]:

$$3\ Fe(CO)_5 + R_3N + 2\ H_2O \longrightarrow [R_3NH]^+[HFe_3(CO)_{11}]^- + 2\ CO_2 + 2\ CO + H_2 \quad (10.31)$$

This equation indicates that, catalyzed by the metal, part of the ligand CO is converted by H_2O to CO_2 and H_2 under mild conditions (*cf.* [2]).

$$CO + H_2O \longrightarrow CO_2 + H_2 \quad\quad\quad\quad (10.32)$$

An X-ray structural analysis of the ionic complex, with $R = C_2H_5$, has been undertaken [52], and the structure of the anionic moiety is shown in Fig. 10.11. The position of the hydrogen could not directly be established from the X-ray work, but was inferred from stereochemical and bonding considerations.

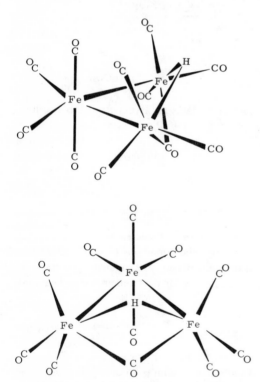

Fig. 10.11. Molecular configuration of $[HFe_3(CO)_{11}]^-$ [52].

The ionic iron cluster reacts, at 100 °C and in the absence of CO, with olefin and water, in a stoichiometric reaction, giving the corresponding alcohol with one carbon atom more than the starting olefin. The complex decomposes in this reaction, giving partly some unidentified Fe(II) compound, some $Fe(CO)_5$, and CO_2 [51]. From these results it was concluded that the trinuclear hydridocarbonyliron ion is the active catalytic species in the alcohol synthesis. Irrespective of the actual bonding we may abbreviate the active species, for convenience, as

$HFe(CO)L_n$, where L_n involves further CO ligands and the other iron centers [Eq. (10.33)]. The first steps in the alcohol synthesis, up to the formation of the acylmetal species, are then straightforward [*cf.* Eqs. (10.27) and (10.28)]. But why does the alcohol form subsequently, instead of the acid, as formulated in Eq. (10.29)? In this context it is interesting that the pentacarbonyliron alone, in the absence of amine, actually catalyzes the formation of carboxylic acids [51].

It is tempting to assume that the hydrogen liberated in the formation of the trinuclear iron complex, Eq. (10.31), is utilized for the hydrogenation of the acylmetal species, giving an intermediate aldehyde, just as in hydroformylation. This proposed step, Eq. (10.34), would reestablish the catalyst. The final formation of the alcohol, Eq. (10.35), is based on experimental evidence: $[R_3NH]^+[HFe_3(CO)_{11}]^-$ hydrogenates aldehydes to give alcohols at elevated temperatures, in the absence of H_2 [51]. Evidently the complex decomposes and we assume that $Fe(CO)_5$ and R_3N are reformed. The whole catalytic cycle may then be represented formally as follows:

$$3\ Fe(CO)_5 + R_3N + 2\ H_2O \longrightarrow [R_3NH]^+[HFe_3(CO)_{11}]^- + 2\ CO_2 + 2\ CO + H_2 \quad (10.31)$$

$$[R_3NH]^+[HFe_3(CO)_{11}]^- \equiv HFe(CO)L_n \quad (10.33)$$

$$HFe(CO)L_n + RCH{=}CH_2 + CO \longrightarrow RCH_2CH_2COFe(CO)L_n \quad (10.27) + (10.28)$$

$$RCH_2CH_2COFe(CO)L_n + H_2 \longrightarrow RCH_2CH_2CHO + HFe(CO)L_n \quad (10.34)$$

$$RCH_2CH_2CHO + [R_3NH]^+[HFe_3(CO)_{11}]^- + 4\ CO$$
$$\longrightarrow RCH_2CH_2CH_2OH + 3\ Fe(CO)_5 + R_3N \quad (10.35)$$

$$RCH{=}CH_2 + 2\ H_2O + 3\ CO \longrightarrow RCH_2CH_2CH_2OH + 2\ CO_2 \quad (10.40)$$

The last equation gives the net effect of the cycle. Evidently this mechanism is somewhat speculative and other pathways might be visualized. Thus, the formation of intermediate aldehyde might be brought about by the hydridometal species $HFe(CO)L_n$ rather than by H_2 (*cf.* hydroformylation, where a similar suggestion has been made [22]). Alternatively, it is also possible that the iron cluster merely catalyzes the conversion of CO with water into CO_2 and H_2, in this way supplying the necessary hydrogen:

$$3\ Fe(CO)_5 + R_3N + 2\ H_2O \longrightarrow [R_3NH]^+[HFe_3(CO)_{11}]^- + 2\ CO_2 + 2\ CO + H_2$$

$$[R_3NH]^+[HFe_3(CO)_{11}]^- + 4\ CO \longrightarrow 3\ Fe(CO)_5 + R_3N + H_2$$

$$2\ CO + 2\ H_2O \longrightarrow 2\ CO_2 + 2\ H_2$$

Rather than the trinuclear iron cluster itself, a mononuclear hydridoironcarbonyl species in equilibrium with the cluster might then be the actual catalyst for the hydrohydroxymethylation. More kinetic mechanistic work is required to solve this problem.

Higher olefins are hydrohydroxymethylated at unsatisfactory rates and yields using the pentacarbonyliron/amine system. A noticeable improvement is obtained with rhodium modified pentacarbonyliron systems, again in the presence of amine [53a]. The rhodium is applied as Rh_2O_3, but under the reaction conditions (CO pressure, presence of H_2O and amine) it is assumed to be transformed to $HRh(CO)_m(R_3N)_n$. Extremely small amounts of rhodium (Rh/Fe = 1/200) have a great effect not only on the reaction rate, but also on the product distribution (see Table 10.7). With the rhodium modified system, at $T < 175\,°C$, a considerable amount of aldehyde is obtained. In all cases, the product contains isomeric mixtures of linear and branched alcohols or aldehydes, but more isomerization is observed with the rhodium modified system. It has been suggested that the rhodium species catalyzes the formation of aldehyde, which is then reduced by the iron species to alcohol.

Table 10.7. Carbonylation of octene-1 with the rhodium modified $Fe(CO)_5$/N-methylpyrrolidine system [53a]. Octene/water $\simeq 2$; $p_{CO} = 200$ atm (cold); excess of amine, Fe/Rh = 200.

Temperature (°C)	Yield*) (%)	Nonanal (%)	Nonanol (%)
100**)	2	0	100
150**)	16	0	100
100	15	91	9
125	36	58	42
150	65	40	60
175	94	0	100
200	97	0	100

* Nonanal + nonanol, relative to octene, after 3 h reaction time.
** Runs without rhodium, for comparison.

The bimetallic iron/rhodium/N-methylpyrrolidine system was also effective in a new development of the carbonylation chemistry, named aminomethylation of olefins [53b]. It involves the formation of tertiary amines from secondary amines, olefin, CO and water, with the stoichiometry:

$$\underset{/}{\overset{\backslash}{C}} = \underset{\backslash}{\overset{/}{C}} + 3\,CO + H_2O + HNR_2 \longrightarrow HC - CCH_2NR_2 + 2\,CO_2$$

The reaction is assumed to proceed *via* interaction of the secondary amine with an acylmetal species formed in the usual way. The exact mechanism is not yet clear.

10.3.3. Carbonylation of Methanol

The synthesis of acetic acid from methanol according to Eq. (10.8) is a process of considerable industrial importance. As in hydroformylation, cobalt and rhodium carbonyl complexes are the most important catalyst precursors; a "halogen promoter" (halogen or halide, preferably iodide) is indispensable. The older cobalt based process [54] operates at $210\,°C$ and 500 atm CO; about 90% of the methanol is converted to acetic acid. The more recently developed rhodium catalyst [7] can be used at much lower pressures and provides reaction rates essentially independent of the CO pressure in the range between 13 and 120 atm. A convenient operating temperature is $175\,°C$. The formation of acetic acid proceeds with a degree of selectivity (99%) which is remarkable even for a homogeneous catalytic process.

We shall consider the rhodium catalyzed carbonylation of methanol in more detail. The catalyst precursor comprises two components, a soluble rhodium complex, and an iodide. Invariant reaction rates have been obtained utilizing a variety of rhodium compounds (*e.g.*

$RhCl_3 \cdot 3\ H_2O$, Rh_2O_3, $[R_4As]^+[Rh(CO)_2I_2]^-$), and several promoter components (*e.g.* aqueous HI, CH_3I, I_2), under comparable operating conditions. These results suggest that ultimately the same active species is formed. The most convenient catalyst system consists of the commonly available $RhCl_3 \cdot 3\,H_2O$ and aqueous hydrogen iodide, in a water acetic acid mixed solvent.

The net reaction for the conversion of methanol to acetic acid is simply:

$$CH_3OH + CO \longrightarrow CH_3COOH$$

However, during the course of the reaction, several equilibria are involved:

$$2\,CH_3OH \rightleftharpoons CH_3OCH_3 + H_2O$$

$$CH_3OH + CH_3CO_2H \rightleftharpoons CH_3CO_2CH_3 + H_2O$$

$$CH_3OH + HI \rightleftharpoons CH_3I + H_2O$$

It has been established that methyl iodide attains a steady state concentration very rapidly which remains constant during most of the reaction. The other two equilibria are shifted during the course of reaction, and ultimately all of the intermediates are converted into acetic acid. Kinetic studies showed that the reaction obeys the following overall rate law:

$$\text{rate} = k\,[CO]^0\,[CH_3OH]^0\,[Rh]\,[I^-] \tag{10.36}$$

In the particular case of the rhodium complex $[Rh(CO)_2I_2]^-$, addition of methyl iodide yields a dinuclear acetyl complex of rhodium, $[Rh_2I_6(CH_3CO)_2(CO)_2]^{2-}$, which was isolated and investigated by X-ray structural analysis [55a]. Its molecular configuration is shown in Fig. 10.12. (The cation is $[R_4As]^+$.)

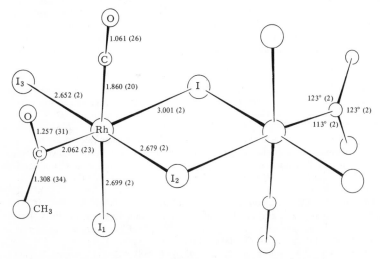

Fig. 10.12. Molecular configuration of $[Rh_2I_6(CH_3CO)_2(CO)_2]^{2-}$ [55a].

Although this observation by no means proves that the dinuclear rhodium cluster is the active catalyst, it indicates the most probable reaction path taking place at a rhodium center: oxidative addition of methyl iodide to the Rh(I) species, with subsequent insertion of a CO ligand into the alkyl-rhodium bond.

The actual catalyst is not yet known, but we may safely assume that it is a square-planar Rh(I) carbonyl species, able to oxidatively add methyl iodide, and to offer a CO ligand for the formation of the acyl ligand. We use the abbreviation $Rh(CO)L_n$ for this active species and the catalytic cycle may then be formulated as [7, 55 b]:

$$CH_3OH + HI \rightleftharpoons CH_3I + H_2O \tag{10.37}$$

$$Rh(I)(CO)L_n + CH_3I \rightleftharpoons CH_3Rh(III)I(CO)L_n \tag{10.38}$$

$$CH_3Rh(III)I(CO)L_n + CO \rightleftharpoons CH_3CORh(III)I(CO)L_n \tag{10.39}$$

$$CH_3CORh(III)I(CO)L_n \rightleftharpoons CH_3COI + Rh(I)(CO)L_n \tag{10.40}$$

$$CH_3COI + H_2O \longrightarrow CH_3CO_2H + HI \tag{10.41}$$

Reductive elimination of CH_3COI reestablishes the catalyst [Eq. (10.40)]. Subsequent hydrolysis of acetyl iodide produces the acetic acid and HI [Eq. (10.41)].

The reaction mechanism postulated is consistent with the kinetic result, Eq. (10.36), if one assumes that the oxidative addition of the methyl iodide to the rhodium complex is rate determining, and that all other steps occur rapidly.

References

[1] O. Roelen, DBP 849548 (1938); Angew. Chem., 60, 62 (1948).   [2] W. Reppe et al., Liebigs Ann. Chem., 582, 1, 38, 72, 87, 116, 133 (1953).   [3] H. Adkins and G. Krsek, J. Amer. Chem. Soc., 71, 3051 (1949).   [4] R. F. Heck and D. S. Breslow, J. Amer. Chem. Soc., 83, 4023 (1961).   [5] R. F. Heck, J. Amer. Chem. Soc., 85, 2013 (1963).   [6] N. v. Kutepow, W. Himmele and H. Hohenschutz, Chem.-Ing.-Tech., 37, 383 (1965).   [7] J. F. Roth, J. H. Craddock, A. Hershman and F. E. Paulik, Chem. Tech., 1971, 600.   [8] M. Orchin, L. Kirch and J. Goldfarb, J. Amer. Chem. Soc., 78, 5450 (1956).   [9] L. H. Slaugh and R. D. Mullineaux, J. Organometal. Chem., 13, 469 (1968).   [10] J. A. Ibers, J. Organometal. Chem., 14, 423 (1968).

[11] K. Noack, Helv. Chim. Acta, 47, 1064 (1964).   [12] C. Gardner Sumner, H. P. Klug and L. E. Alexander, Acta Crystallogr., 17, 732 (1964).   [13] A. Whitaker and J. W. Jeffery, Acta Crystallogr., 23, 977 (1967).   [14] H. J. Plastas, J. M. Stewart and S. O. Grim, Inorg. Chem., 12, 265 (1973).   [15] J. Falbe, Hydroformylation; in: E. G. Hancock, Edit., Propylene and its Industrial Derivatives, Ch. 9, John Wiley, New York, 1973.   [16] B. Heil and L. Markó, Chem. Ber., 102, 2238 (1969).   [17] C. K. Brown and G. Wilkinson, J. Chem. Soc. (A), 1970, 2753, and earlier references therein.   [18] a) J. H. Craddock, A. Hershman, F. E. Paulik and J. F. Roth, Ind. Eng. Chem. Prod. Res. Develop., 8, 291 (1969); b) A. Hershman, K. K. Robinson, J. H. Craddock and J. F. Roth, Ind. Eng. Chem. Prod. Res. Develop., 8, 372 (1969).   [19] C & EN, April 26, 1976, page 25.   [20] D. E. Morris and H. B. Tinker, Chem. Tech. 1972, 554.

[21] R. Whyman, J. Organometal. Chem., a) 66, C23 (1974); b) 81, 97 (1974); c) 94, 303 (1975).   [22] M. van Boven, N. H. Alemdaroğlu, and J. M. L. Penninger, a) Ind. Eng. Chem. Prod. Res. Develop., 14, 259 (1975); b) J. Organometal. Chem., 84, 65 (1975).   [23] F. Ungvary and L. Markó, J. Organometal. Chem., 20, 205 (1969).   [24] G. Natta, Brennst.-Chem., 36, 176 (1955).   [25] K. Nagy-Magos, G. Bor and L. Markó, J. Organometal. Chem., 14, 205 (1968).   [26] G. Csontos, B. Heil and L. Markó, Trans. N.Y. Acad. Sci., 239, 47 (1974).   [27] D. Evans, G. Yagupsky and G. Wilkinson,

J. Chem. Soc. (A), *1968*, 2660.   [28] G. Yagupsky, C. K. Brown and G. Wilkinson, J. Chem. Soc. (A), *1970*, 1392.   [29] D. Evans, J. A. Osborn and G. Wilkinson, J. Chem. Soc. (A), *1968*, 3133.   [30] a) P. Pino, F. Piacenti and P. P. Neggiani, Chem. Ind. (London), *1961*, 1400; b) M. Bianchi, P. Frediani, U. Matteoli, A. Girola and F. Piacenti, Chim. Ind. (Milan), *58*, 223 (1976).

[31] F. Asinger, B. Fell and W. Rupilius, Ind. Eng. Chem. Prod. Res. Develop., *8*, 214 (1969).   [32] V. L. Hughes and I. Kirshenbaum, Ind. Eng. Chem., *49*, 1999 (1957).   [33] a) D. A. v. Bézard, G. Consiglio and P. Pino, Chimia, *28*, 610 (1974); b) D. A. v. Bézard, Ph. D. Thesis, Swiss Federal Institute of Technology, Zürich, May 1976.   [34] E. R. Tucci, Ind. Eng. Chem. Prod. Res. Develop., a) *7*, 32 (1968); b) *7*, 125 (1968); c) *9*, 516 (1970).   [35] B. Fell, E. Müller and J. Hagen, Monatsh. Chem., *103*, 1222 (1972).   [36] B. Fell and J. Shanshool, Chem.-Ztg., *99*, 231 (1975).   [37] C. Y. Hsu and M. Orchin, J. Amer. Chem. Soc., *97*, 3553 (1975).   [38] a) B. Fell and W. Boll, Chem.-Ztg., *99*, 452, 485 (1975); b) B. Fell and W. Rupilius, Tetrahedron Lett., *1969*, 2721.   [39] Belgian Pat. 827–125 (1975).   [40] B. Cornils and R. Payer, Chem.-Ztg., *98*, 596 (1974), and references therein.

[41] J. Falbe, N. Huppes and F. Korte, Chem. Ber., *97*, 863 (1964).   [42] J. Falbe and N. Huppes, Brennst.-Chem., *48*, 46 (1967). [43] P. Pino, G. Consiglio, C. Botteghi and C. Salomon, Advan. Chem. Ser., *132*, 295 (1974), and refs. therein.   [44] H. C. Brown, N. R. Ayyangar and G. Zweifel, J. Amer. Chem. Soc., *86*, 397 (1964).   [45] G. Consiglio, C. Botteghi, C. Salomon and P. Pino, Angew. Chem. Internat. Edit., *12*, 669 (1973).   [46] K. K. Robinson, F. E. Paulik, A. Hershman and J. F. Roth, J. Catal., *15*, 245 (1969).   [47] W. O. Haag and D. D. Whitehurst, Abstr. 2nd. North Amer. Meeting of the Catal. Soc., Houston, Texas, Feb. 1971, p. 16.   [48] A. Rosenthal and I. Wender, in: I. Wender and P. Pino, Edits., *Organic Synthesis via Metal Carbonyls*, Interscience, New York, 1968, vol 1, p. 439.   [49] J. J. Byerley, G. L. Rempel and N. Takebe, Chem. Commun., *1971*, 1482.   [50] a) G. Consiglio and M. Marchetti, Chimia, *30*, 26 (1976); b) G. Consiglio and P. Pino, personal communication.

[51] N. v. Kutepow and H. Kindler, Angew. Chem., *72*, 802 (1960).   [52] L. F. Dahl and J. F. Blount, Inorg. Chem., *4*, 1373 (1965).   [53] A. F. M. Iqbal, a) PhD Thesis, Technische Hochschule Aachen, Germany, 1969; b) Helv. Chim. Acta, *54*, 1440 (1971).   [54] H. Hohenschutz, N. v. Kutepow and W. Himmele, Hydrocarbon Process, *45*, 141 (1966).   [55] a) G. W. Adamson, J. J. Daly and D. Forster, J. Organometal. Chem., *71*, C17 (1974); b) D. Forster, J. Amer. Chem. Soc., *98*, 846 (1976).

Suggested Additional Reading

[56] J. Falbe, *Synthesen mit Kohlenmonoxyd*, Springer-Verlag, Berlin-Heidelberg, 1967; *Carbon Monoxide in Organic Synthesis* (translated by C. R. Adams), Springer-Verlag, New York, 1970.
[57] F. E. Paulik, *Recent Developments in Hydroformylation Catalysis*, Catal. Rev., *6*, 49 (1972).
[58] M. Orchin and W. Rupilius, *On the Mechanism of the Oxo Reaction*, Catal. Rev., *6*, 85 (1972).
[59] L. Markó, *Hydroformylation of Olefins with Carbonyl Derivatives of Noble Metals as Catalysts*, in: R. Ugo, Edit., *Aspects of Homogeneous Catalysis*, D. Reidel Publ. Comp., Dordrecht, Holland, 1974, vol 2.   [60] I. Wender and P. Pino, Edts., *Organic Synthesis via Metal Carbonyls*, Interscience Publ., New York, 1976, vol. 2.

11. Activation of Molecular Oxygen

11.1. General Aspects

Molecular oxygen is a potentially strong oxidizing agent, as judged from its position in the electrochemical series. The standard potential for the redox equilibrium

$$O_2 + 4\,H^+ + 4\,e^- \;\rightleftarrows\; 2\,H_2O \tag{11.1}$$

is $+1.23$ V, which is comparable to the standard potentials of the redox couples I_2/IO_3^- (1.20 V) or $Cr^{3+}/Cr_2O_7^{2-}$ ($+1.33$ V). Nevertheless, reactions of molecular oxygen with most substrates proceed very slowly at room temperature, in the gas phase as well as in homogeneous solution. This relative inertness, which may appear remarkable in view of the biradical character of the ground state of the dioxygen molecule*), is evidently the essential condition for maintaining organic life in an oxygen atmosphere. Two main reasons have been advanced in order to interpret the metastable state of the dioxygen molecule. On the one hand the full redox potential mentioned above would be realized only if the oxygen molecule could be reduced in a four electron step, which is highly improbable under normal conditions. (It may play a role in enzymatic oxidations.) Usually the reduction of dioxygen occurs in a sequence of one-electron steps, and the first electron transfer has an adverse potential ($-0,32$ V), imparting kinetic stability to the molecule [1]. On the other, it has been pointed out that reactions of free triplet molecules (O_2) with singlet molecules (substrate) usually experience high activation energies because of the problem of spin conservation [2].

Complexes of transition metals, containing the dioxygen as a ligand, are able to oxidize certain substrates under unusually mild conditions. Some of these reactions, only a few of which are really catalytic, will be dealt with in Section 11.4. From what is known today, it appears that the metal ion may influence either of the two inhibiting factors mentioned above. This will be discussed in the course of this chapter (particularly in Section 11.3).

Current interest in dioxygen transition metal complexes is also nourished because of a biochemical problem. The respiratory pigments are able to fix oxygen from the atmosphere, to transport it to reacting sites, and there to release it. This, in a sense, is a catalytic process since the natural "oxygen carriers" (e.g. hemoglobin in the red blood cells, myoglobin in the muscle) are not oxidized irreversibly during the process, but retain their activity. It appears well established that transition metals, particularly Fe(II), play an important role. There has been much work and also some speculation as to both the mode of binding of dioxygen to the metal, as well as to why irreversible oxidation to the ferric state does not occur. Investigation of the natural products is complicated by their macromolecular protein component. Therefore, low molecular model substances containing a transition metal center and which are able to combine reversibly with dioxygen, have attracted considerable interest. These model substances will be discussed in the following section.

---

* The dioxygen molecule has two unpaired electrons (cf. Sections 6.2.1 and 11.3.1); its ground state is a triplet state, $^3\Sigma$.

11.2. Models for Natural Oxygen Carriers

11.2.1. Dioxygen Complexes with Metals Other than Iron

In the natural oxygen carriers hemoglobin and myoglobin the metal is surrounded by four nitrogen atoms from the characteristic system of four pyrrol rings (porphyrin). This has been the reason probably why many transition metal complexes with similar and related quadridentate ligands have been investigated with regard to their ability to reversibly fix molecular oxygen. Actually, a considerable number of compounds with this property has been detected during the past years. Mostly, they have cobalt as the central metal, and porphyrins or quadridentate Schiff bases as ligands. The frequent use of cobalt requires some comment. On the one hand it has been shown that replacement of iron in natural hemoglobin and myoglobin by cobalt does not vary essentially the properties most relevant to oxygen fixation of these biological systems [3, 4]. On the other, the synthesis of iron based oxygen carriers met with considerable difficulties until very recently (see Section 11.2.3). Hence cobalt complexes were considered to be suitable models. The reader is referred to a number of reviews which have dealt with various aspects of the problem [5–10]. As an example, we shall discuss one of the most frequently investigated systems, [N,N'-ethylenebis(salicilideneiminato)-Co(II)], abbreviated commonly Co(salen), see Fig. 11.1, as well as several derivatives thereof.

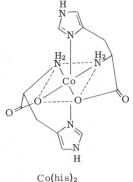

Co(salen)

$R = CH_3$, Co(acacen)

$R = C_6H_5$, Co(bzacen)

Co(his)$_2$

Fig. 11. 1. Some molecular oxygen fixing Co(II) complexes: a) *N,N'*-ethylenebis(salicilideneiminato)-Co(II), b) *N,N'*-ethylenebis(acetylacetoniminato)Co(II) (R = CH$_3$), *N,N'*-ethylenebis(benzoylacetoniminato)Co(II) (R = C$_6$H$_5$), c) bis(histidinato)Co(II).

Early investigations concentrated on reversible oxygenation in the solid state [11]. The compound is perfectly planar, and crystallizes in a layer lattice*). The uptake of dioxygen

---

* There are other crystalline modifications of Co(salen) which are less, or not at all, active in dioxygen uptake.

occurs in the ratio $Co:O_2 = 2:1$, and it is assumed that the dioxygen molecule is bonded between two layers by two cobalt atoms, one in each layer. Dioxygen can be removed from the complex merely by heating. The reversibility is astonishing: up to 3000 oxygenation de-oxygenation cycles have been carried out with the same sample; after 300 cycles 70% activity remained [12], 50% after 3000 cycles [13]. There have even been efforts to develop the Co(salen) system into a technical process for the production of oxygen from air [5, 12], but so far it does not appear that this method would be able to compete with the Linde process.

In the meantime, Co(salen) and its ring substituted derivatives have been found to form reversible dioxygen adducts also in certain aprotic complexing solvents such as dimethyl-formamide (DMF) or dimethylsulfoxide (DMSO), or in noncomplexing solvents in the presence of strong bases such as pyridine or DMF. Depending upon the chelating ligand, the solvent, or the added base, and on the temperature, formation of binuclear complexes $(Co:O_2 = 2)$ or mononuclear complexes $(Co:O_2 = 1)$ has been observed [14–16]. Mostly the $Co:O_2$ ratio has been judged from the uptake of dioxygen. However, some of the adducts have been isolated also in pure form, and it was found that one solvent (or base) molecule is incorporated per Co center. Hence, the related oxygenation equilibria can be formulated as follows, e.g.:

$$Co(3\text{-methoxysalen}) + L + O_2 \rightleftarrows Co(3\text{-methoxysalen})LO_2$$

(adduct isolated with L = pyridine) and

$$2\,Co(salen) + 2\,L + O_2 \rightleftarrows [Co(salen)]_2L_2O_2$$

(adduct isolated with L = DMF or DMSO [14]).

An X-ray structural analysis of the binuclear adduct with DMF has shown that the oxygen molecule forms a bridge between the two cobalt centers:

$$DMF(salen)CoOOCo(salen)DMF$$

The coordinating ligands form distorted octahedra about cobalt, the equatorial positions being occupied by the quadridentate salen ligand. The DMF molecule is bonded in an axial position through its oxygen atom and the other axial position is occupied by one of the oxygen atoms of the dioxygen bridge [17].

Several other Co(II) complexes which are able to fix molecular oxygen reversibly are shown in Fig. 11.1, together with the abbreviations usually found in the literature. Co(acacen) and Co(bzacen) are quite similar to Co(salen) as regards the chelating structure of the equatorial ligand. Both form mononuclear dioxygen complexes $(Co:O_2 = 1:1)$ in the presence of a base, such as pyridine or DMF [18, 19]. Co(his)$_2$ is different since no free coordination site is available; nevertheless a binuclear complex $(Co:O_2 = 2:1)$ is formed in water [20]. Presumably the dioxygen molecule displaces the carboxylate group of one of the histidine ligands. In strongly basic solution even both carboxylate groups appear to be displaced from the metal center which, however, still is able to form a dioxygen complex, with stoichiometry $Co:O_2 = 2:1$ [21]. Even the simple hexamminecobalt(II)ion, $[Co(NH_3)_6]^{2+}$, is oxygenated in aqueous ammonia, the dioxygen displacing one of the ammonia ligands. A brown binuclear complex, with bridging dioxygen, is formed; decrease of the oxygen pressure or increase of the ammonia concentration deoxygenates the complex [5]:

$$2\,[Co(NH_3)_6]^{2+} + O_2 \;\rightleftarrows\; [(NH_3)_5CoO_2Co(NH_3)_5]^{4+} + 2\,NH_3 \tag{11.2}$$

Primary amines or chelating aliphatic polyamines can replace the ammonia, and the complex $[Co(CN)_5]^{3-}$ also reacts reversibly with dioxygen to give a binuclear complex.

A different type of "oxygen carrier" is provided by Vaska's complex, $IrCl(CO)[P(C_6H_5)_3]_2$ (*cf.* Section 7.4.3) which, in benzene solution, reversibly takes up molecular oxygen, forming a 1:1 complex [22]. Numerous related compounds, with different central metals, neutral and halide ligands, have been found subsequently to possess the same property [9].

All these reversible dioxygen complexes, although based on metals other than the actual natural oxygen carriers hemoglobin and myoglobin, are to a certain degree models for the former, and much has been learned from them concerning reversible metal-dioxygen bonding (*cf.* Section 11.3).

Several other transition metal complexes form relatively stable, irreversible mononuclear dioxygen complexes, *e.g.* the iodide analogue of Vaska's complex, or $M[P(C_6H_5)_3]_2$ with $M = Ni(0)$, $Pt(0)$ or $Pd(0)$. The bridged complex $Cl(py)_2Rh-O-O-Rh(py)_2Cl$ (py = pyridine) was reported also to form irreversibly [10]. We shall meet some of these "stable" complexes as active catalysts (in the presence of suitable substrates) in Section 11.4.

11.2.2. Kinetics of the Oxygen Uptake by Co-Complexes

It is now generally agreed that the binuclear, dioxygen bridged complexes are formed in a two step process [14, 15, 20]:

$$Co^{II} + O_2 \; \underset{k_{-1}}{\overset{k_1}{\rightleftharpoons}} \; Co^{II}O_2 \tag{11.3}$$

$$Co^{II}O_2 + Co^{II} \; \underset{k_{-2}}{\overset{k_2}{\rightleftharpoons}} \; Co^{II}O_2Co^{II} \tag{11.4}$$

(Ligands are omitted for the sake of simplicity.)

Wilkins *et al.* accumulated kinetic evidence for the above mechanism from studies on cobalt(II) complexes with various nitrogen containing ligands in water [20, 23, 24]. Although in these cases the mononuclear intermediate has not been observed directly, the rate of oxygen uptake is best accommodated by formation of such an intermediate according to Eq. (11.3). The binuclear bridged species have strong absorption peaks in the 350–420 nm region. Their formation was studied by mixing oxygen free solutions of the cobalt(II) chelates with oxygen saturated water in a stop-flow apparatus, and by following the increase in optical density. A large excess of the cobalt species compared with the concentration of dioxygen ensured pseudo-first order conditions and complete consumption of molecular oxygen (*i.e.* reverse reactions were negligible). Formation of the bridged complex obeyed first order kinetics over several half lives (empirical rate constant, k_{obs}). Assuming steady state conditions for the intermediate, the following rate law applies for the mechanism given by Eqs. (11.3) and (11.4):

$$\frac{d\,[CoO_2Co]}{dt} = \frac{k_1 k_2 \,[Co]^2[O_2]}{k_{-1} + k_2 \,[Co]} = k_{obs} \,[O_2] \tag{11.5}$$

This equation indicates that a nonlinear dependence of the experimental rate constant k_{obs} on the concentration of the starting cobalt(II) compound is required by the assumed mechanism. Eq. 11.5 may be rearranged to give:

$$\frac{[Co]}{k_{obs}} = \frac{1}{k_1} + \frac{k_{-1}}{k_1 k_2 \,[Co]} \tag{11.6}$$

In fact the plot of $[Co]/k_{obs}$ versus $1/[Co]$ was found to be linear, corroborating the two step mechanism.

In the presence of an excess of dioxygen, evidence has been found for a third equilibrium [13]:

$$CoO_2Co + O_2 \underset{k_{-3}}{\overset{k_3}{\rightleftharpoons}} 2\,CoO_2 \tag{11.7}$$

That the 1:1 species resulting from Eqs. (11.3) and (11.7) might be different in their electronic distribution, and that the stable 1:1 complexes might always arise from Eq. (11.7) has been the subject of speculation.

The preponderance of mono or binuclear oxygen complexes appears to be governed by the relative values for the related rate constants, and these seem to be strongly solvent dependent. Thus, the anionic complex $[Co(CN)_5]^{3-}$ reacts with dioxygen in aqueous solution to give the bridged dimer, and a mononuclear 1:1 complex was proposed as an intermediate, based on EPR studies [25]. Reaction of $[Co(CN)_5]^{3-}$ with dioxygen in DMF, on the other hand, leads to the mononuclear species [26]:

$$[Co(CN)_5]^{3-} + O_2 \xrightarrow{\text{water}} [(CN)_5CoO_2Co(CN)_5]^{6-}$$

$$[Co(CN)_5]^{3-} + O_2 \xrightarrow{\text{DMF}} [Co(CN)_5(O_2)]^{3-}$$

11.2.3. „Picket Fence" Fe(II) Porphyrins

Until very recently, it was not possible to isolate more significant models for hemoglobin and myoglobin, having Fe(II) as the central metal ion. In most cases, reaction with dioxygen was found to be irreversible leading, through autoxidation, to Fe(III) species of the form:

$$L_n Fe(III) - O - Fe(III)L_n$$

i.e. with splitting of the $O-O$ bond. In the course of kinetic studies of the autoxidation it was found that an initial 1:1 binding of dioxygen to Fe(II) is followed by a rate-determining bimolecular step including two metal species. The following sequence was suggested (ligands omitted) [27]:

$$Fe(II) + O_2 \ \rightleftarrows \ FeO_2$$
$$FeO_2 + Fe(II) \ \rightarrow \ 2\,Fe(IV)O$$
$$Fe(IV)O + Fe(II) \ \rightarrow \ Fe(III) - O - Fe(III)$$

Nearly simultaneously it occurred to two groups, Baldwin and Huff [28] and Collman and coworkers [27] that, if the second (bimolecular) step could be impeded by the molecular geometry of the ligand, it should be possible to obtain reversible behavior in the first step. Both groups achieved kinetic stabilization of the $1:1$ $Fe(II)O_2$ complex by constructing a macrocyclic quadridentate ligand which forms sort of a pocket, thus shielding the co-ordinated oxygen molecule from reaction with another iron species.

The model ligand of Collman's group, to which the metaphorical name „picket fence por-phyrin" was given, is shown in Fig. 11.2.

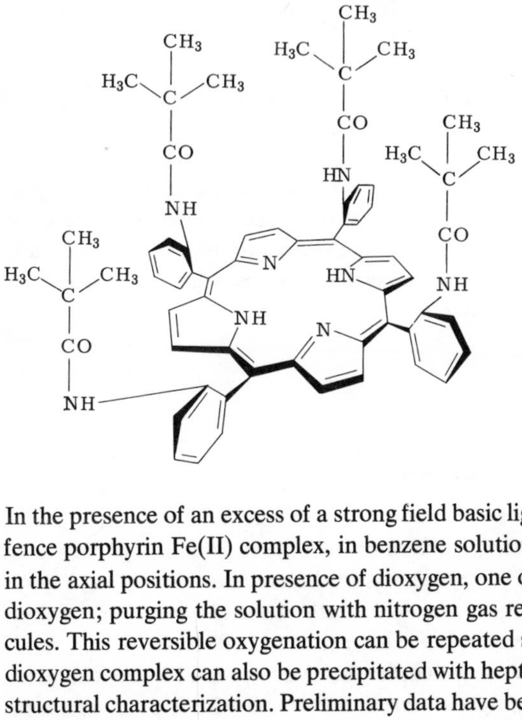

Fig. 11.2. The "picket fence porphyrin" *meso*-tetra($\alpha,\alpha,\alpha,\alpha$-*o*-pivalamido phenyl) porphyrin (Collmann *et al.* [27]).

In the presence of an excess of a strong field basic ligand such as methylimidazole, the picket fence porphyrin Fe(II) complex, in benzene solution, is octahedral with two base molecules in the axial positions. In presence of dioxygen, one of the imidazole molecules is replaced by dioxygen; purging the solution with nitrogen gas restores the complex with two base mole-cules. This reversible oxygenation can be repeated several times, at room temperature. The dioxygen complex can also be precipitated with heptane in crystalline form, permitting X-ray structural characterization. Preliminary data have been reported [27].

Fig. 11.3 shows clearly the position of the coordinated dioxygen molecule in the pocket formed by the picket fence, inaccessible to a second Fe(porphyrin) unit. (There is some statistical disorder in the molecule, the $Fe - O - O$ plane has four possible orientations, and there is also a twofold disorder in the methylimidazole ligand.)

Hitherto, these sterically stabilized $Fe(II)O_2$ complexes are probably the most relevant models for the natural oxygen carriers. They indicate that, although actual oxygen fixation evidently occurs at the metal center, the three-dimensional structure of the ligand environ-

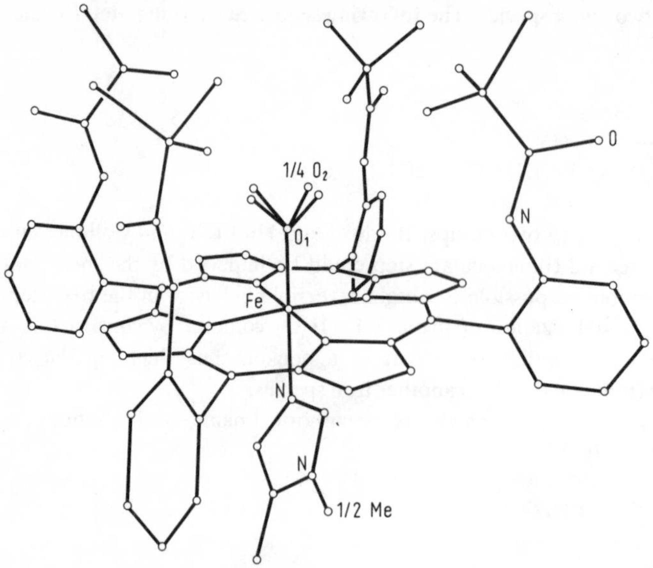

Fig. 11.3. Perspective view of the octahedral Fe(II)O$_2$ complex, having the „picket fence porphyrin" (*cf.* Fig. 11.2) as equatorial ligand, and methylimidazole as one of the axial ligands [27].

ment is undoubtedly of great significance in biological activity. The „picket fence" Fe(porphyrin) also reproduces well the thermodynamic constants (enthalpy and entropy of reaction with oxygen) of the biological systems. This indicates that the apoprotein of the latter does not contribute significantly to the binding of the oxygen, and suggests that the primary role of the protein is to protect the metal from oxidation [27b].

11.3. Bonding of Dioxygen to the Metal

11.3.1. The Electronic Structure of the Oxygen Molecule

The molecular orbital energy diagram for a homonuclear diatomic molecule was shown in Section 6.2.1 (Fig. 6.5). The electronic configuration of the oxygen atom is s^2p^4; hence 12 electrons have to be placed into this scheme, leading to the configuration $(2\,\sigma_g)^2(2\,\sigma_u)^2$ $(1\,\pi_u)^4(3\,\sigma_g)^2(1\,\pi_g)^2$. Since the $1\,\pi_g$ level is degenerate, each of the two electrons occupies one of the two orbitals, *i.e.* the free oxygen molecule has two unpaired electrons, and therefore is paramagnetic. The two unpaired electrons confer the character of a biradical upon the molecule in its ground state. Consequently, the reaction partners with a predilection for dioxygen are radicals and paramagnetic metal ions.

However, the molecular orbital diagram given in Fig. 6.5 is valid only for a free oxygen molecule. For dioxygen under the influence of the electrostatic field of a transition metal ion in a complex, the situation may be somewhat different. It was postulated by Griffith [29] that this influence might remove the degeneracy of the $1\,\pi_g$ level, and that in certain cases the energy difference between the two orbitals may become larger than the energy of pairing, *i.e.* the two electrons will then be located in the more stabilized of the two orbitals, with their spins paired (see Fig. 11.4). Note that this situation would be the same as for the oxygen molecule in its first excited state (singlet oxygen), and that it would relax the problem of spin conservation with singlet molecules.

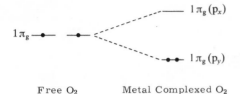

Free O_2 Metal Complexed O_2

Fig. 11. 4. Influence of the transition metal electrostatic field on oxygen molecule electron distribution, in a metal O_2 complex.

Griffith has indicated also that in this situation the oxygen molecule has an electronic configuration comparable to that of ethylene in its ground state. We may visualize, therefore, the electron distribution in terms of two trigonally sp^2 hybridized oxygen atoms. One of the sp^2 hybrid orbitals of each oxygen atom is used for mutual σ bonding and the other two are each doubly occupied; in other words, they form a set of four lone pair orbitals (n_0) oriented in the same way as the four hydrogen atoms in ethylene (Fig. 11.5). In MO language that means that the four lone pair orbitals are linear combinations of 2 σ_u, 3 σ_g, 1 π_u (p_y) and 1 π_g (p_y), so oriented as to minimize interorbital electronic interaction, whereas 2 σ_g is responsible for the σ bond. There is also one effective π bond in the molecule, 1 π_u (p_x); the lowest energy, empty antibonding orbital is 1 π_g (p_x).

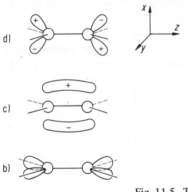

Fig. 11.5. The "valence state" MOs of the oxygen molecule: a) σ bonding MO; b) sp^2 lone pair orbitals (n_0); c) π bonding MO, 1 π_u (p_x); d) π^* antibonding MO, 1 π_g (p_x) [29].

11.3.2. Structure and Bonding in the Complexes

X-Ray structural analysis, optical spectroscopy, magnetic measurements, and theoretical MO considerations have shed some light on the bonding in transition metal-dioxygen complexes. We shall discuss first the mononuclear complexes resulting from the interaction of dioxygen with Schiff base cobalt(II) compounds, such as Co(salen) or Co(acacen). The nonoxygenated starting materials are paramagnetic, having one unpaired electron per cobalt center (low spin d^7 configuration) [30]. As mentioned in Section 11.2.1, these compounds are oxygenated in solution only in the presence of a strongly coordinating axial ligand. The purpose of the axial ligand may be rationalized by considering the d orbital energy levels of cobalt(II) in different environments (Fig. 11.6).

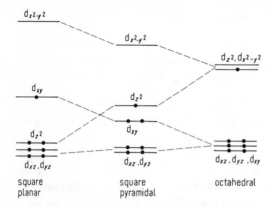

square
planar

square
pyramidal

octahedral

Fig. 11.6. Energy levels of the d orbitals in various environments, for a d^7 system.

A simple ligand field consideration (*cf.* Section 5.1.3) would suggest that the starting square planar cobalt(II) complex has the unpaired electron in the d_{xy} orbital. However, recent convincing evidence, based on EPR [31], and NMR [32] measurements, as well as on theoretical calculations [33], suggests that the unpaired electron is located more appropriately in the d_{yz} orbital. (The x and y axes are in the plane defined by the cobalt center and the Schiff base ligand.) This rather unusual localization of the unpaired electron in an orbital of lower energy arises from electronic repulsion effects [33]. A strong axial fifth ligand leads to a square pyramidal arrangement; it raises the energy of the d_{z^2} level, and the unpaired electron is now found in this orbital [30, 33]. This situation appears to be a necessary prerequisite for oxygenation [30, 34]. (Evidently, the oxygen molecule must approach the complex from the z direction, at the site opposite to that of the basic ligand.)

In the presence of a Lewis base, *e.g.* pyridine, Co(acacen) and related compounds have similar EPR signals. Fig. 11.7 which shows the spectrum of Co(acacen)py in toluene solution frozen at 77 K [30] reveals the interaction of the unpaired electron with the ^{59}Co nucleus

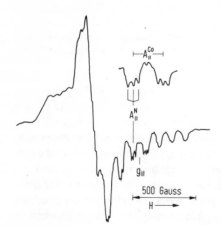

Fig. 11.7. EPR spectrum of Co(acacen)py in frozen toluene solution at 77 K [30].

($I = 7/2$) and with the ^{14}N nucleus of the pyridine ligand ($I = 1$). The signal is interpreted best assuming that the unpaired electron is associated with the 3 d_{z^2} orbital.

After oxygenation, the complexes still have one unpaired electron per cobalt center [30]. Since cobalt(II) has one, and gaseous oxygen has two unpaired electrons, spin pairing must have occurred. One may assume that, upon coordination, the symmetry of the oxygen molecule is sufficiently reduced to split the otherwise degenerate antibonding 1 π_g orbitals and to constrain the two antibonding electrons so that they are paired in the more stable of the two (*cf.* Fig. 11.4). But the EPR signal of the oxygenated complex (Fig. 11.8) indicates that the unpaired electron has only a small spin density at the cobalt nucleus (a_{Co} *ca.* 10 gauss as compared with 80–100 gauss for the nonoxygenated parent compounds, Fig. 11.7). One may conclude therefore that the unpaired electron is delocalized onto the oxygen ligand, thus making it similar to a superoxide ion. This would mean that the complex approaches the formulation Co(III)(O_2^-) [30].

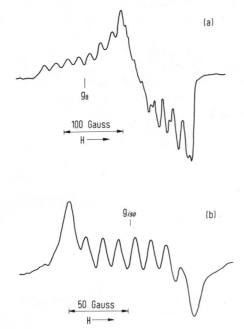

Fig. 11.8. EPR spectra of Co(acacen)pyO$_2$: a) frozen toluene solution at 77 K; b) toluene solution at 215 K [30].

Actually, the same kind of bonding (identical EPR signal) is obtained if a related Co(III) complex is reacted, in solution, with free superoxide ion, O_2^- [35]. Additionally, an X-ray photoelectron spectroscopic investigation (ESCA, *cf.* Section 6.4.4) of the oxygen complexes of Co(salen) and related compounds has shown unambiguously that the cobalt(II) ion is oxidized during the oxygenation process, although the transfer of an electron from cobalt to the oxygen molecule might not be 100% complete [36]. The very fact that the unpaired electron has a definite spin density at the cobalt nucleus (EPR evidence) indicates that the assignment of integral oxidation numbers is not quite correct in this kind of bonding. The situation appears to be described better by stating that the unpaired electron moves in a molecular orbital made up from a 3d atomic orbital of cobalt and the empty antibonding π orbital of dioxygen:

$$\Psi = c_1 \, 3 \, d_{Co} + c_2 \, 1 \, \pi_g(p_x) \quad \text{with } c_2 \gg c_1 \tag{11.8}$$

Further insight comes from EPR measurements with complexes which have been oxygenated with [^{17}O] enriched dioxygen. The mononuclear $Co(bzacen)pyO_2$ (*cf.* Fig. 11.1) shows, in solution, the EPR signal of two equivalent oxygen atoms [37], whereas in a frozen solution, the oxygen atoms are nonequivalent [38]. It was suggested that the O_2^- group is jumping between two conformations in solution, and that this motion is frozen in the solid. (Analogous conclusions have been drawn also from the temperature dependent line width of the dioxygen stretching vibration in a $Fe(II)$ (porphyrin)O_2 complex [27c].)

The X-ray structural analysis of $Co(bzacen)pyO_2$ gives concordant as well as additional information. Fig. 11.9 shows a perspective view of the complex [19]. The cobalt is bonded to only one of the oxygen atoms of the dioxygen ligand. Coordination about cobalt is approximately octahedral with the four Schiff base ligand atoms coplanar with the metal atom. The bond angle $Co-O-O$ is 126° and the $O-O$ distance of 1.26 Å corresponds approximately to that found in a superoxide anion (*cf.* Table 11.1). The same arrangement has been reported also for the complex $[N(C_2H_5)_4]_3[Co(CN)_5O_2] \cdot 5 \, H_2O$, with a $Co-O-O$ bond angle of 153.4° and an $O-O$ distance of 1.24 Å [39], and for the "picket fence porphyrin" $Fe(II)O_2$ complex shown in Fig. 11.3, for which a bond angle $Co-O-O$ of 136°, and a $O-O$ distance of 1.24 Å were given [27].

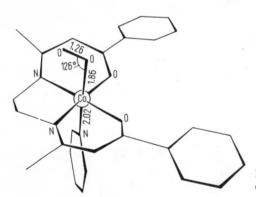

Fig. 11.9. Perspective view of the mononuclear oxygen complex $Co(bzacen)pyO_2$ [19].

The bent bond $Co-O-O$ requires further comment. The molecular orbitals of dioxygen in its „valence state", *i.e.* with the two antibonding π electrons paired under the influence of a transition metal field, have been shown in Fig. 11.5. From this picture two possibilities for coordination arise, one in which the metal lies in a lone pair direction, the metal-$O-O$ angle being approximately 120°, and the other in which the metal forms a coordinative double bond comparable with that between transition metals and olefins:

$$
\begin{array}{ccc}
\overset{\displaystyle O}{\underset{\displaystyle\overset{\Vert}{O}}{}} & \quad & O{=}O \\
\vdots & & \vdots \\
M & \quad M & \quad (M = Metal)
\end{array}
$$

Table 11.1. Oxygen-oxygen bond lengths.

Compound	Type	Structure	O−O (Å)	Ref.
O_2	free molecule	$O=O$	1.2074	[40]
KO_2	superoxide	O_2^-	1.28	[40]
H_2O_2	peroxide	O_2^{2-}	1.453	[41]
Co(bzacen)pyO$_2$	reversible O$_2$ complex	Co−O‚O	1.26	[19]
"Picket fence porphyrin"- Fe(II)(methylimidazole)O$_2$	reversible O$_2$ complex	Fe−O‚O	1.24	[27]
[(NH$_3$)$_5$CoOOCo(NH$_3$)$_5$]$^{4+}$	µ peroxo complex	Co−O‚O−Co	1.47	[42]
[(NH$_3$)$_5$CoOOCo(NH$_3$)$_5$]$^{5+}$	oxidized µ peroxo complex	Co−O‚O−Co	1.31	[43]
(O$_2$)IrCl(CO)[P(C$_6$H$_5$)$_3$]$_2$	reversible O$_2$ complex	O⇌O ‚Ir	1.30	[44]
(O$_2$)IrI(CO)[P(C$_6$H$_5$)$_3$]$_2$	irreversible O$_2$ complex	O⇌O ‚Ir	1.51	[45]

Evidently, the cobalt complexes discussed so far are close to the first type. (Other metals form bonds of the second type, *vide infra*.) As soon as the oxygen molecule has approached the transition metal as the sixth ligand, the d orbital splitting becomes that of an octahedral complex (Fig. 11.6 right-hand side). A coordinative σ bond is established between one of the oxygen lone pairs and the d$_{z^2}$ level of cobalt, and there is also a considerable amount of back donation from one of the doubly occupied d orbitals of cobalt to the antibonding empty 1 π$_g$(p$_x$) of the oxygen molecule. This latter back bonding weakens the O−O bond, as evidenced by the bond distance 1.24–1.26 Å (*cf.* Table 11.1).

Considerable effort has been made to investigate the effect of varying the axial base and the quadridentate chelating ligand on the oxygen uptake ability of Co(II) and Fe(II) in mononuclear complexes [18, 46–50]. Basolo and coworkers [49, 50] found a linear correlation between the equilibrium constants for oxygen complex formation and the ease of oxidation of Co(II) to Co(III), as measured by cyclic voltammetry, with the oxygen free complexes. The authors suggested that this relationship exists because the redox potential of the cobalt chelate is a measure of the electron density on cobalt, which in turn is the most important parameter in determining the oxygen affinity of the cobalt complexes. Thus the ability of the metal center to accept the dioxygen molecule increases with increasing donor strength of the base; it decreases if the chelating ligand tends to delocalize electron density from cobalt into the ligand π-system. These observations are thought to be in line with the general view that the mononuclear cobaltdioxygen complexes approach the formulation Co(III)O$_2^-$.

The binuclear cobaltdioxygen complexes [(NH$_3$)$_5$Co−O−O−Co(NH$_3$)$_5$]$^{4+}$ and L(salen)- Co−O−O−Co(salen)L where L = DMF, have also been investigated by X-ray structural analysis. They possess bent $Co-O-O-Co$ bonds, with $Co-O-O$ bond angles of $110.8°$ [43] and $120.4°$ [17] respectively (see Fig. 11.10).

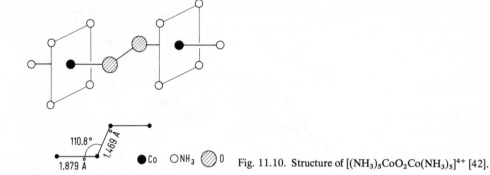

● Co   O NH$_3$ ⊘ O   Fig. 11.10. Structure of $[(NH_3)_5CoO_2Co(NH_3)_5]^{4+}$ [42].

Interestingly, these binuclear cobaltdioxygen complexes are diamagnetic, whereas the parent compounds, *e.g.* $[(NH_3)_6Co]^{2+}$ or Co(salen), are paramagnetic (low spin d^7, one unpaired electron). If we assume again that the unpaired electrons of the free oxygen molecule are paired in one of the $1\pi_g$ orbitals under the influence of the cobalt neighbors, the diamagnetism of the complex implies that the unpaired electrons of the two cobalt centers are now paired in a four-center molecular orbital embracing $Co-O-O-Co$. This hypothesis finds experimental corroboration by EPR. The complex $[(NH_3)_5CoO_2Co(NH_3)_5]^{4+}$ can easily be oxidized to $[(NH_3)_5CoO_2Co(NH_3)_5]^{5+}$; the latter is paramagnetic (one unpaired electron), and has a 15 line EPR signal [51a], showing the interaction of the unpaired electron with two identical Co nuclei (^{59}Co: $I = 7/2$). Similar EPR spectra have been found with other oxidized dioxygen-bridged cobalt complexes. A particularly well resolved signal was observed when the binuclear dioxygen complex of bis(histidinato)cobalt(II) was oxidized with Ce^{4+} [51b], see Fig. 11.11 (*cf.* Fig. 1).

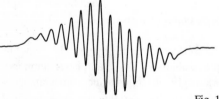

Fig. 11.11. EPR spectrum of the binuclear oxygen complex of bis(histidinato)Co(II) [51b].

These spectra show that the unpaired electron moves in a molecular orbital including the two cobalt nuclei in a symmetric way. A proposal for the four-center MO is given in Fig. 11.12. The four centers are held together by a π bond between the empty antibonding π orbital of the dioxygen, $1\pi_g(p_x)$, and one t_{2g} orbital from each cobalt. From the relatively small EPR coupling constant, $a_{Co} \approx 12$ gauss, it can be concluded that the unpaired electron is mostly localized on the $O-O$ part of the complex. This is in agreement with the $O-O$ bond distance found for the oxidized complex (see Table 11.1), which is similar to that of a superoxide, O_2^-.

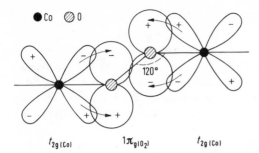

$t_{2g(Co)}$　　$1\pi_{g(O_2)}$　　$t_{2g(Co)}$

Fig. 11.12. Four-center molecular orbital in $[(NH_3)_5CoOOCo(NH_3)_5]^{4+}$ and similar complexes [5].

The bent $Co-O-O-Co$ bond indicates that there is also σ interaction between a lone pair on each oxygen atom and vacant $d(e_g)$ orbitals of the metal on either side.

The starting diamagnetic complexes, such as $[(NH_3)_5CoO_2Co(NH_3)_5]^{4+}$, on the other hand, may be assumed to have two electrons in this same four-center orbital, making the $O-O$ part then similar to a peroxide ion O_2^{2-}. In fact the $O-O$ bond distance is very similar to that of the peroxide ion (see Table 11.1).

Although it is customary to designate these bridge-bonded, binuclear cobalt O_2 complexes as μ-superoxo- and μ-peroxocobalt(III) complexes, it should be reminded that the assignment of integral oxidation numbers is not correct. Evidently the actual „effective" metal oxidation state is determined by the net electron distribution resulting from transfer in the σ-bond to the metal and back-donation to the dioxygen in the π-bond. It has been suggested [8], that the term metaldioxygen complex should apply to all relevant species containing a complexed dioxygen moiety, as is customary with the analogous metaldinitrogen complexes (see Chapter 12).

Vaska's oxygen complex $IrCl(CO)[P(C_6H_5)_3]_2O_2$ (see Section 11.2.1), and some related complexes of iridium and rhodium, have the dioxygen molecule bonded symmetrically. X-Ray studies have revealed equal distances between the metal and either of the two oxygen atoms; the arrangement of the ligands is trigonal bipyramidal, with the phosphines situated at the apices (see Fig. 11.13). The symmetric position of the two oxygen atoms suggests a bonding similar to that of the Dewar-Chatt model for the metalolefin complexes (*cf.* Fig. 7.1). A σ bond is formed between the filled bonding π orbital of the dioxygen and an empty d orbital of the metal. Back donation occurs, to offset the resultant dipole, from a filled d orbital to the antibonding $1\pi_g$ orbital of dioxygen.

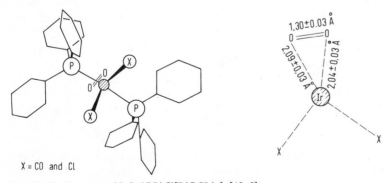

X = CO and Cl

Fig. 11.13. Structure of $IrO_2(CO)Cl[P(C_6H_5)_3]_2$ [43, 5].

Presumably the electron back-donation from the metal to the dioxygen ligand in the π bond is the predominant factor also in these symmetric dioxygen complexes. This is indicated by the observation that higher electron density at the metal center favors complex formation. Thus, as the electronegativity of the halide in the iridium complex (Fig. 11.13) decreases (Cl > I), the strength of the bonding of the oxygen molecule to the metal increases (see Table 11.1; note that dioxygen is bound irrevesibly in the iodide complex). At the same time, the O−O bond is strongly destabilized, *i.e.* the bond length is increased, indicating the higher electron density in the antibonding π orbital.

Variation in donor strength of the phosphine ligand (*cf.* Fig. 11.13) has concordant consequences: the better the donor $[(p-\text{ClC}_6\text{H}_4)_3\text{P} < (\text{C}_6\text{H}_5)_3\text{P} < (p-\text{CH}_3\text{C}_6\text{H}_4)_3\text{P} < (p-\text{CH}_3\text{O-}\text{C}_6\text{H}_4)_3\text{P}]$ the higher the rate of oxygen uptake, and the greater the stability of the dioxygen complex [52].

The importance of π bonding in metaldioxygen complexes is also reflected in the observation that d^{10} metals, which do not possess empty d orbitals, also form complexes with dioxygen but not with hydrogen [2].

After the discussion of the different modes of attachment of the oxygen molecule to the metal center, the question arises as to how dioxygen is bonded in the natural oxygen carriers such as hemoglobin or myoglobin. The macromolecular character of these compounds makes the possibility of a direct X-ray determination of the geometry of the oxygen unlikely. The bent bonding mode, as well as the symmetric attachment are able to provide reversibility (*cf.* Section 11.2), and in this sense both could serve as models for the natural compounds. At present, however, the experience with „picket fence porphyrin" Fe(II) complexes (Section 11.2.3) gives more credance to the end-on attachment of the dioxygen molecule in hemoglobin and myoglobin. Interestingly, *ab initio* calculations of the oxygen binding in iron porphyrin [53] indicate that the dioxygen ligand is *quasi*-neutral, thus supporting the description of the iron-dioxygen unit as Fe(II)-O_2 rather than Fe(III)-O_2^-. (This contrasts with the dioxygen complex of Co(acacen), where the corresponding calculations indicate a net charge of about 0.5 e on the dioxygen ligand [53].)

11.4. Reactions of Coordinated Dioxygen

X-Ray crystallographic studies of the synthetic transition metal dioxygen complexes discussed in the preceding section have indicated that a certain activation of the oxygen molecule takes place upon coordination, as shown by elongation of the O−O bond. This suggests the use of the activated dioxygen for specific oxidations, not possible with free gaseous dioxygen. In fact, several cases of homogeneous catalysis of hydrocarbon autoxidation and epoxidation with dioxygen complexes have been reported. It appears, however, that most of these have the characteristics of radical chain processes [54]. The transition metal complex catalyzes the decomposition of peroxides, as does, for instance, Co(naphthenate). This kind of oxidation, which is generally very unspecific and gives a large amount of acidic byproducts, will not be treated here.

The desired specific oxidation of substrate would require simultaneous coordination of dioxygen and substrate molecule to the same metal center prior to reaction. The catalytic oxidation of phosphine, carbon monoxide, or isocyanide ligands has been reported; olefins and even a saturated hydrocarbon have been oxidized by coordinated oxygen, although with low turnovers (see Section 11.4.2).

11.4.1. Non-Catalytic Reactions

The stoichiometric oxidation of SO_2, CO, NO, and N_2O_4 has been observed repeatedly, particularly with $IrX(CO)(O_2)[P(C_6H_5)_3]_2$ where $X = Cl$, Br, I [55], $PtO_2[P(C_6H_5)_3]_2$ [2, 56], and with $NiO_2(t\text{-}BuNC)_2$ [57].

These reactions involve an increase of the formal valency state of the metal, together with the formation of a stable anion (or two anions) attached to the metal; (see Fig. 11.14). They might, in some cases, be of interest for synthetic purposes. Free dioxygen would not react under mild conditions with SO_2, CO, NO, N_2O_4; but the cited metaldioxygen complexes easily transform these substrates to oxidized, anionic ligands. Since the latter are generally more strongly bonded to the metal than is the starting material, the reaction stops there, and no catalysis is to be expected.

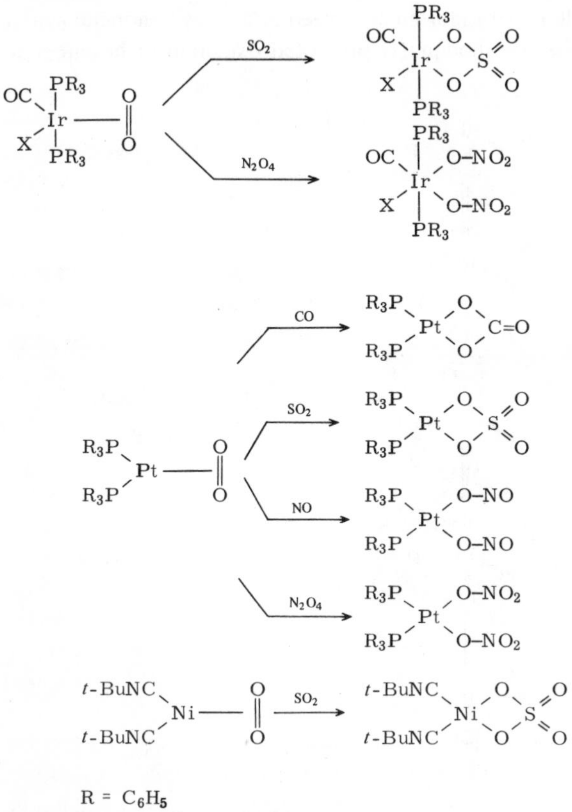

Fig. 11.14. Oxidation of SO_2, CO, NO and N_2O_4, to stable, anionic ligands.

Noncatalytic activation of molecular oxygen with an iron complex was observed when the dimeric anion $[Fe_2(S_2C_2\{CF_3\}_2)_4]^{2-}$ was reacted in solution with triphenylphosphine in the presence of molecular oxygen, and a monomeric anion containing a phosphine oxide ligand,

$[(C_6H_5)_3POFe(S_2C_2\{CF_3\}_2)_2]^-$ was isolated (as the ammonium salt) and characterized by X-ray crystallography [58].

The stoichiometric reduction of coordinated dioxygen with organic reducing agents which would not react with free dioxygen, such as dihydroxybenzene, ascorbic acid, or thiols, was reported for the dioxygen complex of Co(II)-3-methoxysalen in pyridine solution [59]. The resulting cobalt species was identified as a Co(III)aquo complex; the organic products were not characterized. It was postulated that the combined action of Co(II) and the organic reducing agent was able to convert the unfavorable one electron transfer into the more favorable two electron reduction, Co(II) and the organic component supplying one electron each (cf. Section 11.1, and Reference [1]). Evidently the Co(III) species would have to be reduced (e.g. electrolytically) in order to make the oxidation of substrates by such a mechanism catalytic with respect to the metal ion. It was also suggested that similar processes could be involved in the biological activation of molecular oxygen, as in the enzyme cytochrome oxidase, where the metal part is known to be played by iron, and where a second reducing equivalent could be provided by organic redox groups such as a cysteine or tyrosine residue (both present in the protein part of the enzyme).

The first case of oxidation of an olefin with molecular oxygen at a transition metal center under mild conditions appears to be the rhodium(I) promoted oxidation of hexene-1 to

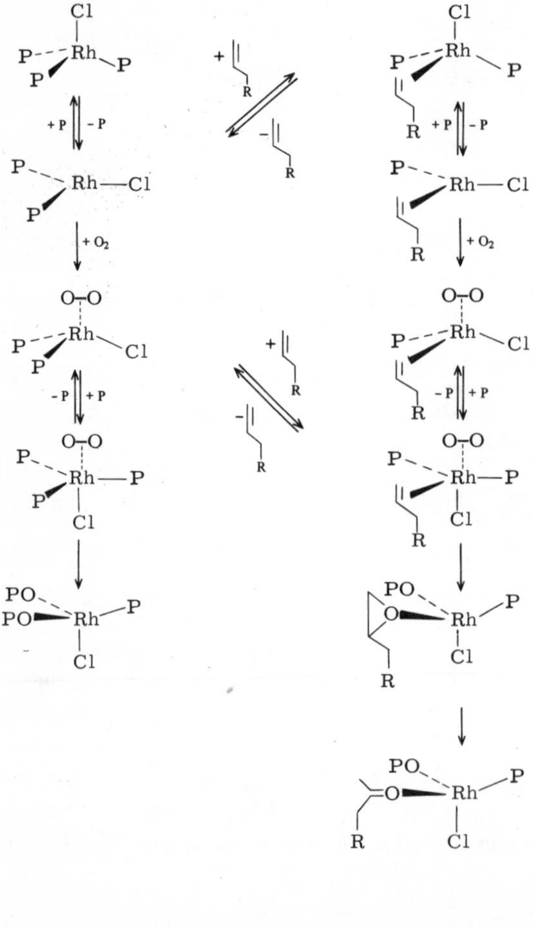

Fig. 11.15. Rhodium(I) catalyzed oxidation of 1-hexene to 2-hexanone. The phenyl groups of the phosphine and phosphine oxide ligands have been omitted; $R = n\text{-}C_3H_7$ [60].

2-hexanone [60, 61]. A free radical type of reaction has been excluded. Although the yield is small (*ca.* 20–30%, with respect to the metal**)), this reaction is remarkable. If RhCl[P-$(C_6H_5)_3]_3$ is brought together with dioxygen at ambient temperature in benzene solution, two of the three phosphine ligands are oxidized to phosphine oxide (see Fig. 11.15, left-hand reaction sequence). If olefin is present it appears able to compete with a phosphine ligand for a coordination site, and it becomes oxygenated instead of one of the phosphine ligands. It seems significant that the system provides an acceptor (the phosphine) for one of the oxygen atoms.

11.4.2. Catalytic Reactions

Triphenylphosphine is oxidized catalytically to triphenylphosphine oxide by several transition metal dioxygen complexes. Thus, Wilke *et al.* [62] found that ML_4, where M = Ni(0), Pd(0), Pt(0), L = $P(C_6H_5)_3$, react with dioxygen, in benzene or toluene solution, to give the complex ML_2O_2. At elevated temperatures (with M = Pd, Pt at $T > 90\,°C$, with M = Ni already at $T > -35\,°C$) the oxygenated complex decomposes to metal and two molecules of phosphine oxide. In the presence of an excess of phosphine, the metal may be scavenged, regenerating the starting complex, *i.e.* the process becomes catalytic. With the Ni(0) compound some 50 mol of phosphine have been oxidized per mol of Ni complex; with Pd(0) a turnover of at least 500 was obtained.

The mechanism of this reaction has been studied, for PtL_4, by Halpern *et al.* [63] on the basis of optical absorption spectroscopy and kinetic measurements. A benzene solution of PtL_4, exhibits strong absorptions in the range 400–560 nm. The same absorptions, with identical intensity, are observed when ethylene is displaced from the complex $PtL_2(C_2H_4)$ (at the same Pt concentration), by adding L in a 1 : 1 ratio. It is concluded convincingly that the predominant species in both cases is PtL_3. Therefore, the first step in the mechanism of the catalytic oxidation of the phosphine with $Pt[P(C_6H_5)_3]_4$ is assumed to be the dissociation of one ligand molecule:

$$PtL_4 \longrightarrow PtL_3 + L \tag{11.9}$$

The subsequent reaction sequence has been established by investigating the two component reactions separately [LO = phosphine oxide, $(C_6H_5)_3PO$]:

$$PtL_3 + O_2 \xrightarrow{k_1} PtL_2O_2 + L \tag{11.10}$$

$$PtL_2O_2 + L \xrightarrow{k_2} PtL_3O_2 \xrightarrow[\text{fast}]{2\,L} PtL_3 + 2\,LO \tag{11.11}$$

Each reaction was followed spectrophotometrically by monitoring the concentration of PtL_3 in the region mentioned above. Reaction (11.10) was studied under conditions ensuring minimum interference from reaction (11.11) ($[O_2] \ll [Pt]$). The kinetic experiments on the reaction (11.11) were performed by adding a large excess of L to a benzene solution of PtL_2O_2 and measuring spectrophotometrically the formation of PtL_3. The resulting rate laws are:

---

* Note added in proof: a turnover of *ca.* 30 has been achieved recently [60b].

$$-d[PtL_3]/dt = k_1[PtL_3][O_2] \tag{11.12}$$

$$+d[PtL_3]/dt = k_2[PtL_2O_2][L] \tag{11.13}$$

with $k_1 = 2.6 \, l \, mol^{-1}s^{-1}$ and $k_2 = 0.15 \, l \, mol^{-1}s^{-1}$, at 25 °C.

The mechanism was checked further under conditions such that both reactions are occurring simultaneously, *i.e.* with a large excess of L, with constant oxygen pressure, and measuring the rate of consumption of dioxygen:

$$-d[O_2]/dt = k_1[PtL_3][O_2] \tag{11.14}$$

Assuming stationary state conditions for PtL_3:

$$d[PtL_3]/dt = k_1[PtL_3][O_2] - k_2[PtL_2O_2][L] = 0 \tag{11.15}$$

and making use of the conservation relation

$$[PtL_3] + [PtL_2O_2] = [Pt]_{total} \tag{11.16}$$

the rate law for dioxygen consumption is:

$$-\frac{d[O_2]}{dt} = \frac{k_1k_2[Pt]_{total}[L][O_2]}{k_1[O_2] + k_2[L]} \tag{11.17}$$

Perfect agreement between the measured rate of dioxygen uptake and values calculated from Eq. (11.17) with the rate constants given above, lends strong support to the assumed reaction sequence, Eqs. (11.10) and (11.11). However, the detailed mechanism of the individual steps, particularly of the transformation of the phosphine into the phosphine oxide at the metal center, is not elucidated from these kinetic data. Intuitively, Halpern *et al.* explained this step in terms of a „dissociative insertion" reaction:

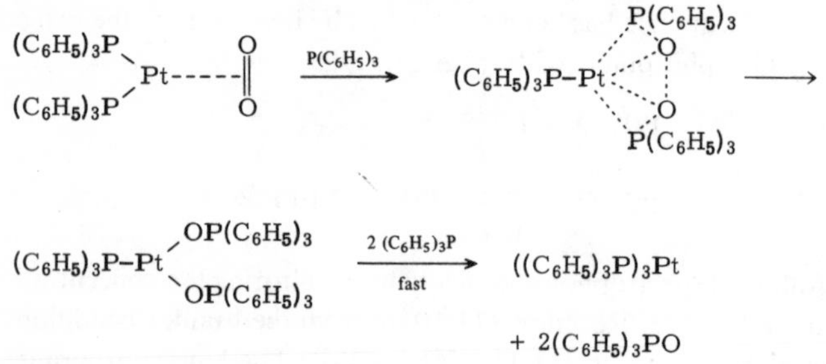

Other complexes which have been reported to oxidize triphenylphosphine catalytically to the oxide include $Ru(O_2)NCS(NO)[P(C_6H_5)_3]_2$ [64], for which a similar reaction mechanism with „dissociative oxygen insertion" has been suggested; furthermore $RuCl_2[P(C_6H_5)_3]_3$ [65] which is relatively active, converts 10 mol of $P(C_6H_5)_3$ into $OP(C_6H_5)_3$ per mol of Ru complex in 15 min at 20 °C, and Vaska's complex, $IrX(CO) [P(C_6H_5)_3]_2$ [66][*] which was, however, found to be reactive only at elevated temperature (110 °C, turnover 4–5 in 9 h).

An interesting simultaneous catalytic (although slow) oxidation of phosphine to phosphine oxide, and of CO to CO_2, was achieved by exposing $RhCl(CO) [P(C_6H_5)_3]_2$ to a slight pressure of CO and O_2 at 100 °C [68].

The catalytic oxidation of *tert*-butyl isocyanide to the corresponding isocyanate has been observed with the complex $[(CH_3)_3CNC]_2NiO_2$ at room temperature [69].

A ligand reaction has also been found for binuclear metaldioxygen complexes, namely $X_2(RNH_2)_3Co-O-O-Co(RNH_2)_3X_2$, with X = organic anion, *e.g.* acetate, and R = organic group, *e.g.* propyl [70]. The complexes are formed if the organic cobalt salt is dissolved in the amine, and dioxygen is introduced into the solution. They are diamagnetic, irreversible, and soluble in alcohols or dimethylformamide. Thermolysis of the dry complexes leads to a homolytic cleavage of the $O-O$ bond with subsequent radical reactions involving amine ligand. When R = propyl, the main products are allylamine and dipropylamine. It is suggested that primary radical formation by cleavage of the $O-O$ bond is followed by dehydrogenation of the propyl groups. The resulting allylic groups exhibit weak $C-N$ (361 kJ/mol) and $C-H$ bonds (345 kJ/mol [71]). Radical cleavage of these bonds leads to the observed alkylation of the primary amine. (Note that the suggested sequence is not a radical chain mechanism.) After thermolysis, the cobalt is present as a paramagnetic, mononuclear Co(II) complex. It dissolves in fresh amine and the resulting solution takes up oxygen at essentially the same rate as during the first oxygenation. Thermolysis leads to the same products (catalysis).

The electrochemically generated dinegative anion of cobalt tetrasulphonate phthalocyanine was reported to catalyze the oxidation of trimethylhydroquinone to trimethylbenzoquinone with molecular oxygen, in a high yield [72].

Finally, the oxidation of saturated hydrocarbons, and of toluene, to alcohols has been achieved, at 20 °C, with a catalyst system consisting of $FeCl_2$ in acetone, in the presence of hydrazobenzene and benzoic acid [73]. Oxygen (1 atm) is consumed for the simultaneous transformation of hydrazobenzene to azobenzene and water, and of hydrocarbon to alcohol:

$$\bigcirc\!\!-NH-NH-\!\!\bigcirc + RH + O_2 \longrightarrow \bigcirc\!\!-N=N-\!\!\bigcirc + ROH + H_2O \quad (11.18)$$

but generally more hydrazobenzene than RH is converted. Cyclohexane has been transformed to cyclohexanol, methylbutane to methylbutanol, and toluene to a mixture of benzyl alcohol and methylphenols. The reaction is slow (with a turnover of *ca.* 3, in three hours).

However, it is remarkable that in this reaction, molecules which are not able to form a coordinative bond with the transition metal (cyclohexane, methylbutane) are activated for reaction with oxygen. Presumably the latter is activated by formation of a $[Fe(III)O_2]^-$ complex; a free radical process was excluded by checking that the addition of a radical scavenger had no influence on the course of the reaction. The process reminds one of a certain type of

---

[*] The catalytic oxidation of styrene to benzaldehyde at 110 °C with Vaska's complex was also reported [66], however, no effort was made to exclude a radical chain mechanism. The free radical alternant copolymerization of styrene and dioxygen is well established [67]; the copolymer decomposes at elevated temperature to benzaldehyde and formaldehyde.

metalloenzyme catalysis. The most common reaction of alkanes in biological systems is the conversion of the alkane to an alcohol, with molecular oxygen as the oxidant. The oxidation is assumed to proceed by reaction of the O_2 and a reducing agent AH_2 with the enzyme to give a very reactive oxygen species; this species then reacts with an alkane molecule. The overall stoichiometry is [cf. Eq. (11.18)]:

$$RH + O_2 + AH_2 \longrightarrow ROH + A + H_2O$$

It is significant that hydroperoxides are not intermediates in these enzyme catalyzed reactions [74]. The difficulty in simulating this enzymatic process with a synthetic catalyst resides partly in the fact that the reducing agent (in the case above the hydrazobenzene) competes with the alkane for the activated oxygen, reducing it to water. In an enzyme such competition might be prevented by stringent steric requirements imposed upon substrate and reducing agent. Moreover, means would be required to regenerate the reducing agent (i.e. azobenzene → hydrazobenzene) continuously, e.g. electrolytically or with hydrogen.

The actual bonding of the dioxygen to metal in the catalytically active (and also in the non-catalytic, but reactive) dioxygen complexes has been studied only in a few cases, e.g. for the dioxygen adduct of Vaska's complex (Fig. 11.13) or for $Pt(O_2)$ $[P(C_6H_5)_3]_2$ [75] and there it is edge-on (olefin-like). The factors controlling the type of bonding appear to be little known. Presumably they are related to the energy match of the metal d orbitals with the available orbitals of the oxygen molecule. For Co(II) and Fe(II) this match might be more favorable for the oxygen lone pair orbitals (cf. Fig. 11.5), whereas for Ir(I), Pt(0) etc. the bonding π orbital of the oxygen molecule appears to be more suitable. It is even possible that for a given complex the energy difference between the two bonding modes is small, and that environmental factors dictate which form is more stable. The degree of bending in the bent end-on complexes has been related recently to the relative importance of σ and π bonding, based on a qualitative MO model [75]. Which type of bonding is more suitable for further reaction remains to be investigated.

Summarizing, we may say that although the scope of truely catalytic reactions of coordinated dioxygen with coordinated substrates is not yet very broad, the present experience nourishes the expectation that transition metal activated molecular oxygen may in future become available for scientifically, or technically, interesting selective oxidations.

References

[1] S. Fallab, Angew. Chem. Internat. Edit., 6, 496 (1967).   [2] J. P. Collman, Accounts Chem. Res. 1, 136 (1958), and ref. therein.   [3] B. M. Hoffman and D. H. Petering, Proc. Natl. Acad. Sci. U.S.A., 67, 637 (1970).   [4] G. S. Hsu, C. A. Spilburg, C. Bull and B. H. Hoffman, Proc. Natl. Acad. Sci. U.S.A., 69, 2122 (1972).   [5] E. Bayer and P. Schretzmann, Struct. Bonding, 2, 181 (1967).   [6] A. G. Sykes and J. A. Weil, Progr. Inorg. Chem., 13, 1 (1970).   [7] R. G. Wilkins, Advan. Chem. Ser. 100, 111 (1971).   [8] V. J. Choy and C. J. O'Connor, Coord. Chem. Rev., 9, 145 (1972).   [9] J. S. Valentine, Chem. Rev. 73, 235 (1973).   [10] G. Henrici-Olivé and S. Olivé, Angew. Chem. Internat. Edit. 13, 29 (1974), and references therein.
[11] A. E. Martell and M. Calvin, Chemistry of Metal Chelate Compounds, Prentice Hall, New York, 1953.   [12] M. Calvin, R. H. Bailes, and W. K. Wilmarth, J. Amer. Chem. Soc., 68, 2254 (1946).   [13] H. Diehl, Iowa State Coll. J. Sci., 22, 271 (1946).   [14] C. Floriani and F. Calderazzo, J. Chem. Soc. A, 1969, 946.   [15] C. Busetto, C. Neri, N. Palladino, and E. Perrotti, Inorg. Chim. Acta, 5, 129

(1971). [16] D. Diemente, B. M. Hoffmann, and F. Basolo, Chem. Commun., *1970*, 467. [17] M. Calligaris, G. Nardin and L. Randaccio, Chem. Commun., *1969*, 763. [18] A. L. Crumbliss and F. Basolo, J. Amer. Chem. Soc., *92*, 55 (1970). [19] G. A. Rodley and W. T. Robinson, Nature, *235*, 438 (1972). [20] J. Simplicio and R. G. Wilkins, J. Amer. Chem. Soc., *89*, 6092 (1967).

[21] K. L. Watters and R. G. Wilkins, Inorg. Chem., *13*, 752, (1974). [22] L. Vaska, Sci., *140*, 809 (1963). [23] F. Miller and R. G. Wilkins, J. Amer. Chem. Soc., *92*, 2687 (1970). [24] F. Miller, J. Simplicio and R. G. Wilkins, J. Amer. Chem. Soc., *91*, 1962 (1969). [25] J. H. Bayston, F. D. Looney and M. E. Winfield, Aust. J. Chem., *16*, 557 (1963). [26] D. A. White, A. J. Solodar and M. M. Baizer, Inorg. Chem., *11*, 2160 (1972). [27] a) J. P. Collman, R. R. Gagne, C. A. Reed, T. R. Halbert, G. Lang, and W. T. Robinson, J. Amer. Chem. Soc., *97*, 1427 (1975), and earlier references therein; b) J. P. Collman, J. I. Brauman and K. S. Suslick, J. Amer. Chem. Soc., *97*, 7185 (1975); c) J. P. Collman, R. R. Gagne, H. B. Gray and J. W. Hare, J. Amer. Chem. Soc., *96*, 6523 (1974). [28] J. E. Baldwin and J. Huff, J. Amer. Chem. Soc., *95*, 5757 (1973). [29] J. S. Griffith, Proc. Roy. Soc., *A235*, 23 (1956). [30] B. M. Hoffmann, D. L. Diemente and F. Basolo, J. Amer. Chem. Soc., 92, 61 (1970); see also B. M. Hoffmann, T. Szymansky and F. Basolo, J. Amer. Chem. Soc., *97*, 673 (1975).

[31] A. von Zelewsky and H. Fierz, Helv. Chim. Acta, *56*, 977 (1973). [32] K. Migita, M. Iwaizumi and T. Isobe, J. Amer. Chem. Soc., *97*, 4228 (1975). [33] P. Fantucci and V. Valenti, J. Amer. Chem. Soc., *98*, 3832 (1976). [34] E. I. Ochiai, J. Chem. Soc., Chem. Commun., *1972*, 489. [35] J. Ellis, J. M. Pratt and M. Green, J. Chem. Soc., Chem. Commun., *1973*, 781. [36] J. H. Burness, J. G. Dillard and L. T. Taylor, J. Amer. Chem. Soc., *97*, 6080 (1975). [37] E. Melamud, B. L. Silver and Z. Dori, J. Amer. Chem. Soc., *96*, 4689 (1974). [38] D. Getz, E. Melamud, B. L. Silver and Z. Dori, J. Amer. Chem. Soc., *97*, 3846 (1975). [39] L. D. Brown and K. N. Raymond, Inorg. Chem., *14*, 2595 (1975). [40] S. C. Abrahams, Quart. Rev., *10*, 407 (1956).

[41] W. R. Busing and H. A. Levy, J. Chem. Phys., *42*, 3054 (1965). [42] F. Fronczek, W. P. Schaefer and R. E. Marsh, Acta Crystallogr., *B30*, 117 (1974). [43] N. G. Vannerberg and G. Brosset, Acta Crystallogr., *16*, 247 (1963). [44] S. J. La Placa and J. A. Ibers, Sci., *145*, 920 (1964); J. Amer. Chem. Soc., *87*, 2581 (1965). [45] J. A. McGinnety, N. C. Payne and J. A. Ibers, J. Amer. Chem. Soc., *91*, 6301 (1969). [46] F. A. Walker, J. Amer. Chem. Soc., *95*, 1154 (1973). [47] D. V. Stynes, H. C. Stynes, B. R. James and J. A. Ibers, J. Amer. Chem. Soc., *95*, 1796 (1973). [48] G. Tauzher, G. Amiconi, E. Antonini, M. Brunori and G. Costa, Nature (NB), *241*, 222 (1973). [49] M. J. Carter, D. P. Rillema and F. Basolo, J. Amer. Chem. Soc., *96*, 392 (1974). [50] Ch. J. Weschler, D. L. Anderson and F. Basolo, J. Amer. Chem. Soc., *97*, 6707 (1975).

[51] a) E. A. V. Ebsworth and J. A. Weil, J. Phys. Chem., *63*, 1890 (1959); b) J. A. Weil and J. K. Kinnaird, Inorg. Nucl. Chem. Lett., *5*, 251 (1969). [52] L. Vaska and L. S. Chen, Chem. Commun., *1971*, 1080. [53] A. Dedieu, M.-M. Rohmer, M. Benard and A. Veillard, J. Amer. Chem. Soc., *98*, 3717 (1976). [54] E. W. Stern, Chem. Commun. *1970*, 736, and ref. therein. [55] R. W. Horn, E. Weissberger and J. P. Collman, Inorg. Chem., *9*, 2367 (1970). [56] J. P. Collman, M. Kubota and J. Hosking, J. Amer. Chem. Soc., *89*, 4809 (1967). [57] S. Otsuka, A. Nakamura, Y. Tatsuno and M. Miki, J. Amer. Chem. Soc., *94*, 3761 (1972). [58] E. F. Epstein, I. Bernal and A. L. Balch, Chem. Commun., *1970*, 136. [59] E. W. Abel, J. M. Pratt, R. Whelan and P. J. Wilkinson, J. Amer. Chem. Soc. *96*, 7119 (1974). [60] a) C. Dudley and G. Read, Tetrahedron Lett., *52*, 5273 (1972); b) G. Read, private communication.

[61] C. W. Dudley, G. Read and P. J. C. Walker, J. Chem. Soc. Dalton Trans., *1974*, 1927. [62] G. Wilke, H. Schott and P. Heimbach, Angew. Chem. Internat. Edit., *6*, 92 (1967). [63] J. P. Birk, J. Halpern and A. L. Pickard, J. Amer. Chem. Soc., *90*, 4491 (1968); J. Halpern and A. L. Pickard, Inorg. Chem., *9*, 2798 (1970). [64] B. W. Graham, K. R. Laing, C. J. O'Connor and W. R. Roper, J. Chem. Soc. Dalton Trans., *1972*, 1237. [65] S. Cenini, A. Fusi and G. Capparella, Inorg. Nucl. Chem. Lett., *8*, 127 (1972). [66] K. Takao, Y. Fujiwara, T. Imanaka and S. Teranishi, Bull. Chem. Soc. Jpn., *43*, 1153 (1970). [67] G. Henrici-Olivé and S. Olivé, Makromol. Chem., *24*, 64 (1957). [68] J. Kiji and J. Furukawa, Chem. Commun., *1970*, 977. [69] S. Otsuka, A. Nakamura and Y. Tatsuno, Chem. Commun., *1967*, 836. [70] G. Henrici-Olivé and S. Olivé, J. Organometal Chem., *52*, C49 (1973).

[71] K. W. Egger and A. T. Cocks, Helv. Chim. Acta, *56*, 1537 (1973). [72] S. Meshitsuka, M. Ichikawa and K. Tamaru, J. Chem. Soc., Chem. Commun., *1975*, 360. [73] H. Mimoun and I. Seree de Roch, Tetrahedron, *31*, 777 (1975). [74] G. W. Parshall and M. L. H. Green, *Hydrocarbon Activation*, in: F. Basolo and R. L. Burwell (Edits.), *Catalysis, Progress in Research*, Plenum Press, New York,

1972.   [75] T. Kashiwagi, N. Yasuoka, N. Kasai, M. Kakudo, S. Takahashi and N. Hagihara, Chem. Commun., *1969*, 743.   [76] B. B. Wayland, J. V. Minkiewicz and M. E. Abd-Elmageed, J. Amer. Chem. Soc., *96*, 2795 (1974).

Suggested Additional Reading

[77] F. Basolo, B. M. Hoffman and J. A. Ibers, *Synthetic Oxygen Carriers of Biological Interest*, Accounts of Chem. Res., *8*, 384 (1975).   [78] G. McLendon and A. E. Martell, *Inorganic Oxygen Carriers as Models for Biological Systems*. Coordin. Chem. Rev., *19*, 1 (1976).

12. Activation of Molecular Nitrogen

12.1. General Aspects

Dinitrogen is generally known to be an inert molecule. Under normal conditions it is so unreactive that Lavoisier named it "azote", meaning without life. It is neither combustible as is hydrogen, nor does it maintain combustion as can oxygen. The German name "Stickstoff" reminds us that not only the flame, but all life would be suffocated (erstickt) in an atmosphere of pure nitrogen. The English expression nitrogen, on the other hand, is deduced from the greek words νιτρον (nitre) and γενναω (generate). Nitre is the name of potassium nitrate, which is found in the form of fine needles or whitish powder on the surface of humid and salt containing soils. Thus the English name points to the fact that nitrogen, once freed from the energetic "cage" of the triply bonded, diatomic molecule, is perfectly capable of forming very important compounds.

Chemically bound nitrogen plays a crucial role in vegetal, animal and human metabolism, however the earth's crust contains only relatively small amounts of nitrogen in combined form, primarily ammonium salts, nitrites and nitrates. Thus mankind is compelled to supply nitrogen to the soil in order to fertilize it for plant growth. The obvious source of the nitrogen is the earth's atmosphere.

Temperatures between 300 and 600 °C and nitrogen pressures of several hundred atmospheres are required in industry to overcome the inertness of the nitrogen molecule and to convert dinitrogen to ammonia (Haber-Bosch process). However, the transformation of atmospheric dinitrogen to fertilizer nitrogen compounds is brought about also, and to a considerable extent, by nature. Microorganisms, living in the soil, convert dinitrogen from the atmosphere into ammonia which, directly or indirectly, may go towards protein synthesis within the plants. This natural nitrogen fixation is probably far more important for the total vegetation on the earth than artificial fertilizing with industrial products. The mechanism by which these microorganisms fix and reduce dinitrogen is one of the most challenging problems of chemistry. Biochemists are trying to probe the secrets of nature by extracting nitrogen assimilating enzymes, the nitrogenases, from the biological cell environment and investigating them separately. They have found that transition metals, particularly iron and molybdenum, are always present in the natural systems, and presumably are involved directly in the process of fixation and reduction of nitrogen. Chemists have been trying for the past decade to imitate nature *in vitro* by fixing nitrogen with the aid of synthetic transition metal complexes. The driving force for these studies was, of course, not only to understand the biocycle, but also to use such knowledge for developing commercial catalytic processes for the production, under mild conditions, of nitrogenated organic compounds of high value. Although we are still far away from this goal, some encouraging progress has been made recently (*cf.* Section 12.4). But even if spectacular success, in the sense of commercial developments, were not to appear before many years, the event of nitrogen fixation by synthetic transition metal complexes has opened a very interesting new field of coordination chemistry.

12.2. Relative Inertness of Dinitrogen as a Consequence of Molecular Properties

The molecular orbital description of the nitrogen molecule has been given in Section 6.2.1 (*cf*. Figs. 6.5 and 6.6). With reference to the inertness of the molecule, the most important features are the low-lying three highest occupied bonding orbitals, and the large energy gap between the HOMO and the LUMO. The energetic position of the HOMO, $3\,\sigma_g$, indicates that the ionization energy of dinitrogen is very high (15.6 eV), approaching that of argon (15.75 eV); this means that *ca*. 1500 kJ/mol are necessary to remove one of the least tightly bound electrons from the molecule in its ground state. The corresponding values for comparable molecules such as acetylene or carbon monoxide (the latter being isoelectronic with dinitrogen) are considerably lower (*ca*. 1130 and 1360 kJ/mol respectively). The LUMO, $1\,\pi_g$, lies 8.3 eV above the HOMO, and so high that it is available only for electrons from alkali metals or other strongly electropositive metals. The electron affinity, defined as the energy released by a neutral gaseous molecule in its lowest energy state when it accepts an electron, is normally determined from a Born-Haber cycle. A recent estimate [1] gives -352 kJ/mol, a rather high negative value. This and the large ionization energy indicate that reactions involving the full transfer of an electron from dinitrogen to an acceptor molecule or from a donor to dinitrogen will be difficult.

Moreover, most conceivable organic substrates have their HOMO's between -15.6 and -7.3 eV where the nitrogen molecule does not offer any orbital (*e.g.* ethylene -10.5 eV, butadiene -9.07 eV, acetylene -11.41 eV). Thus, from the point of view also of bonding criteria of simple MO theory (*cf*. Section 6.1.1), the conditions for effective interaction of dinitrogen orbitals with such substrate orbitals are not available.

The dissociation energy is also extremely large [2], *viz*:

$$N_2(g) \; \rightleftarrows \; 2\,N(g) \qquad\qquad \Delta H = 945 \text{ kJ/mol}$$
$$K_{25\,°C} = 10^{-120}$$

At ordinary pressures there is no appreciable dissociation even at 3000 °C, and the heat of dissociation has been obtained from spectroscopic data.

In Table 12.1, the bond energies of single, double and triple bonds between two nitrogen atoms are compared with those of its neighbors in the Periodic Table. In contrast to the extreme strength of the $N\equiv N$ triple bond, the $N-N$ and $N=N$ bonds are remarkably weak compared with the corresponding $C-C$ bonds. The very pronounced drop in bond energy from $C-C$ to $N-N$ is generally interpreted as a consequence of the appearance of unshared electron pairs [3, 4]. The bond energies for $N-N$ and $N=N$ are low because of the large repulsion energy terms involving these lone pairs. The same is true for $O-O$ and $F-F$. But

Table 12.1. Bond energies in kJ/mol.

$C-C$	$N-N$	$O-O$	$F-F$
347	159	138	155
$C=C$	$N=N$	$O=O$	
611	419	498	
$C\equiv C$	$N\equiv N$		
812	946		

how does the abnormal stability of $N \equiv N$ fit into this picture? Pauling [3] has given a very intuitive description of this phenomenon based on the difference in s character between the orbitals containing the lone pairs of dinitrogen (about 75% s character) and those orbitals forming the single bonds to hydrogen of two carbon atoms connected by a triple bond (acetylene, 25% s character in the $C-H$ bonds). The latter orbitals have relatively pronounced lobes at 180° away from the single bonds, in the direction of the $C \equiv C$ bond (see Fig. 12.1b). Their overlap produces repulsion, decreasing the energy of the carbon-carbon triple bond. On the other hand orbitals with 75% s character have a very small value at 180° away from the direction of their maximum. Fig. 12.1 a shows the approximate shape of the nitrogen orbitals occupied by the lone pairs of electrons in dinitrogen; there is little repulsion between the lone pairs.

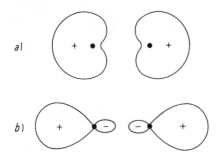

Fig. 12.1. Approximate boundary surfaces of the lone pair orbitals in dinitrogen (a; ca. 75% s character) and of the carbon orbitals forming the $C-H$ bonds in acetylene (b; ca. 25% s character) [3].

12.3. Dinitrogen Complexes of Transition Metals

12.3.1. Preparation

Until 1965 a puzzling feature of inorganic chemistry was that, whereas so many carbon monoxide complexes of low valent transition metals were known, the isoelectronic dinitrogen molecule was apparently completely inert. The discovery by Allen and Senoff [5] of the first metal dinitrogen complex, $[Ru(NH_3)_5(N_2)]^{2+}$, during the attempted synthesis of $[Ru(NH_3)_6]^{2+}$ by the reaction of hydrazine with ruthenium trichloride trihydrate in water, showed that the dinitrogen molecule is well able to become a ligand in a transition metal complex. In this first case, the dinitrogen had been formed *in situ* by disproportionation of hydrazine, with simultaneous formation of ammonia. Soon afterwards, however, Shilov *et al.* [6] showed that free molecular nitrogen can also be taken up as a ligand in Ru(II) complexes. The hope of obtaining information on dinitrogen assimilation in natural systems through the study of this new class of compounds has stimulated an intensive search for further examples.

Several methods for the preparation of dinitrogen complexes were subsequently discovered. Some 50 well defined compounds have been listed in a recent review [38]; a selection of them is given in Table 12.2. Some of the complexes are mononuclear; from X-ray work it is known that the nitrogen molecule is bonded "end-on" to the metal, with a very nearly linear $M-N-N$ arrangement. Other complexes are binuclear or even trinuclear, with a linear $M-N-N-M$ skeleton.

Table 12.2. Molecular data for some dinitrogen complexes (Compounds 11–14 for comparison).

	Compound	Bond length (Å)		Stretching frequency $\nu_{N \equiv N}(cm^{-1})$	Ref.
		M–N	N≡N		
1	$[Ru(NH_3)_5(N_2)]^{2+}$	2.10	1.12	2114	[38]
2	$[Os(NH_3)_5(N_2)]^{2+}$	1.84	1.12	2028	[38]
3	$CoH(N_2)(PR_3)_3$	1.81	1.11	2090	[38]
4	$Mo(N_2)_2(diphos)_2$	2.01	1.10	2040	[38]
5	$[(NH_3)_5RuN_2Ru(NH_3)_5]^{4+}$	1.93	1.124	2100 (Raman)	[38]
6	$(PR_3)_2NiN_2Ni(PR_3)_2$	1.79	1.12		[38]
7	$[\{\eta^5-C_5(CH_3)_5\}_2Ti]_2N_2$	2.02	1.16		[7]
8	$[\{\eta^5-C_5(CH_3)_5\}_2Zr(N_2)]_2N_2$: bridge	2.08	1.18		[8]
	end-on	2.19	1.11	2040	
9	$MoCl_4[(N_2)ReCl(PR_3)_4]_2$	1.75 (Re–N)	1.28	*ca.* 1800	[9]
		1.99 (Mo–N)			
10	$[(C_6H_5Li)_6Ni_2N_2\{(C_2H_5)_2O\}_2]_2$		1.35		[10]
11	$[Ru(NH_3)_6]^{2+}$	2.14			[11a]
12	$[Co(NH_3)_6]^{2+}$	2.11			[11b]
13	$ReCl_3(NCH_3)(PR_3)_2$	1.685			[11a]
14	N≡N		1.0976	2331 (Raman)	[11a]

One of the most straightforward methods of preparing dinitrogen complexes is the replacement of a labile ligand by dinitrogen. In several cases this type of reaction proceeds under mild conditions, and in a reversible way, *e.g.* [12]:

$$CoH_3[P(C_6H_5)_3]_3 + N_2 \underset{C_2H_5OH}{\overset{}{\rightleftharpoons}} CoH(N_2)[P(C_6H_5)_3]_3 + H_2 \qquad (12.1)$$

Even water [13] and ethylene [14] can be displaced by dinitrogen:

$$[Ru(NH_3)_5(H_2O)]^{2+} + N_2 \longrightarrow [Ru(NH_3)_5(N_2)]^{2+} + H_2O \qquad (12.2)$$

$$[Ru(NH_3)_5(H_2O)]^{2+} + [Ru(NH_3)_5(N_2)]^{2+}$$
$$\longrightarrow [(NH_3)_5Ru(N_2)Ru(NH_3)_5]^{4+} + H_2O \qquad (12.3)$$

$$[(C_6H_{11})_3P]_2NiC_2H_4 + P(C_6H_{11})_3 \xrightarrow[-C_2H_4]{100\,°C} [(C_6H_{11})_3P]_3Ni$$
$$\xrightarrow[N_2]{25\,°C} 1/2\,\{[(C_6H_{11})_3P]_2Ni\}_2N_2 + P(C_6H_{11})_3 \qquad (12.4)$$

Frequently, dinitrogen complexes are obtained, if a transition metal compound is treated with dinitrogen under strongly reducing conditions, and in the presence of a tertiary or ditertiary phosphine *e.g.* [15, 39]:

$$MoCl_4[P(CH_3)_2(C_6H_5)]_2 + N_2 \xrightarrow[toluene]{Na/Hg} Mo(N_2)[P(CH_3)_2(C_6H_5)]_4 \qquad (12.5)$$

$$\text{MoCl}_4(\text{THF})_3 + \text{N}_2 + \text{diphos} \quad \xrightarrow[\text{THF}]{\text{C}_2\text{H}_5\text{MgBr}} \quad \text{Mo(N}_2)_2(\text{diphos})_2 \qquad (12.6)$$

$(\text{diphos} = (\text{C}_6\text{H}_5)_2\text{PCH}_2\text{CH}_2\text{P(C}_6\text{H}_5)_2)$

12.3.2. Bonding between Dinitrogen and the Metal

With one exception (complex 10 in Table 12.2, and its analogues) which will be discussed in Section 12.3.4, all isolated dinitrogen complexes have the nitrogen molecule bonded end-on, either to one or two metal centers, *viz:*

$$\text{M} \cdots \text{N} \equiv \text{N} \qquad \text{M} \cdots \text{N} \equiv \text{N} \cdots \text{M}$$

The arrangement is always very close to linear, with $\text{M}-\text{N}-\text{N}$ bond angles in the range from 171,8 to 179° [38]. Evidently the bonding to the metal is comparable to that of carbon monoxide, and may be described as a sort of double bond (*cf.* Fig. 6.13). An occupied σ orbital of the nitrogen molecule (one of the lone pairs of electrons) interacts with an empty dσ orbital of the metal (d_{z^2} or $d_{x^2-y^2}$ in an octahedral complex such as $[\text{Ru(NH}_3)_5(\text{N}_2)]^{2+}$), and electron back-donation takes place *via* an occupied dπ orbital of the metal and one of the empty antibonding $1\,\pi_g$ orbitals of dinitrogen (Fig. 12.2).

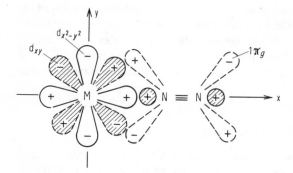

Fig. 12.2. Double bond like interaction between metal and dinitrogen. (Occupied orbitals shaded.)

A comparison of the $\text{Ru}-\text{N}_2$ bond length observed in complexes 1 and 5 (Table 12.1), with the length of a $\text{Ru}-\text{N}$ single bond (complex 11) shows that a certain amount of double bond character in the $\text{M}-\text{N}_2$ bond is actually present. The same argument holds for complex 3 as compared with complex 12. The length of metal-nitrogen double bonds is generally *ca.* 1.70 Å [11c]; an authentic $\text{Re}=\text{N}$ bond is exemplified by complex 13 in Table 12.2. In the dinitrogen complex 9, represented in Fig. 12.3, the $\text{Re}-\text{N}$ bond length is indeed close to that of a double bond.

The electron back-donation described above means that electron density passes, in the π bond, into the antibonding dinitrogen orbital $1\,\pi_g$. This is expected to weaken the bonding between the two nitrogen atoms (*cf.* Fig. 12.2). Table 12.2 shows that the $\text{N}-\text{N}$ bond in the complexes is in fact longer than in the free nitrogen molecule. In most cases the weakening of the bond is not very marked, but in complex 9 (Fig. 12.3) the strong double bond character of the $\text{Re}-\text{N}$ bond has the expected consequence on the length of the $\text{N}=\text{N}$ bond. For

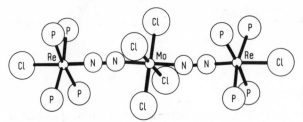

Fig. 12.3. The molecular structure of
trans-[MoCl$_4$((N$_2$)ReCl(PR$_3$)$_4$)$_2$].
PR$_3$ = P(CH$_3$)$_2$(C$_6$H$_5$) [9].

comparison: the length of a N = N double bond (in azomethane) is 1.24 Å, that of a N−N single bond (in hydrazine) is 1.47 Å [38].

The weakening of the N−N bond in the dinitrogen complexes is also manifest in the lowering of the frequency of the N−N stretching vibration, compared with that of the free nitrogen molecule (see Table 12.2). In free dinitrogen, and in complex 5, this vibration is only observable in the Raman spectrum, because of the symmetry of these compounds, but for the mononuclear complexes a strong band can be observed in the infrared region.

A comparison of the bonding of dinitrogen to a metal (Fig. 12.2) with that of carbon monoxide to a metal (Fig. 6.13) immediately indicates why the latter is by far the better ligand. The protruding lone pair orbital located at the carbon atom in carbon monoxide (>50% p character [16]) is much more favorable for bond formation than the lone pair orbitals at each end of the nitrogen atoms in dinitrogen (≈75% s character, *cf.* Fig. 12.1). The unsymmetric shape of the antibonding π orbitals of carbon monoxide (caused by the polarity of the molecule) also contributes to the excellent bonding properties of this ligand. As mentioned in Section 12.2, the nitrogen molecule is additionally disfavored by the extreme energetic positions of its HOMO and LUMO. Thus, it is understandable that the number of stable dinitrogen complexes is still small compared with the hundreds of known transition metal carbonyl complexes.

12.3.3. The Bonding in Dinitrogen-Bridged Complexes

According to the simple picture of double bonding and electron back-donation into the antibonding π orbitals of the nitrogen molecule, one would expect, at first sight, double weakening in complexes of the type M−N$_2$−M, but this is not always so. In complex 5 (Table 12.2), for instance, the N−N bond is only slightly longer than in complex 1. In the zirconium complex 8, on the other hand, where end-on bonding dinitrogen and bridge bonding dinitrogen are both present in the same compound (see Fig. 12.4), the elongation of the N−N bond in the bridge is more significant than in the terminal dinitrogen ligands.

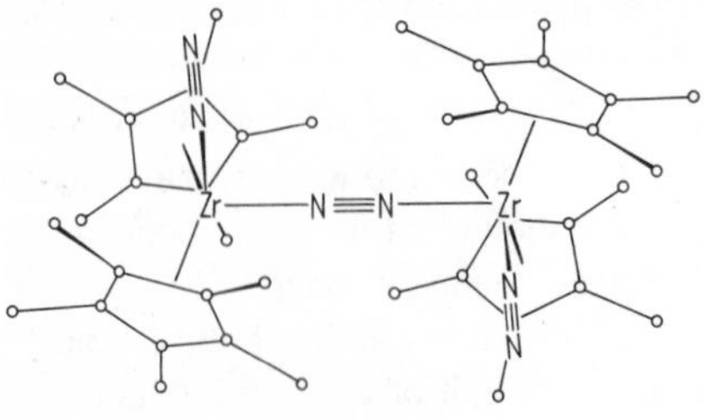

Fig. 12.4. The molecular structure of
[{η^5-C$_5$(CH$_3$)$_5$}$_2$Zr(N$_2$)]$_2$N$_2$ [7].

The varying extent of bond weakening in different dinitrogen bridged binuclear complexes may be understood qualitatively in terms of four-center π molecular orbitals spanning both metals and the bridging nitrogen molecule [38]. From the four π orbitals d_{xz} of metal M, d_{xz} of metal M', and the two p_x orbitals of the two nitrogens, we construct four MO's, by linear combination. (Note that the problem is very similar to that for the construction of the four Hückel MO's for butadiene, described in Section 6.2.3; *cf.* Fig. 6.10.) The four-center MO's are shown in Fig. 12.5. The bond axis is taken as *z* direction. As in the case of butadiene, the energy of the four MO's increases with the number of nodal planes. In addition to the orbitals shown, there exists an equivalent set of four MO's, from the d_{yz} orbitals of the metals and the p_y orbitals of the nitrogens. Hence the energy levels are twofold degenerate (e levels; *cf.* Chapter 4). They are designated as 1e, 2e, *etc.*, with increasing energy. The d_{xy} orbitals at both metal centers are non bonding. Their energy lies probably somewhere between 2e and 3e.

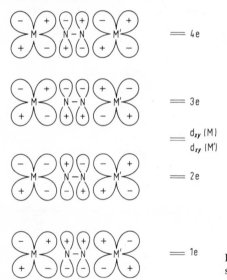

$== 4e$

$== 3e$

$== d_{xy} (M)$
$== d_{xy} (M')$

$== 2e$

$== 1e$

Fig. 12.5. Qualitative four-center molecular orbital scheme for dinuclear dinitrogen complexes [38].

This orbital scheme has to be filled with electrons from the dinitrogen (four electrons, which fill the 1e level), and from the two metal ions. Evidently the strength of the N−N bond (and also of the M−N bond) will depend on the occupancy of the various e levels, and hence on the d electron configuration of the two metal ions. The 2e level is antibonding between the two nitrogen atoms, hence electrons in this level will weaken the N−N bond (and strengthen the M−N bond). Electrons in the 3e level, on the other hand, will contribute to strengthen N−N (and to weaken M−N). Electrons in the non bonding d_{xy} metal orbitals have no effect on the bonding.

In the case of $[(NH_3)_5Ru-N\equiv N-Ru(NH_3)_5]^{4+}$, the two Ru^{2+} ions (d^6 configuration) contribute together twelve electrons. Hence all levels up to 3e are occupied. The antibonding effect of 2e (on N−N) is cancelled by the bonding effect of 3e, and the overall weakening is not very different from that in the mononuclear complex 1 (Table 12.2). In the zirconium complex 8 (*cf.* Fig. 12.4), only four electrons from the two Zr^{2+} ions have to be fed into the scheme, filling the 2e level. In this case the antibonding effect of this orbital (on $N-N$) comes full into play. A similar argument applies also to the trinuclear $Re-N_2-Mo-N_2-Re$ complex 9 (*cf.* Fig. 12.3), if we consider the two bridge bonds separately as made up of four-center MO's. Six electrons from Re^{1+}, and one of the two electrons of Mo^{4+} fill the 2e level and, presumably, d_{xy} of rhenium, leaving d_{xy} of molybdenum with one electron. Since this orbital participates also in the bonding scheme of the other bridge, the two d electrons of Mo^{4+} can pair in this orbital. (Indeed, the compound is diamagnetic [9].) Since the 3e level remains unoccupied, the observed strong destabilization of the $N-N$ bond is, at least qualitatively, plausible. (Evidently there is an additional influence from the polarity of the whole arrangement, $MoCl_4$ being a very strong electron acceptor.)

12.3.4. End-on versus Edge-on Bonding

Complex 10, Table 12.2, differs from all others in the bonding mode of dinitrogen to the metal. The most important part of this complicated compound is represented in Fig. 12.6. The compound is prepared by passing dinitrogen over an ethereal solution of phenyllithium and all-*trans*-1,5,9-cyclododecatrienenickel(0) [10a].

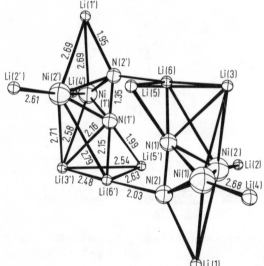

Fig. 12.6. Inner skeleton of the complex $[(C_6H_5Li)_6Ni_2N_2\{(C_2H_5)_2O\}_2]_2$ [10b]. (Phenyl groups and ether molecules are omitted.)

The two nickel atoms and the nitrogen molecule on either side of the compound form a distorted tetrahedron (Fig. 12.7a). The dinitrogen is definitely not end-on bonded to the transition metal; the bonding is more closely related to the edge-on bonding of olefins or acetylenes to metals. Exactly the same type of bonding has been found, for instance, for diphenylacetylene in hexacarbonyldicobalt diphenylacetylene [17] (see Fig. 12.7b).

This particular type of bonding for dinitrogen appears to be stabilized by interaction of the molecule with one or several lithium atoms (Fig. 12.6). A related complex, where lithium is partially replaced by sodium, has been synthesized also [18]. These two compounds represent, up to now, the only isolated examples of edge-on bonded dinitrogen.

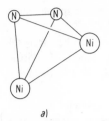
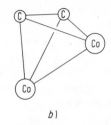

Fig. 12.7. Edge-on bonding of dinitrogen to two nickel centers in the complex represented in Fig. 12.6. (a), and of acetylene in hexacarbonyldicobalt phenylacetylene (b).

The prevalence of end-on bonding may appear amazing in view of the facts that dinitrogen can occupy the same bonding site as ethylene [Eq. (12.4)], and that natural nitrogenases reduce the typical edge-on binder acetylene as well as dinitrogen. However, one should perhaps differentiate between stability and reactivity. That the end-on bonding is more stable than the edge-on bonding, becomes immediately plausible from a qualitative comparison of the molecular orbitals involved in either type of bonding (Fig. 12.8). Whereas in the end-on form two equivalent π bonds between the metal and the nitrogen molecule can be formed, only one such bond is possible in the edge-on form. Thus, from the principle of maximum overlap it can be deduced that the linear configuration should be more stable.

$d_{xz} \rightarrow (1\pi_g)_{xz}$

$d_{xz} \rightarrow (1\pi_g)_{xz}$

$d_{yz} \rightarrow (1\pi_g)_{yz}$

$d_{xy} \rightarrow (1\pi_g)_{xy}$

a)   b)

Fig. 12.8. π Bonding in transition metal dinitrogen complexes; a) end-on structure $M\cdots N\equiv N$,

b) edge-on structure $M\cdots \overset{N}{\underset{N}{|||}}$.

Nevertheless, the more stable end-on form may, in solution, be in equilibrium with the less stable (and potentially more reactive) edge-on form. In one case, certain evidence for such an equilibrium was reported [19]. The ruthenium complex $[Ru(NH_3)_5(N_2)]^{2+}$ was prepared with $^{14}N\equiv^{15}N$, and the species with Ru-$^{14}N\equiv^{15}N$ and with Ru-$^{15}N\equiv^{14}N$ were obtained separately, showing a small, but significant shift (10 cm^{-1}) of the bond maximum of the N$-$N stretching frequency. Interestingly the two species equilibrate in the course of a few hours. This appears to indicate an intermediate state with dinitrogen bonded edge-on:

$$\left[(NH_3)_5Ru\cdots \overset{N}{\underset{N}{|||}} \right]^{2+}$$

From the energetics of this isomerization it was concluded that the edge-on species must have a very short lifetime.

An interesting view of the destabilization of the $N-N$ bond in linear or perpendicular metal dinitrogen complexes comes from a calculation [20] of σ and π bond populations of dinitrogen in the three-center fragments MN_2 for different transition metals M, and for the linear (end-on), as well as for the perpendicular (edge-on) configuration.

The data (Table 12.3) show that the total $N\equiv N$ bond population, $n_{\sigma+\pi}$, in general decreases as compared with that of free dinitrogen. This is in line with experimental evidence for a weakening of this bond ($\nu_{N\equiv N}$ shift, and increased bond distance). But the decrease of bond population is expected to be considerably more pronounced in the perpendicular compounds (*cf.* the long $N-N$ bond in complex 10, Table 12.2). The calculations in Table 12.3 indicate also that the strongest destabilizing effect should be brought about by transition metals on the left of the Periodic Table. Dinitrogen complexes of these metals appear then to be the best candidates for an effective activation of the nitrogen molecule, particularly if the latter is relatively loosely bonded, with possible transitions from end-on to edge-on configuration.

Table 12.3. Calculated σ and π bond population in the $N-N$ bond of the MN_2 fragment [19].

M	$M\cdots N\equiv N$			$M\cdots \overset{N}{\underset{N}{\vert\vert\vert}}$		
	n_σ	n_π	$n_{\sigma+\pi}$	n_σ	n_π	$n_{\sigma+\pi}$
Cr	0.838	0.577	1.415	0.731	0.238	0.969
Mn	0.821	0.634	1.455	0.733	0.384	1.117
Fe	0.809	0.747	1.556	0.738	0.524	1.262
Co	0.807	0.778	1.585	0.797	0.635	1.432
Ni	0.825	0.795	1.620	0.852	0.757	1.609
free N_2	0.796	0.816	1.621	0.796	0.816	1.621

12.4. Reactions of Coordinated Dinitrogen

Whereas the first step in the biological transformation of atmospheric dinitrogen to ammonia, the fixation of dinitrogen to a transition metal center, has been imitated successfully in several cases (Section 12.3), the simulation of the second step, the reduction of dinitrogen to ammonia, has met only limited success, as far as a really catalytic and effective process is concerned. Nevertheless, coordinatively bound dinitrogen has been actually reduced under mild conditions to ammonia as well as to the intermediate state, hydrazine, and also a few other reactions of the "inert" nitrogen molecule have been detected.

12.4.1. Dinitrogen fixation with Ziegler and Related Systems

The reduction of dinitrogen to two nitride anions by strong organometallic reducing agents, in the presence of certain transition metal compounds, was discovered 1964 by Volpin and Shur [21] and has been investigated subsequently by several workers [22]. The reactions proceed at room temperature or below, and at dinitrogen pressures of 1–150 atm. Organic solvents must be used because of the organometallic component, which is generally required in great excess (*e.g.* alkylaluminum, Grignard reagent, alkali metal naphthalide). In most cases the dinitrogen is reduced to nitride which remains attached to the transition metal; NH_3 is obtained by hydrolysis (NH_3/M generally ≤ 2; M = transition metal).

Presumably the reduction of dinitrogen is accomplished by an *in situ* reduced transition metal species which transfers electrons to the coordinatively bonded dinitrogen. At least in one case this view can be corroborated experimentally [22c, 23]. In dinitrogen fixing systems made up from simple transition metal halides in conjunction with lithium naphthalide (LiNp), one dinitrogen molecule is fixed and reduced per transition metal center, under suitable conditions. With the system $VCl_3/LiNp$ it was shown by ESR measurements that the metal is reduced to the zerovalent stage and beyond. On addition of three reduction equivalents (Li/V = 3), the signal of the bis(naphthalene)vanadium(0) sandwich complex is observed in the reaction solution, and at higher Li/V ratios another signal, assigned to a $V(-II)$ species, is observed. In the $CrCl_3/LiNp$ system, as the Li/Cr ratio is increased, the EPR signal of bis(naphthalene)chromium(I) first appears, and then that of bis(napthalene)-chromium($-I$) and bis(napthalene)chromium($-III$) [24]. Only at Li/M ≥ 6, free lithium naphthalide can be detected by its EPR signal. Evidently a further electron transfer occurs from LiNp to M(0), with formation of negatively charged $M(-x)$ species, which are held in solution by the naphthalene ligands. Since, in the particular case of VCl_3, full reduction of one dinitrogen species per vanadium center was achieved at Li/V ≥ 6, it was assumed that $V(-III)$ species are able to transfer the required six electrons to the dinitrogen, with formation of V^{3+} species.

One of the disadvantages of this type of reaction is the fact that the nitrogen combines very strongly to the transition metal in the nitride form. The only method found so far for the removal of the reduced nitrogen from the metal is hydrolysis or alcoholysis. Nevertheless, van Tamelen *et al.* [25] developed a similar process into a cycle of fixation, reduction, and alcoholysis. The fixing and reducing system is a combination of tetraisopropoxytitanium and sodium naphthalide in tetrahydrofuran. The nitride is converted into ammonia with an accurately measured quantity of isopropanol, and is thus separated from the metal. The original combination is regenerated by introduction of fresh sodium naphthalide (or simply of sodium, since the naphthalene is still available in the system). By five such cycles it was possible to convert up to 1.7 molecules of dinitrogen per titanium center into ammonia at normal pressure.

Volpin *et al.* [26] managed to remove the nitride ligand from the transition metal in a batch procedure, so making it available for further fixation. These authors used the reducing combination of aluminum and aluminum bromide, which is well known *e.g.* from the Fischer-Hafner synthesis of aromatic complexes of transition metals [27]. When this combination was used with salts such as $TiCl_4$ or $CrCl_3$, they found that more than 100 nitrogen molecules per transition metal center could be reduced to nitride ion by the use of a large excess of aluminum and Lewis acid (for example, Ti:Al:AlBr_3 = 1:600:1000, 30 hours, 130 °C, 100 atm dinitrogen: 142 molecules of dinitrogen were reduced per titanium atom). The nitrogen is presumably present as nitride ion attached to an aluminum species; it is liberated as ammonia upon hydrolysis.

The fact that the reduction goes generally through to the nitride state, in most systems involving the strong organometallic reducing agents, is not astonishing if one takes into account that opening of the first bond of the nitrogen molecule requires 527 kJ/mol, that of the next 260 kJ/mol, and the last only 159 kJ/mol (*cf.* Table 12.1). Shilov *et al.*, however, detected small amounts of hydrazine after hydrolysis, at $-50\,°C$, in the system $[P(C_6H_5)_3]_2$-$FeCl_3/i$-C_3H_7MgCl/N_2/ether [28].

12.4.2. Reactions with Lewis Acids

In the stable, terminal dinitrogen complexes, the nitrogen molecule suffers considerable electronic polarization, which leads to high intensity of the infrared absorption in the region of the $N-N$ stretching vibration (Table 12.2). The polarization is confirmed by the ESCA spectra of dinitrogen complexes. The nitrogen 1s spectra of *trans*-$RuCl(N_2)(AsR_3)_2$ [29] and of $ReCl(N_2)$ (diphos)$_2$ [30] consist of two peaks, with 1.6 and 2.0 eV separation respectively. This difference in the binding energy of the 1s electrons accounts for a charge difference of approximately 0.3 to 0.4 electrons between the two N atoms, the terminal atom bearing the greater negative charge [38].

Furthermore, the dinitrogen carries an overall negative charge which, in certain cases, is comparable to that of a chlorine ligand. Mononuclear dinitrogen complexes are expected therefore to react as Lewis bases. This is certainly so in complexes of low $N-N$ stretching frequencies [39]. The terminal nitrogen in the rhenium complex *trans*-$ReCl(N_2)L_4$ where $L = P(CH_3)_2(C_6H_5)$ ($\nu_{N_2} = 1925\ cm^{-1}$) is one of the most basic found so far, and adds to a number of Lewis acids such as $TiCl_3(THF)_2$, $CrCl_3(THF)_3$, $MoCl_4$, $Al(C_2H_5)_3$, $TaCl_5$, $NbCl_5$, *etc.*, according to, for instance:

$$ReCl(N_2)L_4 + CrCl_3(THF)_3 \longrightarrow L_4(Cl)Re\text{-}N_2\text{-}CrCl_3(THF)_2 + THF \qquad (12.7)$$

According to the same principle, the trinuclear complex represented in Fig. 12.3 has been prepared [9] [*cf.* also Eq. (12.3)].

12.4.3. Protonation of N_2 in Defined Transition Metal-Dinitrogen Complexes

With most of the isolable dinitrogen complexes all attempts to protonate the dinitrogen (*e.g.* with HCl or HBF_4) led to displacement of dinitrogen from the complex, or to protonation of the metal. Chatt and his group [39], however, discovered that a complex of zerovalent tungsten, containing two nitrogen molecules and four molecules of $P(CH_3)_2(C_6H_5)$ ($= L$) as ligands, reacts at room temperature, in methanol solution, with H_2SO_4 to give ammonia:

$$W(N_2)_2L_4 \xrightarrow[CH_3OH]{H_2SO_4} N_2 + 2\ NH_3 + W^{6+}\ species \qquad (12.8)$$

At least 90% of the starting complex react according to Eq. (12.8); evidently one of the two dinitrogen ligands is liberated, whereas the other is reduced. The metal is assumed to supply the six electrons required for reduction.

In this interesting reaction, the reduction formerly observed in Ziegler type systems (Section 12.4.1) is divided into two clear steps. The starting W(0) complex is obtained reducing WCl_4L_2 with a Grignard reagent in THF, in the presence of a slight excess of phosphine L under an atmospheric pressure of dinitrogen, at room temperature. The complex is isolated and then reacted with H_2SO_4, according to Eq. (12.8). The obvious aim would be to close the cycle reducing back the W^{6+} species, *e.g.* electrolytically, to $W(0)$.

With other $W(0)$ and $Mo(0)$ complexes also containing two dinitrogen ligands, but with two ditertiary phosphine molecules [diphos = $(C_6H_5)_2PCH_2CH_2P(C_6H_5)_2$] instead of the four L above, the protonation stops at an intermediate state, giving a N_2H_2 ligand [39], *e.g.*:

$$Mo(N_2)_2(diphos)_2 \xrightarrow{HCl} MoCl_2(N_2H_2)(diphos)_2 + N_2$$

A whole range of similar complexes, with different phosphines and halogens has been prepared. One of the halide ions in these heptacoordinated Mo complexes can be replaced by noncoordinating ions, such as $[B(C_6H_5)_4]^-$, to give hexacoordinated complexes of the type $[MoCl(N_2H_2)(diphos)_2]$ $[B(C_6H_5)_4]$. X-Ray structural analysis shows that the N_2H_2 ligand has the structure of a formal hydrazide ligand (I) in the heptacoordinated complex, whereas in the hexacoordinated complex, a diazene structure (II) appears more probable:

$$M = N - NH_2 \qquad\qquad M \cdots N \overset{H}{\underset{NH}{<}}$$

$$\text{(I)} \qquad\qquad\qquad \text{(II)}$$

A related reaction was reported by Bercaw *et al.* [31], involving the dinitrogen complex of zirconium shown in Fig. 12.4, which is prepared by reduction of $[\eta^5\text{-}C_5(CH_3)_5]_2ZrCl_2$ with sodium amalgam, under nitrogen. Treatment of this binuclear dinitrogen complex with an excess of HCl, at $-80\,°C$ in toluene yields, after subsequent warming to room temperature, a mixture of N_2, H_2 and $N_2H_4 \cdot 2\,HCl$. Eq. (12.9) consequently is implicated as a major reaction pathway, wherein the four reducing equivalents available in the dimeric complex are utilized in the reduction of one of the three dinitrogen molecules of the complex to N_2H_4:

$$\{[C_5(CH_3)_5]_2Zr(N_2)\}_2N_2 + 6\,HCl$$
$$\longrightarrow\ 2\,[C_5(CH_3)_5]_2ZrCl_2 + 2\,N_2 + N_2H_4 \cdot 2\,HCl \tag{12.9}$$

Contrary to a former report [32], $[\eta^5\text{-}C_5(CH_3)_5]_2TiCl_2$ is believed to react analogously to the zirconium complex [7]. Apparently it is not simply the bridging dinitrogen molecule alone, which is involved in the reduction. This was shown by preparing the starting complex with $^{15}N_2$ and replacing the two terminal dinitrogen ligands, which are substitutionally more labile, by $^{14}N_2$. After decomposition, it was observed that both, dinitrogen and hydrazine contain ^{14}N. These findings require a reaction sequence involving species in which one terminal dinitrogen and the bridging dinitrogen have become equivalent, *e.g.* $[C_5(CH_3)_5]_2\text{-}Zr(N_2H)_2$, which could have formed by protonation of one of the terminal dinitrogens, loss of the other terminal dinitrogen and rearrangement after opening of the bridge.

12.4.4. Acylation and Alkylation of Dinitrogen in Defined Complexes

A further interesting development comprises the reaction of a stable dinitrogen complex with an organic acid chloride, leading to an organonitrogen compound, which was reported by Chatt *et al.* [33a]. Again a bisdinitrogen complex of tungsten(0) is involved:

$$W(N_2)_2(diphos)_2 + CH_3COCl + HCl \longrightarrow WCl_2[N_2H(COCH_3)](diphos)_2 + N_2$$

Traces of moisture are assumed to generate the HCl required by the stoichiometry. With triethylamine it is possible to remove HCl from the complex, leading to $WCl(diphos)_2$-(N_2COR). A chelating structure is suggested for the acyl-azo ligand:

Simple alkyl halides also react, albeit extremely slowly, with the coordinated dinitrogen in the bisdinitrogen complexes of tungsten(0) and molybdenum(0). It was, however, discovered nearly simultaneously by Chatt *et al.* [39] and by Day *et al.* [34] that irradiation (366 nm) speeds up the reaction. Both teams were able to isolate complexes containing a diazenido ligand, *e.g.* $WBr(N_2CH_3)(diphos)_2$ [39] or $MoI(N_2C_6H_{11})(diphos)_2$ [34]. A close to linear structure $M-N-N$ and bond angles $N-N-C$ of $120\,°C$ and $142\,°C$ respectively, were determined by X-ray structural analysis. Irradiation of the tungsten complex in the presence of tetrahydrofuran and methyl bromide leads to reaction of tetrahydrofuran with dinitrogen, whereby the ether ring is opened, and an ω-diazobutanol ligand is formed [33b]:

$$MBr\{N-N=CH(CH_2)_3OH\}(diphos)_2]Br$$

Presumably the coordinated nitrogen molecule is involved in a radical process [33c].

12.4.5. Simulation of Nitrogenase

The dinitrogen fixing and/or reducing systems described in the preceding sections cannot be taken as suitable models for natural nitrogenase, since the environment is more or less abiological. Ligands which are not to be expected in nature (*e.g.* phosphines) and organic solvents are present in most cases.

Fortunately, considerable progress has been made during the past decade in the isolation and analysis of the enzyme nitrogenase, responsible for the biological dinitrogen fixation [35]. The enzyme can be separated into two proteins, a high molecular weight component containing one or, less probably, two molybdenum atoms and about fifteen iron atoms per Mo atom; the smaller molecular weight protein contains two atoms of iron and two ions of "labile" sulphide. The enzyme preparations show maximum activity *in vitro*, if both proteins are present in equimolar amounts. An electron source such as ferredoxin or sodium dithionite is required. Reduction of a substrate (dinitrogen, acetylene) is possible, however, only in the presence of ATP (adenosinetriphosphate).

Schrauzer [40] has tried to approach the biological conditions by copying the natural nitrogenase as far as possible. His nitrogenase model consists of alkali metal molybdate, as molybdenum source, a salt of the complex anion $[Fe_4S_4(SR)_4]^{2-}$ (a model of ferredoxin [36]) as iron source, $Na_2S_2O_4$ as reducing agent, and L-(+) cystein to provide sulfur ligands, in alkaline solution (pH = 7–10). This system is able to copy most of the reactions that preparations of natural nitrogenase would carry out *in vitro*, although with only 0.001–1% of the activity, at comparable Mo concentrations. The active form of the molybdenum compound is assumed to be that shown below [Mo(red)]. The iron containing anion has a quasi-cubic structure, as has natural ferredoxin:

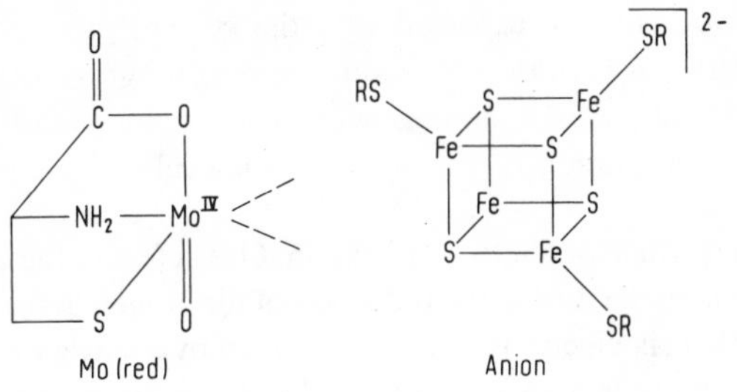

Mo (red)   Anion

Dinitrogen is reduced by this system to ammonia. As in natural nitrogenase the overall stoichiometry is:

$$N_2 + 8\,e + 8\,H^+ \longrightarrow 2\,NH_3 + H_2 \tag{12.10}$$

The iron component is assumed to transfer electrons to Mo(red) which then transfers them to the dinitrogen. The reducing agent acts on the iron component to restore its electron reservoir. The role of ATP is not yet quite clear.

Nitrogenase as well as its model reduce acetylene to ethylene. Hence diimine was suggested as a possible intermediate in the reduction of dinitrogen. Schrauzer [40] prepared diimine *in situ* (from salts of azodicarboxylic acid), in the presence of the nitrogenase model, and found that it was not reduced, but decomposed according to:

$$3\,N_2H_2 \longrightarrow 2\,N_2 + H_2 + N_2H_4 \tag{12.11}$$

From the presence of hydrogen in the reaction products of dinitrogen with the model system, and with nitrogenase [Eq. (12.10)], it was concluded that diimine might well be an intermediate but that it is not further reduced. Instead it decomposes according to Eq. (12.11); the resulting hydrazine is reduced by nitrogenase. Hence the reduction of dinitrogen by nitrogenase and its model was suggested to proceed according to:

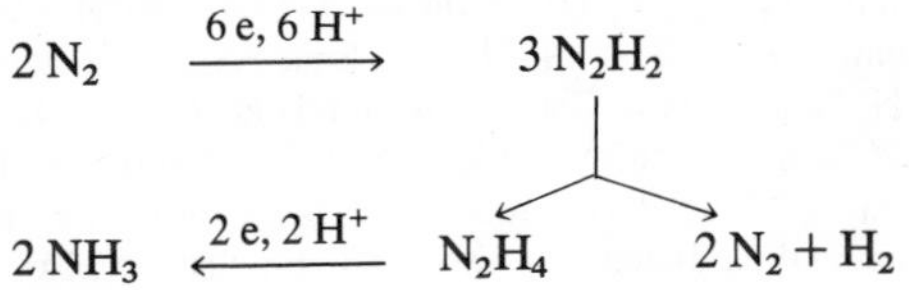

12.5. Outlook

Nitrogen has evidently lost some of its eminence as an inert gas in recent years. On the one hand it is to be expected that the number of new stable complexes with dinitrogen ligands will grow steadily. On the other hand nitrogen has been reduced to nitride ion under mild conditions. One might ask whether this can be expected to lead to competition with industrial processes for the production of ammonia. In view of the present price of NH_3, the answer to this question must be no. However, the question as asked is wrongly phrased, since ammonia is only the starting point for the chemistry of nitrogen. A more sensible question would be whether any competition is to be expected with the current processes for the production of amines, amides, nitriles, *etc.* In view of the outstanding progress that has been made in the course of the last 10–20 years in homogeneous catalysis on transition metal complexes in general, and in the activation of the nitrogen molecule in particular, this possibility must be taken seriously.

Several problems remain open in the field of dinitrogen fixation. One refers to the changes of the oxidation state of the transition metal during the reduction of dinitrogen in the complex. Although it has been suggested that six electrons could be supplied by a single metal center (*cf.* Sections 12.4.1 and 12.4.3), there is a general rule in the theory of electron transfer reactions stating that the probability of simultaneous transfer of several electrons is low [37]. If the reduction is brought about by a cluster of metal ions, a dimer for example, this difficulty is partially overcome. Thus, in the case of the binuclear trisdinitrogenzirconium complex (Fig. 12.4), only a two electron transfer per metal atom is, in principle, required. The intermediate formation of clusters probably cannot be excluded in several of the cases of dinitrogen reduction. Although one may start the reaction with a well defined complex, and be able to write down the overall stoichiometry, not all intermediates are thereby defined. Future kinetic and mechanistic studies of these interesting reactions will certainly help to throw light on this problem. A further aspect refers to the attachment of dinitrogen in the actual active species. Although most isolated complexes have end-on structures, it has repeatedly been postulated that edge-on attachment might be required for activity.

Finally the difficult task remains, that is to transform the so far substoichiometric or, at best, stoichiometric reactions in effective catalytic cycles.

References

[1] Yu. G. Borodko and A. E. Shilov, Usp. Khim., *38*, 761 (1969).   [2] F. A. Cotton and G. Wilkinson, Advanced Inorganic Chemistry. Interscience Publ., 1966.   [3] L. Pauling, Tetrahedron, *17*, 229 (1961). [4] K. S. Pitzer, J. Amer. Chem. Soc., *70*, 2140 (1948).   [5] A. D. Allen and C. V. Senoff, Chem. Commun., *1965*, 621.   [6] A. E. Shilov, A. K. Shilov and Yu. G. Borodko, Kinetika i Kataliz., *7*, 768 (1966).   [7] J. E. Bercaw, private communication.   [8] R. D. Sanner, J. M. Manriquez, R. E. Marsh and J. E. Bercaw, J. Amer. Chem. Soc., in press.   [9] P. D. Cradwick, J. Chatt, R. H. Crabtree and R. L. Richards, J. Chem. Soc. Chem. Commun., *1975*, 351.   [10] a) K. Jonas, Angew. Chem. Internat. Edit. *12*, 997 (1973); b) C. Krüger and Y.-H. Tsay, Angew. Chem. Internat. Edit. *12*, 998 (1973). [11] a) B. R. Davis and J. A. Ibers, Inorg. Chem., *9*, 2768 (1970); b) N. E. Kine and J. A. Ibers, Acta Crystallogr. *B25*, 168 (1969); c) B. R. Davis, N. C. Payne and J. A. Ibers, Inorg. Chem., *8*, 2719 (1969).   [12] A. Yamamoto, S. Kitazume and S. Ikeda, J. Amer. Chem. Soc., *90*, 1089 (1968). [13] D. F. Harrison, E. Weissenberger and H. Taube, Science (Washington) *159*, 320 (1968).   [14] P. W. Jolly, K. Jonas, C. Krüger and Y.-H. Tsay, J. Organometal. Chem., *33*, 109 (1971).   [15] B. Bell, J. Chatt and G. J. Leigh, J. Chem. Soc. Chem. Commun., *1970*, 842.   [16] R. C. Sahni, Trans. Faraday Soc., *49*, 1246 (1953).   [17] W. G. Sly, J. Amer. Chem. Soc., *81*, 18 (1959).   [18] K. Jonas, D. J. Brauer, C. Krüger, P. J. Roberts and Y.-H. Tsay, J. Amer. Chem. Soc., *98*, 74 (1976).   [19] J. N. Armor and H. Taube, J. Amer. Chem. Soc., *92*, 2560 (1970).   [20] K. B. Yatsimirskii, Pure Appl. Chem., *27*, 251 (1971).

[21] M. E. Volpin and V. B. Shur, Nature *209*, 1236 (1966) and ref. therein. [22] Selected Rev.: a) R. Murray and D. C. Smith, Coord. Chem. Rev., *3*, 429 (1968); b) Yu. G. Borodko and A. E. Shilov, Russ. Chem. Rev., *38*, 355 (1969); c) G. Henrici-Olivé and S. Olivé, Angew. Chem. Internat. Edit. *8*, 650 (1969). [23] G. Henrici-Olivé and S. Olivé, Angew. Chem., Internat. Edit. *6*, 873 (1967). [24] G. Henrici-Olivé and S. Olivé, J. Amer. Chem. Soc., *92*, 4831 (1970). [25] a) E. E. van Tamelen, G. Boche and R. H. Greeley, J. Amer. Chem. Soc., *90*, 1677 (1968); b) E. E. van Tamelen, R. B. Fechter, S. W. Schneller, G. Boche, R. H. Greeley and B. Åkermark, J. Amer. Chem. Soc., *91*, 1551 (1969). [26] M. E. Volpin, M. A. Ilatovskaya, L. V. Kosyakova and V. B. Shur, Chem. Commun., *1968*, 1074. [27] E. O. Fischer and W. Hafner, Z. Naturforsch. *10b*, 665 (1955). [28] Yu. G. Borodko, M. O. Broitman, L. M. Kachapina, A. E. Shilov and L. Yu. Ukhin, Chem. Commun. *1971*, 1185. [29] P. Finn and W. L. Jolly, Inorg. Chem. *11*, 1434 (1972). [30] G. J. Leigh, J. N. Murrell, W. Bremser and W. G. Proctor, Chem. Commun. *1970*, 1661.

[31] J. M. Manriquez and J. E. Bercaw, J. Amer. Chem. Soc., *96*, 6229 (1974). [32] J. E. Bercaw, E. Rosenberg and J. D. Roberts, J. Amer. Chem. Soc., *96*, 612 (1974). [33] a) J. Chatt, G. A. Heath and G. J. Leigh, J. Chem. Soc. Chem. Commun., *1972*, 444; b) P. C. Bevan, J. Chatt, R. A. Head, P. B. Hitchcock and G. J. Leigh, J. Chem. Soc. Chem. Commun. *1976*, 509; c) J. Chatt, private communication. [34] V. W. Day, T. A. George and S. D. A. Iske, J. Amer. Chem. Soc., *97*, 4127 (1975). [35] R. W. F. Hardy, R. C. Burns and G. W. Parshall, Advan. Chem. Ser., *100*, 219 (1971). [36] B. A. Averill, T. Herskovitz, R. H. Holm and J. A. Ibers, J. Amer. Chem. Soc., *95*, 3523 (1973). [37] A. J. Thomson, Nature, *253*, 7 (1975).

Suggested Additional Reading

[38] D. Sellmann, *Dinitrogen-Transition Metal Complexes: Synthesis, Properties and Significance.* Angew. Chem. Internat. Edit., *13*, 639 (1974). [39] J. Chatt, *The Reactions of Dinitrogen in its Mononuclear Complexes,* J. Organometal. Chem., *100*, 17 (1975). [40] G. N. Schrauzer, *Non Enzymatic Simulation of Nitrogenase Reactions and the Mechanism of Biological Nitrogen Fixation.* Angew. Chem. Internat. Edit., *14*, 514 (1975). [41] A. E. Shilov, *Nitrogen Fixation with Soluble Transition Metal Complexes.* Usp. Khim. (Russian) *43*, 863 (1974).

Subject Index

Catalyst Formula Index